Marine Pollution and Climate Change

Marine Pollution and Climate Change

Editors

Andrés Hugo Arias

Instituto Argentino de Oceanografía, CCT-CONICET
Universidad Nacional del Sur, Departamento de Química, Area III
Bahía Blanca
Argentina

and

Jorge Eduardo Marcovecchio

Instituto Argentino de Oceanografía, CCT-CONICET
Universidad FASTA, Mar del Plata, Argentina
Universidad Tecnologica Nacional (TN - BHI)
Bahia Blanca
Argentina

CRC Press
Taylor & Francis Group
Boca Raton London New York

CRC Press is an imprint of the
Taylor & Francis Group, an **informa** business
A SCIENCE PUBLISHERS BOOK

Cover illustration: Reproduced by kind courtesy of the editors of the book, Dr. Andrés Hugo Arias and Jorge Eduardo Marcovecchio. The photo is an illustration of Buenos Aires harbor from the sea (Argentina).

CRC Press
Taylor & Francis Group
6000 Broken Sound Parkway NW, Suite 300
Boca Raton, FL 33487-2742

First issued in paperback 2021

© 2018 by Taylor & Francis Group, LLC
CRC Press is an imprint of Taylor & Francis Group, an Informa business

No claim to original U.S. Government works

Version Date: 20170703

ISBN-13: 978-0-367-78191-0 (pbk)
ISBN-13: 978-1-4822-9943-4 (hbk)

Library of Congress Cataloging-in-Publication Data

Names: Arias, Andrâes Hugo, editor. | Marcovecchio, Jorge Eduardo, editor.
Title: Marine pollution and climate change / editors Andrâes Hugo Arias and
     Jorge Eduardo Marcovecchio.
Description: Boca Raton, FL : CRC Press, 2017. | Includes bibliographical
     references and index.
Identifiers: LCCN 2017018990| ISBN 9781482299434 (hardback : alk. paper) |
     ISBN 9781482299441 (e-book)
Subjects: LCSH: Marine pollution. | Ocean-atmosphere interaction. | Marine
     ecology. | Climatic changes.
Classification: LCC GC1085 .M298 2017 | DDC 363.739/4--dc23
LC record available at https://lccn.loc.gov/2017018990

Visit the Taylor & Francis Web site at
http://www.taylorandfrancis.com

and the CRC Press Web site at
http://www.crcpress.com

Preface

As stated by the IPCC, human influence on the climate system is clear, and recent anthropogenic emissions of greenhouse gases are the highest in history. The atmosphere and oceans have warmed, the amounts of snow and ice have diminished, and sea levels have risen. The same sources of greenhouse gases have also generated a variety of pollutants which after releasing and circulating in the air, water streams, soils and organisms eventually end up in the oceans. These concomitant pollutants are driving rapid changes such as decline in sea ice, receding of glaciers and permafrost, increased snow melt and runoff, shifted ranges for plants and animals, changes in populations, timing of many life-cycle events—such as blooms and migration, decoupling of species interactions, damages due to droughts and floods, etc. This book aims to cover the main groups of pollutants impacting the oceans and their effects in the light of the greenhouse gases induced changes.

Marine Pollution and Climate Change presents a comprehensive analysis of marine pollution including contributions from an impressive group of international oceanographers.

The book begins with a look at the state of oceans as they were earlier: what were the environmental and oceanographic conditions of the primitive oceans and what are the main differences/similarities with the modern ocean? Chapter 2 delves into the South Atlantic circulation, throwing light on "El niño" and "La niña" interactions. Chapter 3 tackles the first big family of ocean pollutants: hydrocarbons and greenhouse emissions. In this chapter, there is a review on the state and consequences of the present main energy paradigm: the petroleum combustion. Chapter 4 explores the interaction between the continents and the ocean harbors processes that originate the flux of metals, energy and organisms across those interfaces. Chapter 5 covers the general aspects and features of emerging pollutants in the global change scenario, highlighting the strong need of sustainable developments as the primary goal for reaching healthy ecosystems and environments. Chapter 6 addresses the issue of marine debris: man-made items of debris are now found in marine habitats throughout the world, from the poles to the equator, from shorelines and estuaries to remote areas of the high seas, and from the sea surface to the ocean floor. Chapter 7 combines the study of two ocean hazards: biological pollution and climate change. As the global ship networking is expected to increase in the future, a reduction in natural barriers for the dispersal of aquatic species is expected. This gets worse in the framework of global climate change. Chapter 8 gives to the readers an outstanding view of high-seas fisheries (demersal fishes), covering historical and present trends, vulnerability and actual threats posed by climate change to this activity. Chapter 9 innovatively studies the scarcely covered issue of the Antarctic sea's pollution and the incoming climate changes, giving an updated and accurate overview about climate system, hydrocarbons, metals, sewage and occurrence of persistent organic pollutants in the area. Finally, Chapter 10 examines, international laws pertaining to marine pollution and changes in the oceans and gives a precise overview of the protected and unprotected boundaries of the ocean.

The preparation of this book was significantly facilitated by the collaborative efforts of each of the authors. We are indebted to them, main players in the realization of this book, and the many other colleagues who provided valuable suggestions and support during the entire process of development of the book. An acknowledgement is also given to the main editorial board and all the editorial staff who provided us with the confidence and help to accomplish this project which started in late 2014.

Last but not least, the editors and everyone close to Dr. Peter Menke-Glückert wish to dedicate this book to him *in memoriam*. Peter suddenly passed away during preparation of the first version of Chapter 5 in September 2016. Peter was a totally exceptional person with enormous knowledge paired with an excellent education, rational intelligence and emphatic and human breakthroughs. From scientific, political as well as human point of view, he was one of the most influential person in Germany, Europe and worldwide. For instance, he developed under the leadership of the German former Minister of Foreign Affairs, Hans Dietrich Genscher, the Green and Sustainable developments in the 60s, 70s, 80s and 90s of the past century. In addition, Peter was one of the front men of the Liberal Party in Germany (FDP). To him, our most heartfelt remembrance.

October 2016   **Editors**

Instituto Argentino de Oceanografía (IADO), CONICET
Universidad Nacional del Sur (UNS)
Argentina

Contents

1

Ancient vs. Modern Oceans

Perspectives in a Climate Change Scenario

Jorge E. Marcovecchio,[1,2,3,*] *Silvia G. De Marco*[4,2]
and *Walter D. Melo*[1,5]

Introduction

The history, origin and environmental properties and characteristics of the Oceans, not only the present but also the primitive ones, are an excellent framework which allows to re-build old stages of our planet, as well as the processes which have governed changes between each of them. This step is quite significant considering the Earth is continuously evolving, and the historical sightseeing could be a nice starting point to understand the future of our planet.

---

[1] Instituto Argentino de Oceanografía (IADO–CONICET/UNS), Florida 7000, Edificio E-1, 8000 Bahía Blanca, Argentina.
[2] Facultad de Ingeniería, Universidad FASTA. Gascón 3145, 7600 Mar del Plata, Argentina.
[3] Facultad Regional Bahía Blanca, Universidad Tecnológica Nacional (UTN-BHI), 11 de Abril 461, 8000 Bahía Blanca, Argentina.
[4] Facultad de Ciencias Exactas y Naturales, Universidad Nacional de Mar del Plata (UNMdP), Dean Funes 3350, 3° piso, 7600 Mar del Plata, Argentina.
E-mail: demarco@mdp.edu.ar
[5] Departamento de Geografía y Turismo, Universidad Nacional del Sur (UNS), 12 de Octubre 1198, 4to piso, 8000 Bahía Blanca, Argentina.
E-mail: wdmelo@criba.edu.ar
* Corresponding author: jorgemar@iado-conicet.gob.ar

The Role of the Oceans in the Present Day

The Ocean has historically been recognized as an essential regulator of climate in our planet, playing a significant role in different processes like generation of wind fields and storms, changes in biogeochemical balances, global circulation and heat fluxes (Ganachaud and Wunsch 2000). It is well known that the exchange of latent heat, sensible heat and radiative fluxes between the ocean and the atmosphere represents a significant source for Ocean/Atmosphere interactions, including processes of warming due to increases within atmospheric CO_2 concentration (or other greenhouse causing gases) (Ramanathan 1981).

The Oceans contains about fifty times more carbon than the atmosphere; this fact has been extensively explained through the concept of *biological pump*, which is a mechanism involving either physical or biological processes directed to produce an oceanic carbon sink, balancing CO_2 concentration within the atmosphere (Hernández-León et al. 2010; Lam et al. 2011; Karl et al. 2012). The decrease of pH in seawater responding to absorption of anthropogenic CO_2 has been defined as ocean acidification (i.e., Doney et al. 2004; 2007), and can significantly impact marine ecosystems (Bates and Peters 2007).

Changes in the ocean conditions can directly affect the climate system, not only due to its role on the energy fluxes of the planet, but also by regulating the running and performance of numerous biogeochemical cycles (Rahmstorf 2002). In this sense, the studies by Nguyen et al. (1983) describing the role of the ocean on the global sulfur cycle, or those by Mason and Sheu (2002) concerning the significance of the ocean within the global mercury cycle deserve to be highlighted.

Another characteristic of the oceans that deserves to be remarked is the redox state, which plays a significant role in the regulation of the biological activity at the global scale (Fike 2010). This concept was largely shared by different authors. Canfield et al. (2010) presented an extended overview on the status and evolutionary perspectives of nitrogen cycle in the Earth, pointing out the redox state as one of the most important factors conditioning biological production. In the same viewpoint Rutherford et al. (2012) noted that a redox adequate condition is essential for profitable biological production. Also Raymont (2014) analyzed the conditions which govern the biological production within the marine environment, focusing on that of phytoplankton, and pointing out the significance of the redox state to get a successful season.

One of the topics of greatest concern is the role of the oceans as ecosystem services providers (Doney et al. 2004) with their ability to store inorganic carbon and their activity governing the distribution of major biogeochemical tracers (i.e., oxygen, nutrients, pollutants). Cooley et al. (2009) proposed that ocean acidification process could modify the operation of the marine ecosystem through its effect on all type of ecosystem service categories (i.e., provisioning, regulation, culture and support, according to Millennium Ecosystem Assessment Board 2005). In addition, Constanza et al. (1997) overviewed the importance of the mentioned processes in terms of value of the World's ecosystem services and natural capital. Halpern et al. (2008) pointed out that Humans depend on ocean ecosystems not only in terms of valuable goods but also services; nevertheless, human use has also altered the oceans through direct and indirect means, producing significant and severe damage. Moreover, Barbier et al. (2011) explained that deterioration of ocean ecosystem services due to human activities is in the range of 29–50% worldwide. Marine ecosystems provide numerous services, including food production, wastes discharge and degradation, protection of shorelines against storms, climate and atmosphere dynamics regulation, tourism development, among others (Palumbi et al. 2009).

However, the knowledge on the whole global ocean is uncomplete and remains to be fully considered. Certain parts of present day oceans have still not been adequately studied, and among them the submarine hydrothermal solutions (strongly linked to mid-ocean ridge hot springs) are a good example of this lack of knowledge (Bowers et al. 1985; Mottl et al. 2011). This is a quite remarkable point considering these environments presumably show similar conditions as those from the primitive ocean's ones.

A couple of interesting questions to put on the discussion table are: (*i*) had the ocean always played these environmental roles all along the history of the planet?...(*ii*) What were the environmental and oceanographic conditions of the primitive oceans?...(*iii*) Which are the main differences and similarities between ancient and modern-day oceans?...

How Ancient are The "Ancient Oceans"...???

The original location, distribution and dynamics of primitive oceans within early stages of Earth planet were quite different from those we usually recognize at present times, and a slow but steady evolution has come out along different eons and eras.

Each of them had got their own environmental characteristics and chemical properties, as will be summarily exposed in the next paragraphs. These conditions have been closely linked with different geological processes which have generated the corresponding structural stages of our planet, including continental masses, drifting, tectonic processes, etc. Essential parameters for understanding the physics and chemistry of the present and past oceans exist, i.e., the drainage pattern of the continents, cycles of fresh and hypersaline waters in the oceans, opening and closure of important pathways, and formation and melting of ice sheets. All these parameters influence the climate, the sea-ice distribution, the vertical circulation of the oceans, and the distribution of many elements within their sediments.

First Scenario: The Archaean Ocean

During the first stages of Earth planet consolidation (i.e., Precambrian) the first recorded ocean had appeared, and constitutes the first antecedent within this topic (Fig. 1.1).

In 1968 T.F.W. Barth studies eventually became a classic for researchers dealing with oceanography and geochemistry. In this study the author settled a strong concept directed to understand the possible origin of the ocean within Earth planet: *"the oldest rocks of the continents show distinct marks of having been deposited in water, consequently the Ocean is older than the oldest known rocks"*. Keeping this concept in mind it is reasonable to consider these set of rocks are representing the earliest crustal ones, and this fact suggests that a world-wide primordial oceanic crust arose within the early Precambrian (Glikson 1972).

So, the available information allows sustaining that oceans and continental crust already existed almost four billion years ago, considering the dated gneissic sequences and detrital zircons corresponding to this period, which signed out 4.4–4.2 Gyr age (Nutman 2006).

A large amount of geological evidence has suggested that both the atmosphere and the ocean evolved as a result of outgassing of the Earth, showing that volcanism and associated outgassing have been going on for more than 3 Gyr. The presence of this global shallow ocean was of great importance to the degassing processes. This does not mean that there were continents similar to the modern equivalent, rather that crustal rock with comparatively low density, and thus relatively high buoyancy, must

Figure 1.1. Distribution of the Archaean Ocean. Unnamed continental masses within the primitive World's ocean.

have been accumulating from very ancient times. It could be assumed that degassing of the first ~5% of the late impactors resulted in a proto-ocean with a mass approximating 5% of the present ocean, which might cover most of the Earth with a relatively thin layer of water produced by the mentioned degassing (Shaw 2016). Volatiles from outgassing interacted with the alkaline crust to form an ocean with pH 8–9 as to produce an atmosphere basically consisting of CO, CO_2, N_2 and H_2. The presence of a large shallow proto-ocean during most of the accretion and degassing of the late meteoritic veneer was very significant for the early atmosphere and the degassing processes, recognized as the main source within Earth's volatile inventory. One interesting question on this point is: where did the water come from?... Genda and Ikoma (2008) proposed three possible sources of water within primitive Earth: (*i*) water-containing rocky planetesimals like carbonaceous chondrites (CC's); (*ii*) icy planetesimals like comets; and, (*iii*) the solar nebula. Within this scenario, and considering the occurrence of a massive H_2 + H_2O atmosphere, several consecutive steps may have evolved: (*i*) in a sufficiently hot atmosphere all water would occur in vapor form; (*ii*) When the atmosphere cooled down, water vapor would condense and fall to form an ocean.

While the geological rock register dates back to ~4 Gyr within Earth planet history, a correlated stock of samples from the original proto-ocean seawater does not exist, and this fact fully complicates the understanding of composition and evolution of seawater over time. Fortunately, there is an alternative which can provide a partial record of seawater evolution, even considering that several limitations may exist. So, chemical sediments precipitated from seawater (i.e., limestones, iron formations, phosphorites) can represent the trace element and isotopic characteristics of the water mass from which they form, allowing inferring several chemical features of the ancient oceans (Derry and Jacobsen 1990). In addition, Komiya et al. (2008) have proposed three types of methods to estimate seawater composition: (1) from the composition of fluid inclusions in evaporate minerals, quartz and halite (Foriel et al. 2004), (2) from the mineral parageneses in chemical sediments of carbonate rocks and evaporites (Hardie 2003), and (3) from the composition of the carbonate rocks and banded iron formation (Kato et al. 2006).

Biogeochemical signatures preserved in ancient sedimentary rocks provide clues to identify the magnitude, distribution trend and evolution of main seawater chemical parameters, and so the reconstruction of palaeo-environmental conditions within the Primitive Ocean is possible (Scott et al. 2008). In this sense, that of ocean palaeo-temperatures has received huge efforts during the last decades, and the obtained results allows to sustain that between 3.5–1.2 Gyr ago the ocean was hot, with temperatures varying among 55 to 85°C (De la Rocha 2006). The same range of temperature values has been determined by Blake et al. (2010) working with studies on oxygen and silicon isotope compositions of cherts as well as on protein evolution during the early Palaeo-Archaean era (3.5 Gyr ago).

Simultaneously, salinity values within the Archaean Ocean was estimated to be ~1.5–2 times the modern values (~72–75‰), remaining with very high levels throughout the Archaean, and it has been attributed to the absence of long-lived continental cratons required to sequester giant halite beds and brine derived from evaporating seawater (Knauth 2005). This information fully agreed with that previously reported by De Ronde et al. (1997), who have assessed Archaean Ocean's salinity values between 180–490% modern ones.

What happened with nutrient salts from seawater during this period? Globally, nitrogen and phosphorus have been recognized as potentially limiting of the biologically mediated carbon assimilation by photoautotrophs within the oceans (Falkowski 1997). During planetary accretion nitrogen was delivered to the protoplanet as solid NH_3 (ice), amino acids, and other simple organics; these reduced forms were slowly oxidized via high-temperature reactions in the upper mantle to form atmospheric N_2, which was outgassed from volcanoes (Yokochi et al. 2009), and consequently entered into the Archaean Ocean. Godfrey and Falkowski (2009) proposed a very simple nitrogen cycle which could be occurred within the Archaean Ocean, including biological reduction of N_2 to NH_4^+, and release of NH_4^+ to water column via bacteria respiration and grazing.

Precambrian concentrations of dissolved silicon were extreme, not far from the point of saturation, and were largely controlled by abiotic reactions (Siever 1991). Silica concentrations in Precambrian oceans would have been much higher in the absence of silica-secreting organisms, and its precipitation would have been induced by evaporative supersaturation or coprecipitation with solid-phase iron minerals (Hamade et al. 2003). In addition, van den Boorn et al. (2007) proposed both seawater saturated as well as hydrothermal activity as primary sources of silica within the Archaean ocean.

Phosphorus on primitive Earth was originally trapped in igneous rock, mainly as calcium orthophosphate, which was slowly leached from the surface rocks and carried to the seas in run-off water. It has been proposed that ~3 Gyr were necessary to decompose enough rocks to saturate the seas with respect to apatites (Griffith et al. 1977). After this, the amount of P available to organisms rapidly increased based on rich sedimentary deposits exposed to weathering and leaching.

In this way, a quiet aggressive scenario for the development of any kind of life within Primitive Ocean has been described: an extremely hot, saline and alkaline environment is present during Precambrian (i.e., T up to 85°C, S up to 75‰, pH up to 9). These physical-chemical conditions fully ruled the emergence, distribution and concentration of a significant element from the view point of life evolution: oxygen. The natural occurrence of this gas was scarce during Precambrian, considering its extremely little concentration within primitive atmosphere, very low solubility due to high seawater temperature (Henry's law) and small proportion of photochemical dissociation of water vapor (Berkner and Marshall 1965). Consequently, the declining solubility of oxygen gas with temperature would have magnified ocean hypoxia under the low oxygen levels of the Precambrian atmosphere, virtually voiding the development of multicellular animal life (De la Rocha 2006). Nevertheless, different evidences from several lines of investigation confirmed that the oxygen level in the atmosphere and oceans was very low during the Archaean and Early Palaeoproterozoic. This conclusion was also confirmed by the discovery of large degrees of Mass-Independent Fractionation (MIF) of the sulfur isotopes in sulfides and sulfates in pre-2.45 Gyr sedimentary rocks (Gaucher et al. 2008). So, it is possible that the inferred trend in palaeotemperature reflects an ecological trajectory as ancient bacteria made the transition from hot springs and thermal vents to the open ocean. Marine life was limited to microbes (including cyanobacteria) that could tolerate the hot, saline early ocean. Because O_2 solubility decreases strongly with increasing temperature and salinity, the Archaean ocean was anoxic and dominated by anaerobic microbes even if atmospheric O_2 were somehow as high as 70% of the modern level (Knauth 2005).

Within this palaeo-environmental framework organic geochemical data, mainly those from carbon isotopic composition of kerogen (insoluble organic matter) pointed out carbonaceous material within these sediments as probably produced by photoautotrophs, such as photosynthetic bacteria or blue-green algae (Schopf 1974). So, evidence indicates that living systems were present and widespread as early as 3.2 Gyr ago, and photoautotrophs would presumably be the dominant one (Schopf 2011; Farquhar et al. 2011). The earliest attributable microfossils of possible cyanobacteria occurred in the 3.5 Gyr Apex Chert of Western Australia, at the same time as the first evidence for oxygen in the Archaean oceans (Schopf 1993; Hoashi et al. 2009). The function of this ocean system during at least 1 billion years has built the new scenario which will displayed along the Proterozoic.

Great changes of Ocean during Proterozoic (2.5–0.543 Gyr ago)

The Archaean-Proterozoic boundary presumably represents the most important transition time within the Earth's crust and atmosphere evolution (Watanabe et al. 1997). In this sense, Poulton et al. (2004) have properly described that *"The Proterozoic aeon (2.5 to 0.54 Gyr ago) marks the time between the largely anoxic world of the Archaean (>2.5 Gyr ago) and the dominantly oxic world of the Phanerozoic (<0.54 Gyr ago)"*. It is a transcendental event considering that during this aeon the adequate scenario which would allow the explosive biological development from Paleozoic has been settled.

Approximately 2 Gyr ago a significant oxidation of the Earth's surface occurred, probably driven by increasing input of oxygen into the atmosphere linked to the raise within sedimentary burial of organic matter among 2.3 and 2.0 Gyr (Karhu and Holland 1996; Canfield 1998). This was a slow and complex process, and the geochemical evidence suggests that a delay of several hundred million years between the evolution of oxygenic photosynthesis and the accumulation of oxygen in Earth's atmosphere must have occurred. In addition, the deep ocean would have remained euxinic for several hundred million years after the atmosphere became oxygenated (Fennel et al. 2005). In this sense, Holland (2006), in a very detailed overview on *The oxygenation of the atmosphere and oceans*, has proposed that the last 3.85 Gyr of Earth history should been divided into five stages; *(i) Stage 1* (3.85–2.45 Gyr): the oxygen level in the atmosphere and oceans was very low during the Archaean and Early Palaeoproterozoic, and measurements indicate that the O_2 content of the Archaean atmosphere was generally less than ~10^{-5} Present Atmospheric.

Level (PAL) (2 p.p.m.v.; Kasting et al. 2001; Pavlov and Kasting 2002). *(ii) Stage 2* (2.45–1.85 Gyr): atmospheric O_2 clearly appeared between 2.41 and 2.32 Gyr; the lack of Banded Iron Formation (BIF) deposition reduced or eliminated a minor O_2 sink, contributing to the transition of the atmosphere to an oxygenated state. A decrease in the hydrothermal flux of H_2 and H_2S, and changes in the biosphere and in the nutrient flux to the oceans should have been other potentially important changes within the Earth's redox system. The period between 2.4 and 2.0 Gyr has become known as the Great Oxidation Event (GOE). *(iii) Stage 3* (1.85–0.85 Gyr): the O_2 content of the atmosphere was 10–20% PAL. Consequently, the O_2 content of seawater would become exhausted during its passage from the sea surface downward and along the oceanic conveyor belt. The deep ocean would therefore have become anoxic or euxinic, if all other parameters of the system had remained unchanged. *(iv) Stage 4* (0.85–0.54 Gyr): The largest three ice ages impacted the Earth and may have been followed by unusually hot climates (Hoffman and Schrag 2002). High positive $\delta^{13}C$ excursions of marine carbonates occurred during this period, pointing out the burial of excess carbon and the generation of excess O_2. Atmospheric O_2 and SO_4^- in seawater should have reached levels probably not much lower than those of the present day. *(v) Stage 5* (the last 0.54 Gyr): atmospheric O_2 might have probably significantly varied during the Phanerozoic. Berner (2004) indicated that atmospheric O_2 may have reached values as high as 0.35 atm during the Permo–Carboniferous. The surface oceans must have been oxygenated throughout the Phanerozoic, but the oxidation state of the deeper oceans had widely fluctuated.

Oxygen deficiency has demonstrated to be a major driver of evolution and extinction throughout Earth history. The evolution and extinction of life are strongly linked to the oxygen state of the ocean, and particularly to the presence of anoxic and/or euxinic water on a global scale (Lyons et al. 2009).

Oceans chemical composition has significantly changed with the oxidation of Earth's surface, and both evolutionary and biological history of life are strongly related with this process (Cloud 1972). The early Earth was characterized by a reducing ocean-atmosphere system, whereas the Phanerozoic Eon (< 0.543 Gyr) represented a stably oxygenated biosphere including complex ecosystems with large biological diversity (Reinhard et al. 2013). The ocean started to be ventilated at ~1.8 Gyr linked to BIF duck-out; nevertheless, it has been commonly assumed that the mid-Proterozoic Earth presented a globally euxinic ocean supporting the corresponding marine systems (Brocks et al. 2005).

So, significant changes occurring on the Earth's surface from 2.5–2.0 Gyr included increase in atmospheric O_2, SO_4^- in seawater, and accumulation of great amount of organic-rich sediments. This scenario led to the *Great Oxidation Event* (GOE), characterized by the transition from deposition of iron formation to the deposition of red beds (Rogers and Santosh 2009). In this sense, Archaean ocean seawater could be characterized as hot (> 50°C), saline and anoxic. After this (~1.2 Gyr ago) ocean salinity decreased and O_2 level increased, presumably mediated by the appearance of large areas of shallow water on continental crust (Knauth 2005). Evidence suggested that seawater was enriched in carbonate before 1.8 Gyr, whereas sulfate was more abundant after that time (Ohmoto et al. 2004; Johnston et al. 2006). Thereby, and even taking into account that organic evolution presents significant uncertainties, it could be sustained that ~1.8 Gyr ago both the atmosphere and oceans contained enough oxygen to support eukaryotic life (Rogers and Santosh 2009a). In fact, evidence pointed out that the mentioned oxidation within Earth's environment has occurred in two steps: a first one (~2.4–2.2 Gyr) described as surface oxidation; and a second one (~0.8–0.58 Gyr) described as a biospheric oxygenation, turning up just before large animals appeared within the fossil record (Canfield and Teske 1996; Shen et al. 2003). This two-staged oxidation condition allowed proposing unique ocean chemistry for much of the Proterozoic eon, which would have been neither completely anoxic and iron-rich as hypothesized for Archaean seas, nor fully oxic as supposed for most of the Phanerozoic eon (Canfield 1998).

The starting times of the Earth were characterized by the presence of supercontinents, which have played a significant role within the planet history (Meert 2012). The properties and characteristics of the supercontinents have conditioned the development and magnitude of the associated processes (i.e., global circulation, terrestrial weatherability, etc.), as well as their corresponding breakups have originated not only new continents and large continental landmasses but also the oceans included around them (Kheraskova et al. 2010) (Fig. 1.2). There are at least three periods in Earth history during which most (> 75%) of the Earth's

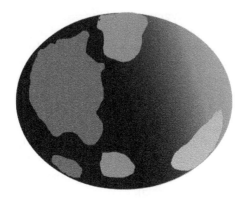

Figure 1.2. Distribution of supercontinents during Proterozoic era. In the image Columbia supercontinent is observed.

continental crust was assembled in a rigid (or quasi-rigid) supercontinent. These three supercontinents are named Columbia (during the Archaean-Paleoproterozoic, up to ~1.8 Gyr) (Rogers and Santosh 2009b), Rodinia (during Neoproterozoic, ~1.1 Gyr) (McMenamin and McMenamin 1990), and Pangaea (during Paleozoic, ~0.3 Gyr) (Wegener 1912; 1915).

Rodinia supercontinent's breakup represented a significant environmental change during the Neoproterozoic Era (~1.0–0.54 Myr). It was a period plenty of ice sheets, and global temperatures dropped to –50°C during two long-lived 'snowball' events (Gernon et al. 2016). During Neoproterozoic Snowball Earth glaciations, the oceans' alkalinity significantly increased, generating massive cap carbonates deposition on deglaciation. Continental breakup led to acute increases in riverine runoff and silicate weathering, producing a high CO_2 depletion, leading to a 'snowball' state (Donnadieu et al. 2004).

Maruyama (1994) suggested that supercontinent breakup was initiated by a superplume situated in what at present is the South Pacific. A triple-point junction over the superplume could have resulted in the opening of the North Pacific, South Pacific and Paleo-Asian Oceans (Maruyama 1994). Two stages of the Rodinia break-up have been suggested by Dobretsov et al. (1995): (1) the first stage (0.9–0.8 Gyr) resulted in the opening of the Paleo-Asian ocean; and (2) the second stage (0.75–0.7 Gyr) resulted in the opening of the Paleo-Pacific. The breakup of the latest-Mesoproterozoic supercontinent Rodinia and its transformation into the end-Neoproterozoic to Paleozoic supercontinent Gondwana is recorded in the life cycle of four main ocean basins and their margins: the Mirovoi, Mozambique, Pacific and Iapetus Oceans (Fig. 1.3).

Summarizing, the Proterozoic aeon (2.5 to 0.54 billion years (Gyr) ago) marks the time between the largely anoxic world of the Archaean (> 2.5 Gyr ago) and the dominantly oxic world of the Phanerozoic (< 0.54 Gyr ago). The

Figure 1.3. Distribution of oceans within Proterozoic: the breakup of the latest-Mesoproterozoic Rodinia supercontinent.

course of ocean chemistry through the Proterozoic has traditionally been explained by progressive oxygenation of the deep ocean in response to an increase in atmospheric oxygen around 2.3 Gyr ago. Sulfidic conditions may have persisted until a second major rise in oxygen between 0.8 to 0.58 Gyr ago, possibly reducing global rates of primary production and arresting the pace of algal evolution. The redox chemistry of Proterozoic oceans has important implications for biological evolution. A significant point which deserves to be highlighted was that increasing terrestrial weatherability during the late Neoproterozoic may explain low temperature, increases in ocean phosphate, ocean sulfate, and atmospheric oxygen concentration at this time.

The Life Revolution within Earth: the Paleozoic Oceans (543–248 Myr)

The late Precambrian and Cambrian world experienced explosive evolution of the biosphere agreeing with deep changes in climate, atmospheric and oceanic conditions over this span of time (Brasier 1992). The largely documented early adaptive radiation of large energetic metazoans along the Proterozoic-Phanerozoic transition was necessarily based in a solid source of primary productivity, which should have been the phytoplankton. In this sense Butterfield (1997) proposed that the first significant shift in phytoplankton diversity was therefore the rapid radiation of small acanthomorphic acritarchs in the Early Cambrian, as well as the occurrence of a 'trophic cascade scenario' accountable for a strong top-down ecological motor supporting the Cambrian radiation. The Proterozoic-Phanerozoic transition is noted for secular fluctuations in the $\delta^{13}C$ values of sedimentary carbon as a consequence of differential carbon burial and/or net primary productivity presumably linked to major biotic events (Brasier et al. 1994).

Secular changes in seawater chemistry also appear to have influenced phytoplankton evolutionary trajectories, i.e., strongly redox state dependent trace elements (Fe, Cu, Zn, Mn) played essential roles in mediating critical biochemical reactions in all phytoplankton (Katz et al. 2004). In addition, Quigg et al. (2003) reported that historical changes in the redox state of the ocean had a critical role in determining the evolutionary trajectory of different heterogeneous groups of photosynthetic organisms, like dinoflagellates among others.

The Precambrian-Cambrian boundary was a time of dramatic changes in global environmental conditions, including glaciations (even extended to tropical and equatorial latitudes), intense carbonates depositions, and significant modifications in the chemistry of ocean waters (Dalziel 1997). Sea level globally rose during Cambrian time, starting the first order eustatic cycle which developed along the whole Paleozoic Era. It is necessary to fully consider the ever-changing distribution of continents and ocean basins to properly understand that Earth's environment (Fig. 1.4).

Strong changes in carbonate mineralogy have been reported across the Precambrian-Cambrian transition. It could be supported on two geochemical hypothesis: increasing ratio of Ca:Mg ions in seawater or increasing pCO_2 due to high volcanic activity. Major episodes of evaporation close to the Precambrian-Cambrian and Lower-Middle Cambrian boundaries were also recorded. Warm, saline bottom waters could also explain numerous indications of anoxia in latest Precambrian and Cambrian strata. During rising sea levels, such waters may result in oxygen depletion, leading to the formation of black shales (Brassier 1992).

It has been shown that glacial conditions were widespread in the late Precambrian Varangian epoch. The change towards 'greenhouse' conditions possibly began with the development of deep, saline bottom waters, produced in evaporitic rift basins. Different transgressions brought

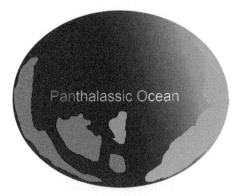

Figure 1.4. Oceans distribution along Paleozoic era: the dominant Panthalassic Ocean.

salinity stratified waters onto the shelf, laying down extensive metalliferous black shales and phosphorites. Reduced rates of nutrient recycling during transgressions may account for lowered primary productivity in surface waters. Briefly, and in the words by Brassier (1992) *"the Precambrian-Cambrian boundary interval perhaps indicates what happens when a salinity-stratified ocean is perturbed by climatic change and/or sea level rise; a rich broth of nutrients flooded the carbonate platforms and eutrophication arguably led to a huge but temporary rise in the biomass of primary producers"*. In this sense, it is important to keep in mind that cycles of changing sea level with very different periodicity have been recognized during the whole Phanerozoic: from the 100–200 m.y. cycles of Sloss (1963; 1972) up to the oscillations during the Pleistocene which may have been as short as 10,000 years (McKerrow 1979).

Links between the biogeochemical cycles of carbon and sulfur are expressed in the evolving stable isotope composition of the ocean. Carbonate rocks record the inorganic carbon isotope composition of the oceanic reservoir through geological time, along with the sulfate sulfur isotope composition preserved as Carbonate-Associated Sulfate (CAS). Throughout the Phanerozoic there is a general first order inverse relationship between the carbon and sulfur isotope records (Veizer et al. 1980). This relationship has been linked to the mass-balance between the oxidized and reduced reservoirs of both elements. Removal of sulfur from the ocean occurs through two major pathways: precipitation of sulfate minerals during evaporite deposition and the burial of pyrite. The biogeochemical cycles of sulfur and carbon are coupled through a network of input and output fluxes that are linked to the environmental conditions. The carbon and sulfur isotope records of the Paleozoic track an evolving oceanic reservoir and atmosphere. A progressive decoupling of the short-term carbon and sulfur isotope systems over the duration of the Paleozoic may have recorded an increasing oceanic sulfur reservoir against a backdrop of generally low DIC in the Paleozoic ocean (Gill et al. 2007). $\delta^{13}C$ data for Palaeozoic carbonates of the Great Basin, USA showed to be extremely 'spikey' (Saltzman 2005), suggesting that between 299 and 513 Myr ago (Ma) the rate of organic carbon burial varied rapidly in this part of the oceans and possibly in the world ocean.

Widespread anoxia in the ocean is frequently invoked as a primary driver of mass extinction as well as a long-term inhibitor of evolutionary radiation on early Earth. Gill et al. (2011) reported a large and rapid excursion in the marine carbon isotope record (SPICE), which is recognized as indicative of a global carbon cycle perturbation (Saltzman et al. 2000; 2004). These results identify the SPICE interval as the best characterized ocean anoxic event in the pre-Mesozoic ocean and an extreme example of oxygen deficiency in the later Cambrian ocean. Consequently, the environmental challenges presented by widespread anoxia may have been a dominant influence on animal evolution in Cambrian oceans. If anoxic water masses occurred widely in the subsurface of the later Cambrian ocean the high rates of biological turnover (Bambach et al. 2004) and repeated trilobite extinctions documented for later Cambrian fossils can be at least partially explained by episodic expansion of oxygen-depleted waters (Gill et al. 2011). In this way, broad patterns of Cambrian animal evolution may reflect persistent oxygen deficiency in subsurface waters of Cambrian oceans. Black shales are common in Proterozoic marine successions (particularly during much of the Paleozoic; Arthur and Sageman 1994), which indicates oxygen depletion beneath surface water masses in the oxygen-minimum zone (Shen et al. 2002; 2003). Deep-water anoxia may have been particularly pronounced near the end of the Permian (Isozaki 1997b). Ocean anoxia increases the availability of Fe, Mn, P, and NH_4^+, and decreases the availability of Cd, Cu, Mo, Zn, and NO_3^-, favoring some phytoplankton lineages over others when subsurface reducing conditions prevailed (Katz et al. 2004).

A rise in atmospheric O_2 has been linked to the Cambrian explosion of life. For the plankton and animal adaptive radiation that began some 40 Myr later and continued through much of the Ordovician (*Great Ordovician Biodiversification Event*), the search for an environmental trigger has remained elusive. Carbon and sulfur isotope mass balance model for the latest Cambrian time interval spanning the globally recognized Steptoean Positive Carbon Isotope Excursion (SPICE), indicating a major increase in atmospheric O_2. The SPICE is followed by an increase in plankton diversity that may have been related to changes in macro- and micronutrient abundances in increasingly oxic marine environments, representing a critical initial step in the trophic chain (Saltzman et al. 2011). Highest phytoplankton diversity of the Palaeozoic presumably occurred due to palaeogeography (greatest continental dispersal) and major orogenic and volcanic activity, which provided maximum ecospace and large amounts of nutrients. With its warm climate and high atmospheric CO_2 levels, the Ordovician was a period when phytoplankton diversity was at its maximum during the Paleozoic. With increased phytoplankton availability in the Late Cambrian and Ordovician a radiation of zooplanktonic organisms took place at the same time as a major diversification of suspension feeders (Servais et al. 2008).

The first suggestion that a wide ocean was present in the early Palaeozoic between America and Europe (Wilson 1966) was based primarily on faunal differences; this ocean was named Iapetus (Harland and Gayer 1972) after the father of Atlas (from whom the Atlantic Ocean takes its name) (McKerrow 1988). During the Late Proterozoic and Early Palaeozoic the supercontinent Rodinia broke up, in a process that was the origin of Iapetus Ocean which emerged by continental rifting, progressing to seafloor spreading in approximately the same position as the Atlantic Ocean (which will be formed later during the Early Tertiary) (Fig. 1.4). This process could have occurred in a short period around 620–605 Myr (Svenningsen 2001; McCausland et al. 2007). This continental separation may have played a role—together with the ecological drivers that govern biological diversity—in promoting Cambro-Ordovician phytoplankton and marine invertebrate expansion (Katz et al. 2004). In fact, the Iapetus Ocean has been described as a paleogeographic key change which aided the diversification of life (Grotzinger et al. 1995). McKerrow (1988) have steadily proposed that Iapetus Ocean did not close finally until the Silurian (in Greenland and Norway) or the Devonian (in the northern Appalachians), but many exposed ophiolites have been dated as Cambrian or Ordovician; clearly there were different parts of Iapetus which 'closed' at different times, and many of the Ordovician closures were related to marginal basins.

High concentrations of vanadium, molybdenum, uranium, arsenic, antimony with low concentrations of manganese, iron and cobalt were recorded in coeval black shales in the Saint John, New Brunswick area. Gee (1981) and Wilde et al. (1989) suggested that this geochemical signature is a feature of eastern Iapetus on the basis of these results as well as on palaeo-oceanographic reconstructions of the mentioned ocean. The described distinctive geochemical signature resulted from the coincidence of anoxic waters transgressing the shelf at latitudes of high organic productivity at the polar Ekman planetary divergence. V, U and Mo were concentrated in the shales within these conditions. Furthermore, metal enriched anoxic bottom waters produced either by leaching of volcanic or through hydrothermal activity may presumably be the source of the other enhanced signature elements such as As and Sb (Wilde et al. 1989). The major Cambrian transgression of ocean water across the continents was probably driven by the tectonic processes responsible for the fragmentation of Pannotia, which would include a significant decrease in the generally rising $^{87}Sr/^{86}Sr$ ratio of global seawater, pointing out hydrothermal activity along the newly formed mid-Iapetus spreading ridge (Dalziel 1997).

Wilde et al. (1989) described a Primary Productivity (PP) cycle with different support processes within Iapetus Ocean, and could briefly be summarized as follows: PP would be enhanced in the vicinity of 0° and 60°S due to upwelling (Ekman pumping) or P, N and Si-rich waters along the major planetary Ekman divergence zone. Upwelling also would have been produced by entrainment and off shore advection because of the equatorward flow at the eastern meridional boundary. At midlatitudes, the eastern shore of Iapetus would have had seasonal spring-summer upwelling caused by Ekman Transport produced by seasonal non-zonal winds. Such seasonal upwelling, occurring during maximum insolation, would have extended the enhanced productivity northward from the main divergence at 60°S.

Another Paleozoic ocean which deserves to be mentioned and briefly characterized is the Panthalassic Ocean (Fig. 1.4), which encompassed more than half the Earth's surface at 252 Myr. An upward transition from gray organic-poor cherts to black siliceous mudstones at both sites occurred in conjunction with increased primary productivity, intensified euxinia within the Oxygen-Minimum Zone (OMZ), and decimation of the radiolarian zooplankton community. Euxinia in the OMZ of the equatorial Panthalassic Ocean developed episodically for a ~200–250 kyr interval during the Late Permian, followed by an abrupt intensification and lateral expansion of the OMZ around the Permian–Triassic boundary. Throughout the study interval, bottom waters at both sites remained mostly suboxic, a finding that counters hypotheses of development of a "superanoxic" Permo-Triassic deep ocean as a consequence of stagnation of oceanic overturning circulation (Algeo et al. 2011). The Panthalassic Ocean, covering some 70% of the globe, was the largest surficial feature of the Permian and Triassic world, and had the potential to exercise fundamental control of global biogeochemistry (Schoepfer et al. 2012). Latest Permian deposits display distinctive enrichments in ^{15}N and $^{13}C_{org}$, suggesting a period of elevated productivity and anoxia before the final termination of the western Panthalassic upwelling zone, the loss of benthic siliceous sponges, and the Permian/Triassic extinction. This productivity spike may have been due to the synergistic effects of nutrient upwelling and rapid warming and may have contributed to photic zone euxinia by increasing biological oxygen demand.

The Permian/Triassic Boundary (PTB) mass extinction at ~252 Ma was the largest biotic catastrophe of the Phanerozoic Eon, resulting in the disappearance of ~90% of marine species as well as a large fraction of terrestrial taxa (Bambach et al. 2004). The cause of this event has long been debated, and various hypotheses have been proposed: i.e., a meteorite impact, flood basalt volcanism, global oceanic anoxia, and long-term climate change (Hallam and Wignall 1997; Wignall 2007). Lately it has been demonstrated that many shallow-marine platforms of Late Permian and Early Triassic age experienced euxinic conditions, i.e., a lack of dissolved oxygen along with free H_2S in the water column (Grice et al. 2005; Riccardi et al. 2006; Algeo et al. 2008). Environmental conditions in the larger Panthalassic Ocean, comprising 85–90% of the area of the Permian–Triassic global ocean, are fully defined on the basis of the only surviving marine strata from the Panthalassic Ocean now located within accretionary terranes in Japan, New Zealand and western Canada (Kojima 1989; Isozaki 1997a). Results from these studies on Permo-Early Triassic Panthalassic structures allowed recording a transient interval of environmentally hostile conditions (Sano and Nakashima 1997; Musashi et al. 2001).

Environmental conditions in the deep Panthalassic Ocean during the Permo-Triassic have included controversial interpretations of the evidences. On the one hand, strongly reducing (probably euxinic) conditions in Panthalassic bottom waters have been described (Suzuki et al. 1998; Matsuo et al. 2003), and such results have been used to infer widespread deep-ocean anoxia during the Late Permian to Early Triassic interval (Wignall and Twitchett 1996; 2002). On the other hand, the persistence and duration of deep-water anoxia recorded by Japanese abyssal sections on the basis of ichnofabric and other data have been questioned by several authors (Kakuwa 2008; Algeo et al. 2010).

Summarizing, (1) major changes in marine primary productivity rates and plankton community composition in conjunction with the PTB boundary crisis, were reported and (2) the most pronounced changes in redox conditions of the Panthalassic Ocean occurred within the OMZ rather than in the deep ocean (Algeo et al. 2011).

The last particular type of marine environments characterizing the Paleozoic was the epeiric seas, which were extensively distributed along most palaeocontinents (Fig. 1.5) (Ziegler et al. 1977).

Epicontinental or epeiric seas represent shallow oceanic bodies resulting from the flooding of continental interiors. Their geographic position and temporal distribution is dominantly controlled by two major factors: sea level and continental elevation. Epeireic seas are virtually absent at the present time not only because the continental regions in today's world are generally quite elevated, but also due to the fact that current tectonic activity is fairly slow (Harries 2009).

Although the flooded continental area is a small percentage of the total oceanic area suitable for phytoplankton, the shallow seas appear to have contributed proportionally more to niche space because of high nutrient input, high rates of primary production, and habitat heterogeneity.

Figure 1.5. Distribution of Epeireic seas during Paleozoic.

Flooded continental area provides variable, high-nutrient habitat by creating additional upwelling zones and increasing turbulence and nutrient suspension from below the thermocline. In addition, terrestrial nutrient input likely increases because nutrients that were previously sequestered in the large supercontinent interior are more readily transported to the newly opened, nearby oceans (Katz et al. 2004).

The Permian–Triassic Boundary (PTB: ~251.0–253.0 Myr) is known as a period when the most profound collapse both of marine and terrestrial ecosystems and the global environmental devastation occurred (Erwin 1994). Palaeontological studies on fossil records have showed that approximately 90% of marine and terrestrial species became extinct at the end of the Permian (Yin and Song 2013; Clarkson et al. 2015), being the most abrupt biotic turnover of the Phanerozoic (Chen and Benton 2012; Payne and Clapham 2012). Sano et al. (2012) in an extensive review have reported numerous causes which could have been the promoters of this catastrophe: i.e., changes in crustal weathering patterns (Kidder and Worsley 2004), changes in paleosoils in non-marine sections (Retallack et al. 2003), aridification, vegetation loss, and fluvial drainage alteration, greenhouse conditions and warming of the global ocean due to massive volcanic eruptions and increased atmospheric CO_2, as well as possible methane clathrate releases (Benton and Twitchett 2003; Winguth et al. 2002). There is also growing evidence that between the Middle Permian and Early Jurassic the oxygen content of the Earth's atmosphere may have varied markedly with important biological consequences (Berner and Kothavala 2001; Berner 2004; Retallack et al. 2003; Huey and Ward 2005). Dramatic changes in inorganic and organic $\delta^{13}C$ records characterize Permian/Triassic Boundary (PTB) geochemical records in the available rock record (Retallack et al. 2003; Payne et al. 2004). Negative shifts have been explained by addition of ^{13}C-depleted CO_2 to the global atmosphere–ocean reservoir by catastrophic methane release and oxidation (Retallack et al. 2003), exposure and oxidation of buried organic matter (Berner 2004), anoxic ocean overturn (Knoll et al. 1996), and CO_2 release during massive volcanism (Erwin et al. 2002; White 2002). Positive shifts in the organic $\delta^{13}C$ of marine sediments, which have been observed in other Panthalassic deep water sections (Isozaki 2009) have proven more difficult to explain, potentially reflecting both changes in the productivity and composition of marine planktonic communities (Butterfield 1997; Katz et al. 2004), and variation in the degree to which various organic components (such as proteins) are preferentially degraded in a stratified versus pervasively anoxic ocean (Isozaki 2009). As primary productivity and associated biological oxygen demand may have fueled Permian marine anoxia, it is important to understand the interplay between Panthalassic biology and oxygen availability (Schoepfer et al. 2012).

After this environmental cataclysm as occurred within the Permian-Triassic Boundary (PTB) a new scenario is coming for the evolving Earth planet. ...

A new oceanographical, biogeochemical and biological outlook within Mesozoic Oceans (248–65 My)

The Phanerozoic has presented secular oscillation in eustatic sea-level with levels of 300 m, showing highest sea-level at Early-to-Mid Paleozoic, which produced flooding extents of 20–50% of total cratonic area (Ridgwell 2005). Particularly during Mesozoic a different Earth landscape had appeared, which included the break-up of megacontinents and the occurrence (in both, temporary or permanent ways) of different scales seas and oceans. Regional oceans like Tethys (and its different stages: Palaeo Tethys, Neo Tethys), pelagic epeiric or epicontinental facies (large intra-continental water bodies produced at transgression peak), as well as the rising new oceans (Atlantic, Pacific, Indian) should clearly be included within this pack (Fig. 1.6).

A strong set of evidences exist which indicates that Mesozoic ocean water temperature was generally higher than at present, particularly in identified mid- to high-latitudes environments (Hudson and Anderson 1989). These palaeo-temperature studies have applied both the oxygen isotope method (measurement of the $^{18}O/^{16}O$ ratio on calcium carbonate- or calcium phosphate-, considering this precipitation is temperature dependent) (Buchardt 1978), as well as corresponding biotic distribution.

When the oceanic isotopic record of the Mesozoic was carefully analyzed it could be observed that a recognized warmth and homogeneously

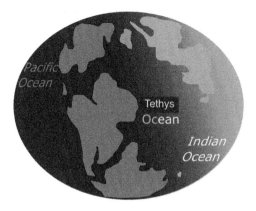

Figure 1.6. Oceans distribution along Mesozoic era: the central position of Tethys Ocean, and the rising Pacific and Indian Oceans.

distributed Cretaceous oceanic climates, indicating a mid-Cretaceous thermal maximum which declined towards the Campanian-early Maastrichtian (Hudson and Anderson 1989). In addition, the isotopic compositions suggest that earlier Mesozoic oceans were not greatly different from those of the Cretaceous.

The new rising Mesozoic oceans (i.e., Atlantic, Pacific, Indian) had received an important impact on their chemical and isotopic composition and evolution, showing that their deep waters were chemically and thermally rejuvenated owing to hot spots, seafloor hydrothermal vents and mid-ocean ridge hot springs activity (Bowers et al. 1985), continued apace during Mesozoic (De Ronde et al. 1997).

On the other hand, a chain of events occurred along Mesozoic which determined particular conditions characterizing this time. Therefore oceans received large deposition of organic matter linked to flooding of land-masses which transported much terrestrial plant material seawards. This fact has stimulated two processes: the production and widespread of marine plankton, as well as bacterial consumption of this organic matter (Jenkyns 1980). These scenarios favored the development of poorly oxygenated mid- to late Cretaceous waters, and simultaneously the sedimentary record pointed out very low oxygen concentrations in much of the world's oceans at those times (Leckie et al. 2002). These periods, so-called Oceanic Anoxic Events (OAEs), presented global marine waters relatively depleted in oxygen, and with high deposition of organic matter derived from both terrestrial and planktonic sources (Jenkyns 2010), and were one of the characteristics of Mesozoic oceans (Pancost et al. 2004). In addition, the described characterization of these OAEs has also been endorsed through the marine isotope geochemistry of molybdenum (Mo) which strongly suggested near total anoxia during those Mesozoic events (Archer and Vance 2008).

Several studies have reported significant differences in the nature of the two major OAE's within Mesozoic oceans stages; i.e., Arthur and Natland (1979) informed a significant terrestrial component in the organic material from Aptian-Albian levels, as well as a more enriched planktonic one from Cenomanian-Turonian age. Jenkyns (2010) proposed that these two OAEs correlated with equivalent climates/transgressive pulses periods, which agreed within their occurrence. In the modern oceans (since Cenozoic and up to present) areas of high organic productivity are predominantly localized within Mediterranean, marginal and shelf seas, and the ratio of the surface of such seas to the global ocean surface is ~1:30; nevertheless, during the peak of the Cretaceous transgression this ratio was at least 1:6 (Schlanger and Jenkyns 1976; Jenkyns 1980; 2010). Thus, assuming no limiting factors such as primary nutrient supply from restricted landmasses, it is likely that at that time the total amount of organic carbon produced in the ocean had been vastly greater than it is today. It has been assumed that the late Cretaceous epeiric and marginal seas had produced enough organic material as to generate, via bacterial oxidation, significant sources of oxygen-depleted waters. So, they were generally laid down in environments sufficiently shallow to feel the effect of algal photosynthesis and interchange with atmospheric oxygen. Summarizing, the general picture is of fertile marginal and epeiric seas, produced by marine transgressions, acting as oxygen sinks through bacterial consumption of organic matter, and delivering tongues of anoxic waters spreading out from these shelf seas across continental margins into ocean basins (Jenkyns 2010).

Another important instance in the interpretation of palaeo-environmental conditions of the oceans is the consideration of the carbon cycle. The relationships between the global carbon cycle and palaeo-climates on short and long time scales have been based on studies of accumulation rate of the two main components of the sedimentary carbon reservoir: organic carbon and carbonate carbon; variations in the rate and proportion of carbonate burial through Phanerozoic time have been attributed to the effects of tectonics on eustasy, atmospheric CO_2 concentration, Mid-Ocean Ridge (MOR) hydrothermal flux, and weathering and riverine flux (Locklair and Lerman 2005). The long-term carbon cycle, which characteristic times are $\sim 10^8$–10^9 years, involves a set of concatenated steps: i.e., accumulation of oceanic sediments, their partial subduction and partial incorporation in the continental crust, volcanic and metamorphic degassing, continental uplift, erosion and chemical weathering (Berner 2003). Oceanic reservoir of carbon and environmental conditions in that reservoir are important components of long-term carbon cycling.

The Tertiary is characterized by relatively strong mechanical erosion of the continents and deposition of $CaCO_3$ formed to a large extent by planktonic and benthonic oceanic organisms, and coral reefs and possibly calcareous algae on continental shelves (Berner 2004; Berner and Kothavala 2001). Moreover, an extraordinary change had occurred during Mesozoic with the proliferation of planktonic calcifiers, in the so called 'Mid-Mesozoic Revolution' when the modern mode of carbonate cycling had started (Ridgwell 2005). The biologically-driven $CaCO_3$ depositional 'mode' changes along with geochemical and tectonic variations in boundary conditions such as sea-level and calcium ion concentrations all affected the carbonate chemistry of the ocean. The available data indicated that Mesozoic oceans' alkalinity would have been to be greater (by a factor of 2.5–3.5) than that of present ones (Erba 2004). Locklair and Lerman (2005) proposed that the increase in surface ocean alkalinity were more linked to weathering and runoff, presumably due to a correlative increase in riverine transport of major ionic species, instead of other typical processes conditioning it (i.e., interaction between ocean water and mid-ocean ridge system). In addition, the results also pointed out that changes in the concentration of atmospheric CO_2 and oceanic Ca^{2+} determined the surface pH (Ridgwell 2005). Precisely this author indicated that the Mid Mesozoic Revolution heralded the development of a responsive deep-sea carbonate sink that introduced a new and powerful negative feedback to the Earth system.

On the other hand, Roth (1987) reported that major anoxic events coincided with high sea level stands while times of high extinction rates and thus high turnover rates coincided with times of major regressions along Mesozoic. This could be sustained by facts linked to changes in the sea level, whereas they involve variations in significant environmental conditions (i.e., flooding surface, occurrence of thermal gradients, latent heat transport, circulation/ventilation/stratification of ocean waters) which govern stability and mixing rate within the oceans.

These new scenarios might have allowed that during Mesozoic, the three principal phytoplankton clades that would have come to dominate the modern seas (dinoflagellates, coccolithophores and diatoms) fully consolidated their ecological prominence (Falkowski et al. 2004). It is a central highlighted point considering—even phytoplankton represents less than 1% of the Earth's photosynthetic biomass—these organisms are responsible for more than 45% of our planet's annual net primary production (Ito 2011). The Mesozoic adaptive radiation of dinoflagellates, coccolithophores and diatoms paralleled a long-term increase in sea level with an accompanying expansion of flooded continental shelf area (Falkowski et al. 2004). The diversity trends of the phytoplankton in the Mesozoic indicate that the increase was relatively slow, and a maximum was only reached after about hundred million years (Servais et al. 2008). Several authors have commented that a bolide impact at the Cretaceous/Tertiary boundary would have removed a major part of the phytoplankton diversity, which would be followed by a recovery of the coccolithophores and dinoflagellates up to the Eocene, before the long-term decline that continues to the present (Falkowski et al. 2004).

Towards the end of the Cretaceous (~70–72 Myr) multiple large environmental changes in both climate and ocean chemistry had occurred, creating a highly variable environment several millions of years prior to the Cretaceous-Tertiary (K-T) boundary (Barrera 1994). The $\delta^{18}O$ results suggest accelerated cooling, resulting in the lowest marine temperatures of the Late Cretaceous. Carbon isotopic ratios from planktonic and benthic microfossils across the K-T boundary reveal not only a breakdown in the normal surface-water to deep-water gradient of $^{13}C/^{12}C$, but also a reversal at the boundary (Ivany and Salawitch 1993).

A $^{87}Sr/^{86}Sr$ spike in seawater strontium with amplitude 2×10^{-4} and duration of order 2 Myr is superimposed on longer-term variations at the Cretaceous-Tertiary boundary, anomaly which has been attributed to increased continental runoff due either to meteorite impact-related acid rain or sea-level regression (MacDougall 1988), even though other authors (i.e., Javoy and Courtillot 1989) also included explosive acid volcanism preceding the development of the Deccan traps as partially responsible of that spike.

The shape of the Sr isotope curve near the KTB should constitute a tool for distinguishing between rapid vs. gradual causes for the KT extinctions. That is climate change, sea-level regression or a period of intense volcanism (specifically Deccan volcanism), are the most popular gradual extinction mechanisms (Officer et al. 1987; Vajda et al. 2001; Keller et al. 2002). In contrast, a bolide impact (i.e., Schulte et al. 2010) would have resulted in geologically instantaneous environmental destruction. If the Sr isotope change is related to any of these processes, it should be possible to distinguish the bolide case from the others due to its very short time scale, even if the volcanism occurred over as short a time interval as 0.5 Myr as has been proposed for the Deccan basalt eruptions (Hooper et al. 2010).

Whether this environmental crisis was caused by one of the aforementioned factors or by the sum of them all, a completely different scenario was generated for the following historical instance. ...

The Cenozoic Oceans: the gateway to present ones

Changes in oceans conditions and configuration as occurred between 55 and 14 Ma (including opening and closures of interoceanic connections) are expected to have deeply modified the ocean thermohaline circulation, as well as the corresponding heat transport (Bice et al. 2000), providing a similar look to the current (Fig. 1.7). These changes would have been recorded from early Eocene through middle Miocene oceans scenario. Decreasing high latitude surface and deep ocean temperatures during the early Eocene through Pliocene interval are well-documented by both continental and marine records (i.e., Lear et al. 2000; Zachos et al. 2001).

Figure 1.7. Oceans distribution along Cenozoic era: the present oceans have already appeared.

By the early Oligocene, an extensive ice-sheet existed on Antarctica, which subsequently fluctuated (Zachos et al. 1993), with benthic and planktonic foraminiferal $\delta^{18}O$ values diverging since the early Oligocene, reflecting the development of a thermally well-stratified ocean (Wade and Pearson 2008). Since the late Cretaceous a global sea level decrease could have been recorded, presumably sustained by a combination of decreased global ocean ridge volume, ice-sheet construction and thermal contraction of sea water with cooling (Miller et al. 2005).

Long-term late Paleocene-Eocene $\delta^{18}O$ and $\delta^{13}C$ records suggest a comparatively stable water column structure with a deep mixed layer and less seasonal variability (Bralower et al. 1995). Peak sea surface temperatures of 24–25°C were inferred in the lowermost Eocene, and this warm interval had reached the end by a sharp cooling event of 4°C in the latest middle Eocene; vertical thermal gradients had decreased dramatically in the late Paleocene and had reached a minimum in the latest part of this period (Zachos et al. 2003). A strong $\delta^{18}O$ increase near the Eocene/Oligocene boundary was gathered, which indicated the occurrence of a sea level drop and cooler climates, and allowed to sustain the development of cooling and glaciation beginning about 32 Ma and reaching maximum levels 29 Ma (Lear et al. 2008). In addition, Miller and Fairbanks (1983) proposed that the increased $\delta^{13}C$ in North Atlantic benthonic foraminifera of earliest and latest Oligocene age reflected increased production of northern-source deep water which would have not accumulated large amounts of CO_2 derived from the degradation of organic matter, unlike Keigwin and Keller (1984) who suggested lowered $\delta^{13}C$ values in middle Oligocene time might have resulted from lowered sea level exposing organic matter trapped in estuaries and on continental shelves to erosion and oxidation. Which ever it is the right one, the change within carbon cycle on this period is evidenced (Corliss and Keigwin 2013).

The Paleogene (~65–23.7 Myr) was characterized by cooling high latitude temperatures and the development of greater latitudinal thermal contrast that eventually led to the predominantly glacial mode of the Neogene (~23.7–1.8 Myr). According to one model, during the Paleogene, the mode of deep water formation changed from predominantly in the low and mid latitudinal marginal seas, producing warm, saline bottom water, that was characteristic of the late Cretaceous, to predominantly in the high latitudinal areas, producing cold, dense bottom water, characteristic of the present time (Haq 1981). Paleogene avaliable data (i.e., planktonic biogeographic reports, oxygen-isotopic results) showed cooling events in middle Paleocene, middle Eocene and near the Eocene/Oligocene boundary, as well as a major warming event that culminated in peak warming during the early Eocene (Lear et al. 2000; Pearson and Palmer 2000). In the history of the oceans, the Paleogene interval is perceived as an intermediate phase, characterized by changing thermal patterns in the world ocean and a transition from a predominantly thermospheric circulation to predominantly thermohaline circulation, as well as from a non-glacial to a glacial mode. The late Paleocene global warming trend culminated in the early Eocene in the warmest interval of the entire Cenozoic, which has been the best possible scenario for the optimal spread and growth of temperate marine biota. Increased productivity led to high carbonate (and total sediment) accumulation rates in the later part of early Eocene and early middle Eocene (Gibbs et al. 2006). The Eocene/Oligocene (E/O) boundary event represents the most dramatic step-like cooling episode within the larger cooling trend on the Cenozoic causing wide-spread climatic deterioration (Zanazzi et al. 2007). The sharp drop in the bottom and surface water temperatures at the E/O boundary have produced adverse effects on marine biota, with strong declination on both the phyto- and zooplankton biodiversity groups (Ivany et al. 2000).

The change in the Cenozoic climate from relatively warm and certainly ice free in the Paleocene to massive ice-sheets in both the southern and northern hemispheres in the Pliocene has been ascribed to a number of causes, i.e., opening of circum-Antarctic seaways, declining of atmospheric CO_2, shallow seaways versus deep seaways, and development of a vigorous circum-Antarctic current may be mentioned among others (DeConto and Pollard 2003). Within this climate change framework deep-sea temperatures might have decreased from ~12°C during Eocene up to ~8°C along Oligocene, less than one million years later (Zachos et al. 2001). In the same way, atmospheric CO_2 concentration might have declined from 2000–>3000 ppm for the earliest Eocene (~52 Ma) to less than 800 ppm by 40 Ma (Pagani et al. 2005). In addition, during the Cenozoic a series of tectonic events impacted ocean circulation, i.e., the closure of a low-latitude Tethyan seaway to circulation, the mid-Cenozoic opening of a southern seaway around Antarctica, the middle Miocene restriction of the Southeast Asian seaway, and finally the closure of the Pliocene Panamanian seaway (Lawver and Gahagan 2003).

These systems were developed until the end of the Pliocene (~1.8 Ma), when the Quaternary Period (i.e., early Pleistocene) started. Within this time oceans distribution agreed with the modern (present) one, and so entered up the age that we are living.

Modern oceans: from Pleistocene up to the present

During the Quaternary (the last 1.8 Myr) the oceans with which we live at present (i.e., Atlantic, Pacific, Indian) were not only in the geographic position but also within the configuration as we recognize today (Fig. 1.8).

During this time thermohaline circulation demonstrated to be the dominant one, and included among the main drivers of the planetary

Figure 1.8. Quaternary/modern oceans distribution.

climate. In this sense, salinity has a profound influence on thermohaline circulation (Ekwurzel et al. 2001). Even though the distribution and behavior of salinity within modern oceans is relatively stable and with a quite bounded variation, its fluctuations along the Quaternary have been significant. In this sense, Dickson et al. (1988) reported this as the 'Great Salinity Anomaly' (GSA) at the northern North Atlantic; Schlosser et al. (1991) documented a large salinity reduction on Greenland Sea Deep Water. Finally, Steele and Boyd (1998) demonstrated a significant decrease in the extent of the Cold Halocline Layer in the Eurasian Basin.

By the way, the levels of dissolved oxygen within ocean water from the past can be used to improve the knowledge on the history of ocean circulation, climate variations and associated biological processes (Kaiho 1994). Oxygen Minimum Zones (OMZs) are potential traces of a primitive ocean within the present ones, which were systems where Archaean bacteria lived and where numerous reduced chemical anomalies were recorded (Paulmier and Ruiz Pino 2009). OMZs have been mainly known for playing an essential role in the global nitrogen cycle, including various chemical species according to their degree of oxidation (i.e., NH_4^+; NO_2^-; NO_3^-; N_2O; N_2), and are also involved in the cycle of very important climatic gases (i.e., production of the oceanic N_2O, of H_2S, of CH_4, among others) (Paulmier et al. 2006). Moreover, OMZs also have influence on the limitation of atmospheric CO_2 sequestration by the ocean (Falkowski 1997), as well as potential dimethyl-sulphonates DMS consumption due to higher bacterial activity (Kiene and Bates 1990). At present the deep ocean is well oxygenated due to a conveyor-belt circulation which carries cold, oxygen-rich waters from high latitudes to the distant reaches of the abyssal zone (Meyer and Kump 2008). Despite the efficient functioning of this system, its condition is less robust than it may be thought, i.e., half of the oxygen injected into the deep sea within the North Atlantic deep-water formation regions is consumed during the decomposition of organic detritus settled through the water column (Holland 2006). Unlike long intervals of Earth history when euxinic conditions may have been ubiquitous (Canfield 1998), the present ocean is strongly stratified but largely oxygenated (Jaccard et al. 2014).

Another interesting proxy applied to better understand the functioning of oceans is Neodymium (Nd), which is transported by water masses and marine particles. Dissolved Nd is removed from the surface water by scavenging, which is released back to seawater by remineralization in deeper waters (Tachikawa et al. 2003). Continental materials are the principal Nd sources to the ocean, and seawater Nd isotopic ratio ($^{143}Nd/^{144}Nd$) varies according to the water masses, which allows to characterize both the mixing of water masses and the dissolved/particulate exchanges, through the "dissolved/particulate exchange ratio" ($\varepsilon_{Nd(0)}$), which can be measured within seawater (Pearce et al. 2013).

The Quaternary is a period characterized by strong climate variability in relatively short periods, and this fact has also driven to variations in ocean conditions. So, alternation of cool and warm periods generated variations in circulation patterns, dissolved oxygen levels, changes in CO_2–$CO_3^=$ equilibrium, etc. By the way, periods of glacier growth ruled out the occurrence of slightly warm surface waters temperature, minimal ice drift, and increasing moisture load and salinity (Ruddiman et al. 1980). On the other hand, warm (deglaciation) periods direct to decrease in salinity due to large inputs of freshwater, decreasing surface waters temperature and DO levels, and maximal loss of ice to the oceans by calving (Rahmstorf 2002).

Once again the biological pump has been identified as the main regulator within the modulation of atmospheric pCO$_2$ over glacial-interglacial timescales, increasing the oceanic nitrate budget and biological production in surface waters as well as the pool of inorganic carbon stored in the glacial ocean, drawdowning glacial CO_2 levels (Ganeshram et al. 2000). At present, oceans are causing seawater acidification due to anthropogenic carbon dioxide accumulation which direct to lowering both the concentration of carbonate ions and the pH (Feely et al. 2004). This process decreases the saturation state of carbonate minerals within seawater, thus causing difficulty in precipitation of carbonate (including biological ones) (Erez 2003).

Anthropogenic emissions of carbon dioxide (i.e., primarily due to fossil fuel burning, land use change and cement manufacturing) have increased its atmospheric concentration by ~40% since the beginning of the industrial era. These CO_2 increased levels are causing changes in the radiative forcing of the Earth's climate (Zeebe 2012). In addition, ocean acidification has been recognized as the major impact of CO_2 emissions commonly described as the ongoing decrease in ocean pH (Orr et al. 2005). Zeebe (2012) reported that over the period from 1750 to 2000 the oceans have absorbed approximately one-third of the CO_2 the ocean's uptake of anthropogenic CO_2 emitted by humans; this absorption has caused a decrease of surface-ocean pH by ~0.1 units from ~8.2 to ~8.1. The pair 'increasing CO_2 and decreasing pH' determines the declining of carbonate ion (CO_3^{2-}) and the corresponding increase of bicarbonate ion (HCO_3^-), which strongly diminish the stability of calcium carbonate ($CaCO_3$), mineral which used many marine organisms to build up shells and skeletons (Bijma et al. 2002). Unlike the mentioned current anthropogenic disruption which generates a large and rapid carbon perturbation, the Holocene ocean carbonate chemistry appeared to be nearly constant, including its remarkable carbon cycle stability and absence of large variations or acidification events (Caldeira and Wickett 2003).

Furthermore, the study of variations in the ratio of magnesium to calcium (Mg/Ca) in fossil ostracodes from the deep North Atlantic allowed understanding that bottom water temperatures changed by 4.5°C during the last two 100,000-year glacial-to-interglacial climatic cycles within the Quaternary, determining that Mg/Ca ratios increase during interglacial stages and decrease during glacial stages (Dwyer et al. 1995). This result is quite significant considering that deep-ocean circulation affects the storage and transfer of heat and nutrients in the ocean, as well as atmospheric CO_2 (Sigman et al. 2010).

Finally, the Holocene (~11,500 yr B.P. to the present) has been characterized by many (no less than six), short periods of significant rapid climate change which included alternative conditions of polar cooling, tropical aridity, major atmospheric circulation changes or polar cooling accompanied by increased moisture in some parts of the tropics (Mayewski et al. 2004).

Conclusion

The present *status*, configuration and behavior of the Earth's oceans fully govern the scenario where life develops within this planet. This fact occurs not only through the structure and conditions of the own marine ecosystems and associated environments (i.e., coastal, oceanic, abyssal, estuaries, mangroves, salt marshes, etc.) but also due to ocean influence on global climate, and consequently on the whole planetary system. As it was reflected within the previous paragraphs, even extremely changeable, this scenario has been performed along the whole Earth history, and numerous of the well studied cases we know at present were strongly related to those past ocean conditions. The future evolution of the oceans is underway, and certainly in the next few million years the planet will have a different image, and changes the oceans generate will determine how life should continue on Earth.

References

Algeo, T.J., Shen, Y., Zhang, T., Lyons, T., Bates, S., Rowe, H. and Nguyen, T.K.T. 2008. Association of ^{34}S-depleted pyrite layers with negative carbonate δ^{13}C excursions at the Permian–Triassic boundary: evidence for upwelling of sulfidic deep ocean water masses. Geochemistry, Geophysics, Geosystems 9: Q04025.

Algeo, T.J., Hinnov, L., Moser, J., Maynard, J.B., Elswick, E., Kuwahara, K. and Sano, H. 2010. Changes in productivity and redox conditions in the Panthalassic Ocean during the latest Permian. Geology 38: 187–190.

Archer, C. and Vance, D. 2008. The isotopic signature of the global riverine molybdenum flux and anoxia in the ancient oceans. Nature Geoscience 1: 597–600.

Arthur, M. and Natland, J.H. 1979. Carbonaceous Sediments in the North and South Atlantic: the role of salinity in stable stratification of early cretaceous Basins. pp. 375–401. *In*: Talwani, M., Hay, W. and Ryan, W.B.F. (eds.). Deep Drilling Results in the Atlantic Ocean: Continental Margins and Paleoenvironment. Maurice Ewing Ser. 3, American Geophysical Union.

Arthur, M. and Sageman, B. 1994. Marine black shales: depositional mechanisms and environments of ancient deposits. Annual Review on Earth Planetary Science 22: 499–451.

Bambach, R.K., Knoll, A.H. and Wang, S.C. 2004. Origination, extinction, and mass depletions of marine diversity. Paleobiology 30: 522–542.

Barbier, E.B., Hacker, S.D., Kennedy, C., Koch, E.W., Stier, A.C. and Silliman, B.R. 2011. The value of estuarine and coastal ecosystem services. Ecological Monographs 81(2): 169–193.

Barrera, E. 1994. Global environmental changes preceding the Cretaceous-Tertiary boundary: Early-late Maastrichtian transition. Geology 22: 877–880.

Barth, T.F.W. 1968. The geochemical evolution of continental rocks: a model. pp. 587–597. *In*: Ahrens, L.H. (ed.). Origin and Distribution of the Elements, Pergamon Press, Oxford (UK).

Bates, N.R. and Peters, A.J. 2007. The contribution of atmospheric acid deposition to ocean acidification in the subtropical North Atlantic Ocean. Marine Chemistry 107: 547–558.

Benton, M.J. and Twitchett, R.J. 2003. How to kill (almost) all life: the end-Permian extinction event. Trends in Ecology & Evolution 18(7): 358–365.

Berkner, L.V. and Marshall, L.C. 1965. On the origin and rise of oxygen concentration in the Earth's atmosphere. Journal of the Atmospheric Sciences 22(3): 225–261.

Berner, R.A. and Kothavala, Z. 2001. GEOCARB III: a revised model of atmospheric CO_2 over Phanerozoic time. American Journal of Science 301: 182–204.

Berner, R.A. 2003. The long-term carbon cycle, fossil fuels and atmospheric composition. Nature 426: 323–326.

Berner, R.A. (ed.). 2004. The Phanerozoic Carbon Cycle: CO_2 and O_2. Oxford University Press, Oxford (UK), 150 pp. ISBN 0-19-517333-3.

Bice, K.L., Scotese, C.R., Seidov, D. and Barron, E.J. 2000. Quantifying the role of geographic change in Cenozoic ocean heat transport using uncoupled atmosphere and ocean models. Palaeogeography, Palaeoclimatology, Palaeoecology 161: 295–310.

Bijma, J., Hönisch, B. and Zeebe, R.E. 2002. Impact of the ocean carbonate chemistry on living foraminiferal shell weight: comment on "Carbonate ion concentration in glacial-age deep waters of the Caribbean Sea" by W.S. Broecker and E. Clark. Geochemistry, Geophysics, Geosystems 3(11): 1064.

Blake, R.E., Chang, S.J. and Lepland, A. 2010. Phosphate oxygen isotopic evidence for a temperate and biologically active Archaean ocean. Nature 464: 1029–1033.

Bowers, T.S., Von Damm, K.L. and Edmond, J.M. 1985. Chemical evolution of mid-ocean ridge hot springs. Geochimica et Cosmochimica Acta 49: 2239–2252.

Bralower, T.J., Zachos, J.C., Thomas, E., Parrow, M., Paull, C.K., Kelly, D.C., Premoli Silva, I., Sliter, W.V. and Lohmann, K.C. 1995. Late Paleocene to Eocene paleoceanography of the equatorial Pacific Ocean: stable isotopes recorded at Ocean Drilling Program Site 865, Allison Guyot. Paleoceanography 10(4): 841–886.

Brasier, M.D. 1992. Global ocean-atmosphere change across the Precambrian-Cambrian transition. Geological Magazine 129(2): 161–168.

Brasier, M.D., Rozanov, A.Y., Zhuravlev, A.Y., Corfield, R.M. and Derry, L.A. 1994. A carbon isotope reference scale for the Lower Cambrian succession in Siberia: report of IGCP Project 303. Geological Magazine 131: 767–783.

Brocks, J.J., Love, G.D., Summons, R.E., Knoll, A.H., Logan, G.A. and Bowden, S.A. 2005. Biomarker evidence for green and purple sulphur bacteria in a stratified Palaeoproterozoic sea. Nature 437: 866–870.

Buchardt, B. 1978. Oxygen isotope palaeotemperatures from the Tertiary period in the North Sea area. Nature 275: 121–123.

Butterfield, N.J. 1997. Plankton Ecology and the Proterozoic-Phanerozoic Transition. Paleobiology 23(2): 247–262.

Caldeira, K. and Wickett, M.E. 2003. Anthropogenic carbon and ocean pH. Nature 425: 365–367.

Canfield, D.E. and Teske, A. 1996. Late Proterozoic rise in atmospheric oxygen concentration inferred from phylogenetic and sulphur-isotope studies. Nature 382: 127–132.

Canfield, D.E. 1998. A new model for Proterozoic ocean chemistry. Nature 396: 450–453.

Canfield, D.E., Glazer, A.N. and Falkowski, P.G. 2010. The Evolution and Future of Earth's Nitrogen Cycle. Science 330: 192–196.

Chen, Z.-Q. and Benton, M.J. 2012. The timing and pattern of biotic recovery following the end-Permian mass extinction. Nature Geosciences 1475: 1–9.

Clarkson, M.O., Kasemann, S.A., Wood, R.A., Lenton, T.M., Daines, S.J., Richoz, S., Ohnemueller, F., Meixner, A., Poulton, S.W. and Tipper, E.T. 2015. Ocean acidification and the Permo-Triassic mass extinction. Science 348(6231): 229–232.

Cloud, P. 1972. A working model of the primitive Earth. American Journal of Sciences 272(6): 537–548.

Constanza, R., d'Arge, R., de Groot, R., Farber, S., Grasso, M., Hannon, B., Limburg, K., Naeem, S., O'Neill, R.V., Paruelo, J., Raskin, R.G., Sutton, P. and van den Belt, M. 1997. The value of the world's ecosystem services and natural capital. Nature 387: 253–260.

Cooley, S.R., Kite-Powell, H.L. and Doney, S.C. 2009. Ocean Acidification's Potential to Alter Global Marine Ecosystem Services. Oceanography 22(4): 172–181.

Corliss, B.H. and Keigwin Jr., L.D. 2013. Eocene-Oligocene paleoceanography. Mesozoic and Cenozoic Oceans, Geodynamics Series, Vol. 15: 101–118 (American Geophysical Union).

Dalziel, I.W.D. 1997. Neoproterozoic-Paleozoic geography and tectonics: Review, hypothesis, environmental speculation. Geological Society of America Bulletin 109(1): 16–42.

De La Rocha, C.L. 2006. In hot water. Nature 443: 920–921.

De Ronde, C.E.J., Channer, D.M.D., Faure, K., Bray, C.J. and Spooner, E.W.C. 1997. Fluid chemistry of Archaean seafloor hydrothermal vents: Implications for the composition of circa 3.2 Ga seawater. Geochimica et Cosmochimica Acta 61(19): 4025–4042.

Derry, L.A. and Jacobsen, S.B. 1990. The chemical evolution of Precambrian seawater: evidence from REEs in banded iron formations. Geochimica et Cosmochimica Acta 54: 2965–2977.

Dickson, R.R.J., Meincke, J., Maimberg, S.-A. and Lee, A.J. 1988. The "great salinity anomaly" in the northern North Atlantic 1968–1982. Progress in Oceanography 20: 103–151.

Dobretsov, N.L., Benin, N.A. and Buslov, M.N. 1995. Opening and tectonic evolution of the Paleo-Asian Ocean. International Geological Review 37: 335–360.

Doney, S.C., Lindsay, K., Caldeira, K., Campin, J.M., Drange, H., Dutay, J.C., Follows, M., Gao, Y., Gnanadesikan, A., Gruber, N., Ishida, A., Joos, F., Madec, G., Maier-Reimer, E., Marshall, J.C., Matear, R.J., Monfray, P., Mouchet, A., Najjar, R., Orr, J.C., Plattner, G.K., Sarmiento, J., Schlitzer, R., Slater, R., Totterdell, I.J., Weirig, M.F., Yamanaka, Y. and Yool, A. 2004. Evaluating global ocean carbon models: the importance of realistic physics. Global Biogeochemical Cycles 18(3): 3. GB3017.

Doney, S.C., Mahowald, N., Lima, I., Feely, R.A., Mackenzie, F.T., Lamarque, J.-F. and Rasch, P.J. 2007. The impact of anthropogenic atmospheric nitrogen and sulfur deposition on ocean acidification and the inorganic carbon system. Proceedings of the National Academy of Sciences of the United States of America 104(37): 14580–14585.

Donnadieu, Y., Goddéris, Y., Ramstein, G., Nédélec, A. and Meert, J. 2004. A "snowball Earth" climate triggered by continental break-up through changes in runoff. Nature 428: 303–306.

Dwyer, G.S., Cronin, T.M., Baker, P.A., Raymo, M.E., Buzas, J.S. and Correge, T. 1995. North Atlantic deepwater temperature change during Late Pliocene and Late Quaternary climatic cycles. Science 270: 1347–1351.

Ekwurzel, B., Schlosser, P., Mortlock, R.A. and Fairbanks, R.G. 2001. River runoff, sea ice meltwater, and Pacific water distribution and mean residence times in the Arctic Ocean. Journal of Geophysical Research 106(C5): 9075–9092.

Erba, E. 2004. Calcareous nannofossils and Mesozoic oceanic anoxic events. Marine Micropaleontology 52: 85–106.

Erez, J. 2003. The source of ions for biomineralization in foraminifera and their implications for paleoceanographic proxies. Reviews on Mineralogy & Geochemistry 54: 115–149.

Erwin, D.H. 1994. The Permo-Triassic extinction. Nature 367: 231–236.

Erwin, D.H., Bowring, S.A. and Jin, Y.G. 2002. End-Permian mass extinctions: a review. Boulder Geological Society of America Special Paper 356: 363–384.

Falkowski, P.G. 1997. Evolution of the nitrogen cycle and its influence on the biological sequestration of CO_2 in the ocean. Nature 387: 272–275.

Falkowski, P.G., Katz, M.E., Knoll, A.H., Quigg, A., Raven, J.A., Schofield, O. and Taylor, F.J.R. 2004. The evolution of modern eukaryotic phytoplankton. Science 305: 354–360.

Farquhar, J., Zerkle, A.L. and Bekker, A. 2011. Geological constraints on the origin of oxygenic photosynthesis. Photosynthesis Research 107: 11–36.

Feely, R.A., Sabine, C.L., Lee, K., Berelson, W., Kleypas, J., Fabry, V.J. and Millero, F.J. 2004. Impact of anthropogenic CO_2 on the $CaCO_3$ system in the ocean. Science 305: 362–366.

Fennel, K., Follows, M. and Falkowski, P.G. 2005. The co-evolution of the nitrogen, carbon and oxygen cycles in the Proterozoic Ocean. American Journal of Science 305: 526–545.

Fike, D. 2010. Earth's redox evolution. Nature Geosciences 3: 453–454.

Foriel, J., Philippot, P., Rey, P., Somogyi, A., Banks, D. and Ménez, B. 2004. Biological control of Cl/Br and low sulfate concentration in a 3.5-Gyr-old seawater from North Pole, Western Australia. Earth and Planetary Science Letters 228: 451–463.

Ganachaud, A. and Wunsch, C. 2000. Improved estimates of global ocean circulation, heat transport and mixing from hydrographic data. Nature 408(23): 453–457.

Ganeshram, R.S., Pedersen, T.F., Calvert, S.E., McNeill, G.W. and Fontugne, M.R. 2000. Glacial-interglacial variability in denitrification in the world's oceans: causes and consequences. Paleoceanography 15(4): 361–376.

Gaucher, E.A., Govindarajan, S. and Ganesh, O.K. 2008. Palaeotemperature trend for Precambrian life inferred from resurrected proteins. Nature 451: 704–708.

Gee, D.G. 1981. The *Dictyonema*-bearing phyllites at Nordaunevoll, eastern Trondelag, Norway. Norsk geologisk Tidsskrift 61: 93–95.

Genda, H. and Ikoma, M. 2008. Origin of the ocean on the Earth: early evolution of water D/H in a hydrogen-rich atmosphere. Icarus 194: 42–52.

Gibbs, S.J., Bralower, T.J., Bown, P.R., Zachos, J.C. and Bybel, L.M. 2006. Shelf and open-ocean calcareous phytoplankton assemblages across the Paleocene-Eocene Thermal Maximum: Implications for global productivity gradients. Geology 34(4): 233–236.

Gill, B.C., Lyons, T.W. and Saltzman, M.R. 2007. Parallel, high-resolution carbon and sulfur isotope records of the evolving Paleozoic marine sulfur reservoir. Palaeogeography, Palaeoclimatology, Palaeoecology 256: 156–173.

Gill, B.C., Lyons, T.W., Young, S.A., Kump, L.R., Knoll, A.H. and Saltzman, M.R. 2011. Geochemical evidence for widespread euxinia in the Later Cambrian ocean. Nature 469: 80–83.

Glikson, A.Y. 1972. Early Precambrian evidence of a primitive Ocean crust and island nuclei of sodic granite. Geological Society of America Bulletin 83: 3323–3344.

Godfrey, L.V. and Falkowski, P.G. 2009. The cycling and redox state of nitrogen in the Archaean ocean. Nature Geoscience 2: 725–729.

Grice, K., Cao, C., Love, G.D., Böttcher, M.E., Twitchett, R.J., Grosjean, E., Summons, R.E., Turgeon, S.C., Dunning, W. and Jin, Y. 2005. Photic zone euxinia during the Permian–Triassic superanoxic event. Science 307: 706–709.

Griffith, E.J., Ponnamperuma, C. and Gabel, N.W. 1977. Phosphorus, a key to life on the primitive Earth. Origins of Life 8: 71–85.

Grotzinger, J.P., Bowring, S.A., Saylor, B.Z. and Kaufman, A.J. 1995. Biostratigraphic and geochronologic constraints on early animal evolution. Science 270: 598–604.

Hallam, A. and Wignall, P.B. 1997. Mass Extinctions and Their Aftermath. Oxford University Press, Oxford (UK). 320 pp.

Halpern, B.S., Walbridge, S., Selkoe, K.A., Kappel, C.V., Micheli, F., D'Agrosa, C., Bruno, J.F., Casey, K.S., Ebert, C., Fox, H.E., Fujita, R., Heinemann, D., Lenihan, H.S., Madin, E.M.P., Perry, M.T., Selig, E.R., Spalding, M., Steneck, R. and Watson, R. 2008. A Global Map of Human Impact on Marine Ecosystems. Science 319: 948–952.

Hamade, T., Konhauser, K.O., Raiswell, R., Goldsmith, S. and Morris, R.C. 2003. Using Ge/Si ratios to decouple iron and silica fluxes in Precambrian banded iron formations. Geology 31(1): 35–38.

Haq, B.U. 1981. Paleogene paleoceanography: Early Cenozoic oceans revisited. Oceanologica Acta N°SP, Proceedings 26th International Geological Congress, Geology of oceans symposium, Paris (France), July 7–17, 1980, pp. 71–82.

Hardie, L.A. 2003. Secular variation in Precambrian seawater chemistry and the timing of Precambrian aragonite seas and calcite seas. Geology 31: 785–788.

Harland, W.B. and Gayer, R.A. 1972. The Arctic Caledonides and earlier oceans. Geological Magazine 109: 289–314.

Harries, P.J. 2009. Earth System: History and Natural Variability. Epeiric Seas: a continental extension of shelf biotas. Encyclopedia of Life Support Systems, Vol. IV, UNESCO Electronic Library (http://www.eolss.net.ebooklib).

Hernández-León, S., Franchy, G., Moyano, M., Menéndez, I., Schmoker, C. and Putzeys, S. 2010. Carbon sequestration and zooplankton lunar cycles: Could we be missing a major component of the biological pump? Limnology & Oceanography 55(6): 2503–2512.

Hoashi, M., Bevacqua, D.C., Otake, T., Watanabe, Y., Hickman, A.H., Utsunomiya, S. and Ohmoto, H. 2009. Primary haematite formation in an oxygenated sea 3.46 billion years ago. Nature Geoscience 2: 301–306.

Hoffman, P.F. and Schrag, D.P. 2002. The snowball Earth hypothesis: testing the limits of global change. Terra Nova 14: 129–155.

Holland, H.D. 2006. The oxygenation of the atmosphere and oceans. Philosophical Transactions of the Royal Society B 361: 903–915.

Hooper, P., Widdowson, M. and Kelley, S. 2010. Tectonic setting and timing of the final Deccan flood basalt eruptions. Geology 38(9): 839–842.

Hudson, J.D. and Anderson, T.F. 1989. Ocean temperatures and isotopic compositions through time. Transactions of the Royal Society of Edinburgh: Earth Sciences 80(3-4): 183–192.

Huey, R.B. and Ward, P.D. 2005. Hypoxia, global warming, and terrestrial Late Permian extinctions. Science 308: 398–401.

Isozaki, Y. 1997a. Jurassic accretion tectonics of Japan. The Island Arc 6: 25–51.

Isozaki, Y. 1997b. Permo-Triassic boundary superanoxia and stratified superocean: records from lost deep sea. Science 276: 235–238.

Isozaki, Y. 2009. Integrated "plume winter" scenario for the double-phased extinction during the Paleozoic–Mesozoic transition: The G-LB and P-TB events from a Panthalassan perspective. Journal of East Asian Earth Sciences 36: 459–480.

Ivany, L.C. and Salawitch, R.S. 1993. Carbon isotopic evidence for biomass burning at the K-T boundary. Geology 21: 487–490.

Ivany, L.C., Patterson, W.P. and Lohmann, K.C. 2000. Cooler winters as a possible cause of mass extinctions at the Eocene/Oligocene boundary. Nature 407: 887–890.

Ito, A. 2011. A historical meta-analysis of global terrestrial net primary productivity: are estimates converging? Global Change Biology 17: 3161–3175.

Jaccard, S.L., Galbraith, E.D., Frölicher, T.L. and Gruber, N. 2014. Ocean (de)oxygenation across the last deglaciation. Insights for the future. Oceanography 27(1): 26–35.

Javoy, M. and Courtillot, V. 1989. Intense acidic volcanism at the Cretaceous-Tertiary boundary. Earth and Planetary Science Letters 94: 409–416.

Jenkyns, H.C. 1980. Cretaceous anoxic events: from continents to oceans. Journal of the Geological Society 137: 171–188.

Jenkyns, H.C. 2010. Geochemistry of oceanic anoxic events. Geochemistry, Geophysics, Geosystems 11(3): Q03004.

Johnston, D.T., Poulton, S.W., Fralick, P.W., Wing, B.A., Canfield, D.E. and Farquhar, J. 2006. Evolution of the oceanic sulfur cycle at the end of the Paleoproterozoic. *In*: Canfield, D.E., Lyons, T.W. and Morse, J.W. (eds.). A Special Issue Dedicated to Robert A. Berner. Geochimica et Cosmochimica Acta 70: 5723–5739.

Kaiho, K. 1994. Benthic foraminiferal dissolved-oxygen index and dissolved-oxygen levels in the modern ocean. Geology 22: 719–722.

Kakuwa, Y. 2008. Evaluation of palaeo-oxygenation of the ocean bottom across the Permian–Triassic boundary. Global & Planetary Change 63: 40–56.

Karhu, J.A. and Holland, H.D. 1996. Carbon isotopes and the rise of atmospheric oxygen. Geology 24: 867–870.

Karl, D.M., Church, M.J., Dore, J.E., Letelier, R.M. and Mahaffey, C. 2012. Predictable and efficient carbon sequestration in the North Pacific Ocean supported by symbiotic nitrogen fixation. Proceedings of the National Academy of Sciences of the United States of America 109(6): 1842–1849.

Kasting, J.F., Pavlov, A.A. and Siefert, J.L. 2001. A coupled ecosystem–climate model for predicting the methane concentration in the Archaean atmosphere. Origins of Life & Evolution of Biosphere 31: 271–285.

Kato, Y., Yamaguchi, K.E. and Ohmoto, H. 2006. Rare earth elements in Precambrian banded iron formations: secular changes of Ce and Eu anomalies and evolution of atmospheric oxygen. pp. 269–289. *In*: Kesler, S.E. and Ohmoto, H. (eds.). Evolution of Early Earth's Atmosphere, Hydrosphere, and Biosphere—Constraints from Ore Deposits. Geological Society of America Memoir, Geological Society of America, Boulder, Colorado (USA).

Katz, M.E., Finkel, Z.V., Grzebyk, D., Knoll, A.H. and Falkowski, P.G. 2004. Evolutionary Trajectories and Biogeochemical Impacts of Marine Eukaryotic Phytoplankton. Annual Review of Ecology, Evolution, and Systematics 35: 523–556.

Keller, G., Adatte, T., Stinnesbeck, W., Luciani, V., Karoui-Yaakoub, N. and Zaghbib-Turki, D. 2002. Paleoecology of the Cretaceous-Tertiary mass extinction in planktonic foraminifera. Palaeogeography, Palaeoclimatology, Palaeoecology 178: 257–297.

Keigwin, L. and Keller, G. 1984. Middle Oligocene cooling from equatorial Pacific DSDP Site 77B. Geology 12: 16–19.

Kheraskova, T.N., Bush, V.A., Didenko, A.N. and Samygin, S.G. 2010. Breakup of Rodinia and Early Stages of Evolution of the Paleoasian Ocean. Geotectonics 44(1): 3–24.

Kidder, D.L. and Worsley, T.R. 2004. Causes and consequences of extreme Permo-Triassic warming to globally equable climate and relation to the Permo-Triassic extinction and recovery. Palaeogeography, Palaeoclimatology, Palaeoecology 203: 207–237.

Kiene, R.P. and Bates, T.S. 1990. Biological removal of dimethylsulphide from seawater. Nature 345: 702–705.

Knauth, L.P. 2005. Temperature and salinity history of the Precambrian ocean: implications for the course of microbial evolution. Palaeogeography, Palaeoclimatology, Palaeoecology 219: 53–69.

Knoll, A.H., Bambach, R.K., Canfield, D.E. and Grotzinger, J.P. 1996. Comparative Earth history and late permian mass extinction. Science 273(5274): 452–457.

Kojima, S. 1989. Mesozoic terrane accretion in northeast China, Sikhote–Alin and Japan regions. Palaeogeography, Palaeoclimatology, Palaeoecology 69: 213–232.

Komiya, T., Hirata, T., Kitajima, K., Yamamoto, S., Shibuya, T., Sawaki, Y., Ishikawa, T., Shu, D., Li, Y. and Han, J. 2008. Evolution of the composition of seawater through geologic time, and its influence on the evolution of life. Gondwana Research 14: 159–174.

Lam, P.J., Doney, S.C. and Bishop, J.K.B. 2011. The dynamic ocean biological pump: Insights from a global compilation of particulate organic carbon, $CaCO_3$, and opal concentration profiles from the mesopelagic. Global Biogeochemical Cycles 25: GB3009.

Lawver, L.A. and Gahagan, L.M. 2003. Evolution of Cenozoic seaways in the circum-Antarctic región. Palaeogeography, Palaeoclimatology, Palaeoecology 198: 11–37.

Lear, C.H., Elderfield, H. and Wilson, P.A. 2000. Cenozoic deep-sea temperatures and global ice volumes from Mg/Ca in benthic foraminiferal calcite. Science 287: 269–272.

Lear, C.H., Bailey, T.R., Pearson, P.N., Coxall, H.K. and Rosenthal, Y. 2008. Cooling and ice growth across the Eocene-Oligocene transition. Geology 36: 251–254.

Leckie, R.M., Bralower, T.J. and Cashman, R. 2002. Oceanic anoxic events and plankton evolution: Biotic response to tectonic forcing during the mid-Cretaceous. Paleoceanography 17(3): 1–29.

Lyons, T.W., Anbar, A.D., Severmann, S., Scott, C. and Gill, B.C. 2009. Tracking Euxinia in the Ancient Ocean: A Multiproxy Perspective and Proterozoic Case Study. Annual Review on Earth & Planetary Sciences 37: 507–534.

MacDougall, J.D. 1988. Seawater Strontium Isotopes, Acid Rain, and the Cretaceous-Tertiary Boundary. Science 239: 485–487.

Maruyama, Sh. 1994. Plume tectonics. Journal of the Geological Society of Japan 100: 24–34.

Mason, R.P. and Sheu, G.-R. 2002. Role of the ocean in the global mercury cycle. Global Biogeochemical Cycles 16(4): 1093–1107.

Matsuo, M., Kubo, K. and Isozaki, Y. 2003. Mössbauer spectroscopic study on characterization of iron in the Permian to Triassic deep-sea chert from Japan. Hyperfine Interaction (C)5: 435–438.

Mayewski, P.A., Rohling, E.E., Stager, J.C., Karlén, W., Maasch, K.A., Meeker, L.D., Meyerson, E.A., Gasse, F., van Kreveld, S., Holmgren, K., Thorp, J.L., Rosqvist, G., Rack, F., Staubwasser, M., Schneider, R.R. and Steig, E.J. 2004. Holocene climate variability. Quaternary Research 62: 243–255.

McCausland, P.J.A., Van der Voo, R. and Hall, R.C.M. 2007. Circum-Iapetus paleogeography of the Precambrian–Cambrian transition with a new paleomagnetic constraint from Laurentia. Precambrian Research 156: 125–152.

McKerrow, W.S. 1979. Ordovician and Silurian changes in sea level. Journal of the Geological Society of London 136: 137–145.

McKerrow, W.S. 1988. The development of the Iapetus Ocean from the Arenig to the Wenlock. pp. 405–412. *In*: Harris, A.L. and Fettes, D.J. (eds.). The Caledonian-Appalachian Orogen, Geological Society Special Publication N°. 38, Washington DC (USA).

McMenamin, M.A.S. and McMenamin, D.L.S. 1990. The Emergence of Animals: The Cambrian Breakthrough. Columbia University Press, New York (USA), 217 pp.

Meert, J.G. 2012. What's in a name? The Columbia (Paleopangaea/Nuna) supercontinent. Gondwana Research 21: 987–993.

Meyer, K.M. and Kump, L.R. 2008. Oceanic euxinia in Earth history: causes and consequences. Annual Review of Earth and Planetary Sciences 36: 251–288.

Millennium Ecosystem Assessment Board. 2005. Ecosystems and human well-being: current state and trends, Volume 2. Millennium Ecosystem Assessment Board, 47 pp.

Miller, K.G. and Fairbanks, R.G. 1983. Evidence for Oligocene-Middle Miocene abyssal circulation changes in the western North Atlantic. Nature 306: 250–253.

Miller, K.G., Kominz, M.A., Sugarman, P.J., Browning, J.V., Cramer, B.S., Wright, J.D., Christie-Blick, N., Mountain, G.S., Pekar, S.F. and Katz, M.E. 2005. The Phanerozoic record of global sea-level change. Science 310: 1293–1298.

Mottl, M.J., Seewald, J.S., Wheat, C.G., Tivey, M.K., Michael, P.J., Proskurowski, G., McCollom, T.M., Reeves, E., Sharkey, J., You, C.-F., Chan, L.-H. and Pichler, T. 2011. Chemistry of hot springs along the Eastern Lau Spreading Center. Geochimica et Cosmochimica Acta 75(4): 1013–1038.

Musashi, M., Isozaki, Y., Koike, T. and Kreulen, R. 2001. Stable carbon isotope signature in mid-Panthalassa shallow-water carbonates across the Permo-Triassic boundary: evidence for ^{13}C-depleted ocean. Earth & Planetary Science Letters 196: 9–20.

Nguyen, B.C., Bonsang, B. and Gaudry, A. 1983. The Role of the Ocean in the Global Atmospheric Sulfur Cycle. Journal of Geophysical Research 88(C15): 10903–10914.

Nutman, A.P. 2006. Antiquity of the Oceans and Continents. Elements 2: 223–227.

Officer, C.B., Hallam, A., Drake, C.L. and Devine, J.D. 1987. Late Cretaceous and paroxysmal Cretaceous/Tertiary extinctions. Nature 326: 143–149.

Ohmoto, H., Watanabe, Y. and Kumazawa, K. 2004. Evidence from massive siderite beds for a CO_2-rich atmosphere before approximately 1.8 billion years ago. Nature 429: 395–399.

Orr, J.C., Fabry, V.J., Aumont, O., Bopp, L., Doney, S.C., Gnanadesikan, A., Gruber, N., Ishida, A., Joos, F., Key, R.M., Maier-Reimer, E., Matear, R., Monfray, P., Mouchet, A., Feely, R.A., Lindsay, K., Najjar, R.G., Plattner, G.-K., Rodgers, K.B., Sabine, C.L., Sarmiento, J.L., Slater, R.D., Totterdell, I.J., Weirig, M.-F., Yamanaka, Y., Schlitzer, R. and Yool, A. 2005. Anthropogenic ocean acidification over the twenty-first century and its impact on calcifying organisms. Nature 437: 681–686.

Pagani, M., Zachos, J.C., Freeman, K.H., Tipple, B. and Bohaty, S. 2005. Marked decline in atmospheric carbon dioxide concentrations during the Paleogene. Science 309: 600–603.

Palumbi, S.R., Sandifer, P.A., Allan, J.D., Beck, M.W., Fautin, D.G., Fogarty, M.J., Halpern, B.S., Incze, L.S., Leong, J.-A., Norse, E., Stachowicz, J.J. and Wall, D.H. 2009. Managing for ocean biodiversity to sustain marine ecosystem services. Frontiers in Ecology and the Environment 7(4): 204–211.

Pancost, R.D., Crawford, N., Magness, S., Turner, A., Jenkyns, H.C. and Maxwell, J.R. 2004. Further evidence for the development of photic-zone euxinic conditions during Mesozoic oceanic anoxic events. Journal of the Geological Society, London 161: 353–364.

Paulmier, A., Ruiz-Pino, D., Garçon, V. and Farias, L. 2006. Maintaining of the East South Pacific oxygen minimum zone (OMZ) off Chile. Geophysical Research Letter 33: L20601.

Paulmier, A. and Ruiz-Pino, D. 2009. Oxygen minimum zones (OMZs) in the modern ocean. Progress in Oceanography 80: 113–128.

Pavlov, A.A. and Kasting, J.F. 2002. Mass-independent fractionation of sulfur isotopes in Archaean sediments: strong evidence for an anoxic Archaean atmosphere. Astrobiology 2: 27–41.

Payne, J.L., Lehrmann, D.J., Wei, J., Orchard, M.J., Schrag, D.P. and Knoll, A.H. 2004. Large perturbations of the carbon cycle during recovery from the end-Permian extinction. Science 305: 506–509.

Payne, J.L. and Clapham, M.E. 2012. End-Permian Mass Extinction in the Oceans: An Ancient Analog for the Twenty-First Century? Annual Review on Earth Planetary Sciences 40: 89–111.

Pearson, P.N. and Palmer, M.R. 2000. Atmospheric carbon dioxide concentrations over the past 60 million years. Nature 406: 695–699.

Pearce, C.R., Jones, M.T., Oelkers, E.H., Pradoux, C. and Jeandel, C. 2013. The effect of particulate dissolution on the neodymium (Nd) isotope and Rare Earth Element (REE) composition of seawater. Earth and Planetary Science Letters 369-370: 138–147.

Poulton, S.W., Fralick, P.W. and Canfield, D.E. 2004. The transition to a sulphidic ocean, 1.84 billion years ago. Nature 431: 173–177.

Quigg, A., Finkel, Z.V., Irwin, A.J., Rosenthal, Y., Ho, T.-Y., Reinfelder, J.R., Schofield, O., Morel, F.M.M. and Falkowski, P.G. 2003. The evolutionary inheritance of elemental stoichiometry in marine phytoplankton. Nature 425: 291–294.

Ramanathan, V. 1981. The role of Ocean-Atmosphere interactions in the CO_2 climate problem. Journal of the Atmospheric Sciences 38: 918–930.

Rahmstorf, S. 2002. Ocean circulation and climate during the past 120,000 years. Nature 419: 207–214.

Raymont, J.E.G. 2014. Plankton and productivity in the Oceans. Vol. 1: Phytoplankton. 2nd Edition, Pergamon Press, Oxford (UK), 482 pp. ISBN: 0-08-021552-1.

Reinhard, C.T., Planavsky, N.J., Robbins, L.J., Partin, C.A., Gill, B.C., Lalonde, S.V., Bekker, A., Konhauser, K.O. and Lyons, T.W. 2013. Proterozoic ocean redox and biogeochemical stasis. Proceedings of the National Academy of Sciences 110(14): 5357–5362.

Retallack, G.J., Smith, R.M.H. and Ward, P.D. 2003. Vertebrate extinction across Permian–Triassic boundary in Karoo Basin, South Africa. Geological Society of America Bulletin 115: 1133–1152.

Riccardi, A., Arthur, M.A. and Kump, L.R. 2006. Sulfur isotopic evidence for chemocline upward excursions during the end-Permian mass extinction. Geochimica et Cosmochimica Acta 70: 5740–5752.

Ridgwell, A. 2005. A Mid Mesozoic Revolution in the regulation of ocean chemistry. Marine Geology 217: 339–357.

Rogers, J.J.W. and Santosh, M. 2009a. Configuration of Columbia, a Mesoproterozoic Supercontinent. Gondwana Research 5(1): 5–22.

Rogers, J.J.W. and Santosh, M. 2009b. Tectonics and surface effects of the supercontinent Columbia. Gondwana Research 15: 373–380.

Roth, P.H. 1987. Mesozoic calcareous nannofossil evolution: kelation to paleoceanographic events. Paleoceanography 2(6): 601–611.

Ruddiman, W.F., McIntyre, A., Niebler-Hunt, V. and Durazzi, J.T. 1980. Oceanic evidence for the mechanism of rapid northern hemisphere glaciation. Quaternary Research 13: 33–64.

Rutherford, W., Osyczka, A. and Rappaport, F. 2012. Back-reactions, short-circuits, leaks and other energy wasteful reactions in biological electron transfer: Redox tuning to survive life in O_2. FEBS Letters 586: 603–616.

Saltzman, M.R., Ripperdan, R.L., Brasier, M.D., Lohmann, K.C., Robison, R.A., Chang, W.T., Peng, S., Ergaliev, E.K. and Runnegar, B. 2000. A global carbon isotope excursion (SPICE) during the Late Cambrian: relation to trilobite extinctions, organic-matter burial and sea level. Palaeogeography, Palaeoclimatology, Palaeoecology 162: 211–223.

Saltzman, M.R., Cowan, C.A., Runkel, A.C., Runnegar, B., Stewart, M.C. and Palmer, A.R. 2004. The Late Cambrian SPICE ($\delta^{13}C$) event and the Sauk II-SAUK III Regression: new evidence from Laurentian basins in Utah, Iowa and Newfoundland. Journal of Sedimentary Research 74: 366–377.

Saltzman, M.R. 2005. Phosphorus, nitrogen, and the redox evolution of the Paleozoic oceans. Geology 33: 573–576.

Saltzman, M.R., Young, S.A., Kump, L.R., Gill, B.C., Lyons, T.W. and Runnegar, B. 2011. Pulse of atmospheric oxygen during the late Cambrian. Proceedings of the National Academy of Sciences 108(10): 3876–3881.

Sano, H. and Nakashima, K. 1997. Lowermost Triassic (Griesbachian) microbial bindstone–cementstone facies, southwest Japan. Facies 36: 1–24.

Sano, H., Wada, T. and Naraoka, H. 2012. Late Permian to Early Triassic environmental changes in the Panthalassic Ocean: Record from the seamount-associated deep-marine siliceous rocks, central Japan. Palaeogeography, Palaeoclimatology, Palaeoecology 363–364: 1–10.

Schlanger, S.O. and Jenkyns, H.C. 1976. Cretaceous oceanic anoxic events: causes and consequences. Geologie en Mijnbouw 55: 179–184.

Schlosser, P., Bönisch, G., Rhein, M. and Bayer, R. 1991. Reduction of deepwater formation in the Greenland Sea during the 1980s: Evidence from tracer data. Science 251: 1054–1056.

Schoepfer, S.D., Henderson, C.M., Garrison, G.H. and Ward, P.D. 2012. Cessation of a productive coastal upwelling system in the Panthalassic Ocean at the Permian–Triassic Boundary. Palaeogeography, Palaeoclimatology, Palaeoecology 313-314: 181–188.

Schopf, J.W. 1974. The development and diversification of Precambrian life. Origins of Life 5: 119–135.

Schopf, J.W. 1993. Microfossils of the Early–Archaean Apex Chert: new evidence of the antiquity of life. Science 260: 640–646.

Schopf, J.W. 2011. The paleobiological record of photosynthesis. Photosynthesis Research 107: 87–101.

Schulte, P., Alegret, L., Arenillas, I., Arz, J.A., Barton, P.J., Bown, P.R., Bralower, T.J., Christeson, G.L., Claeys, P., Cockell, C.S., Collins, G.S., Deutsch, A., Goldin, T.J., Goto, K., Grajales-Nishimura, J.M., Grieve, R.A.F., Gulick, S.P.S., Johnson, K.R., Kiessling, W., Koeberl, C., Kring, D.A., MacLeod, K.G., Matsui, T., Melosh, J., Montanari, A., Morgan, J.V., Neal, C.R., Nichols, D.J., Norris, R.D., Pierazzo, E., Ravizza, G., Rebolledo-Vieyra, M., Reimold, W.U., Robin, E., Salge, T., Speijer, R.P., Sweet, A.R., Urrutia-Fucugauchi, J., Vajda, V., Whalen, M.T. and Willumsen, P.S. 2010. The Chicxulub Asteroid Impact and Mass Extinction at the Cretaceous-Paleogene Boundary. Science 327: 1214–1218.

Scott, C., Lyons, T.W., Bekker, A., Shen, Y., Poulton, S.W., Chu, X. and Anbar, A.D. 2008. Tracing the stepwise oxygenation of the Proterozoic ocean. Nature 425: 456–459.

Servais, T., Lehnert, O., Li, J., Mullins, G.L., Munnecke, A., Nützel, A. and Vecoli, M. 2008. The Ordovician Biodiversification: revolution in the oceanic trophic chain. Lethaia 41: 99–109.

Shaw, G.E. 2016. Ocean basin and continental evolution. *In*: Shaw, G.E. (ed.). Earth's Early Atmosphere and Oceans, and the Origin of Life. Springer International Publishing, Heidelberg (Germany), 113 pp. Ch. 3: 19–24. ISBN: 978-3-319-21971-4.

Shen, Y., Canfield, D.E. and Knoll, A.H. 2002. The chemistry of mid-Proterozoic oceans: evidence from the McArthur Basin, northern Australia. American Journal of Sciences 302: 81–109.

Shen, Y., Knoll, A.H. and Walter, M.R. 2003. Evidence for low sulphate and anoxia in a mid-Proterozoic marine basin. Nature 423: 632–635.

Siever, R. 1991. Silica in the oceans: biological-geochemical interplay. pp. 287–295. *In*: Scheneider, S.H. and Boston, P.J. (eds.). Scientists on Gaia, MIT Press, Cambridge (MA, USA).

Sigman, D.M., Hain, M.P. and Haug, G.H. 2010. The polar ocean and glacial cycles in atmospheric CO_2 concentration. Nature 466: 47–55.

Sloss, L.L. 1963. Sequences in the cratonic interior of North America. Bulletin of the Geological Society of America 74: 93–114.

Sloss, L.L. 1972. Synchrony of Phanerozoic sedimentary-tectonic events of the North American craton and the Russian platform. 24th International Geological Congress, Montreal 6: 24–32.

Steele, M. and Boyd, T. 1998. Retreat of the cold halocline layer in the Arctic Ocean. Journal of Geophysical Research 103(10): 419–435.

Suzuki, N., Ishida, K., Shinomiya, Y. and Ishiga, H. 1998. High productivity in the earliest Triassic ocean: black shales, Southwest Japan. Palaeogeography, Palaeoclimatology, Palaeoecology 141: 53–65.

Svenningsen, O.M. 2001. Onset of seafloor spreading in the Iapetus Ocean at 608 Ma: precise age of the Sarek Dyke Swarm, northern Swedish Caledonides. Precambrian Research 110: 241–254.

Tachikawa, K., Athias, V. and Jeandel, C. 2003. Neodymium budget in the modern ocean and paleo-oceanographic implications. Journal of Geophysical Research 108(C8): 3254.

van den Boorn, S.H.J.M., van Bergen, M.J., Nijman, W. and Vroon, P.Z. 2007. Dual role of seawater and hydrothermal fluids in Early Archaean chert formation: evidence from silicon isotopes. Geology 35(10): 939–942.

Veizer, J., Holser, W.T. and Wilgus, C.K. 1980. Correlation of $^{13}C/^{12}C$ and $^{34}S/^{32}S$ secular variations. Geochimica et Cosmochimica Acta 44: 579–587.

Vajda, V., Raine, J.I. and Hollis, C.J. 2001. Indication of Global Deforestation at the Cretaceous-Tertiary Boundary by New Zealand Fern Spike. Science 294: 1700–1702.

Wade, B.S. and Pearson, P.N. 2008. Planktonic foraminiferal turnover, diversity fluctuations and geochemical signals across the Eocene/Oligocene boundary in Tanzania. Marine Micropaleontology 68: 244–255.

Watanabe, Y., Naraoka, H., Wronkiewicz, D.J., Condie, K.C. and Ohmoto, H. 1997. Carbon, nitrogen, and sulfur geochemistry of Archaean and Proterozoic shales from the Kaapvaal Craton, South Africa. Geochimica et Cosmochimica Acta 61(16): 3441–3459.

Wegener, A. 1912. Die Entstehung der Kontinente. Geologische Rundschau 3: 276–292.

Wegener, A. 1915. Die Entstehung der Kontinente und Ozeane. *In*: Skerl, J.G.A. (ed.). (1924). On the Origin of Continents and Oceans, English Translation of 3rd edition, Methuen Ed., London (UK), 212 pp.

Wignall, P.B. 2007. The End-Permian mass extinction—how bad did it get? Geobiology 5: 303–309.

Wignall, P.B. and Twitchett, R.J. 1996. Oceanic anoxia and the end-Permian mass extinction. Science 272: 1155–1158.

Wignall, P.B. and Twitchett, R.J. 2002. Extent, duration, and nature of the Permian–Triassic superanoxic event. pp. 395–413. *In*: Koeberl, C. and MacLeod, K.G. (eds.). Catastrophic Events and Mass Extinctions: Impacts and Beyond, Geological Society of America Special Paper 356.

Wignall, P.B., Newton, R. and Brookfield, M.E. 2005. Pyrite framboid evidence for oxygen poor deposition during the Permian–Triassic crisis in Kashmir. Palaeogeography, Palaeoclimatology, Palaeoecology 216: 183–188.

White, R.V. 2002. Earth's biggest 'whodunnit': unravelling the clues in the case of the end-Permian mass extinction. Philosophical Transactions of the Royal Society of London, Series B 360: 2963–2985.

Wilde, P., Quinby-Hunt, M.S., Berry, W.B.N. and Orth, C.J. 1989. Palaeo-oceanography and biogeography in the Tremadoc (Ordovician) Iapetus Ocean and the origin of the chemostratigraphy of *Dictyonema flabelliforme* black shales. Geology Magazine 126(1): 19–27.

Wilson, J.T. 1966. Did the Atlantic close and then re-open? Nature, London 211: 676–81.

Winguth, A.M.E., Heinze, C., Kutzbach, J.E., Maier-Reimer, E., Mikolajewicz, U., Rowley, D., Rees, A. and Ziegler, A.M. 2002. Simulated warm polar currents during the middle Permian. Paleoceanography 17(4): 1057–1065.

Yin, H.-F. and Song, H.-J. 2013. Mass extinction and Pangea integration during the Paleozoic-Mesozoic transition. Science China: Earth Sciences (doi: 10.1007/s11430-013-4624-3).

Yokochi, R., Marty, B., Chazot, G. and Burnard, P. 2009. Nitrogen in peridotite xenoliths: lithophile behavior and magmatic isotope fractionation. Geochimica et Cosmochimica Acta 73: 4843–4861.

Zachos, J.C., Lohmann, K.C., Walker, J.C.G. and Wise, S.W., Jr. 1993. Abrupt climate change and transient climates during the Paleogene: a marine perspective. Journal of Geology 101: 191–213.

Zachos, J., Pagani, M., Sloan, L., Thomas, E. and Billups, K. 2001. Trends, rhythms, and aberrations in global climate 65 Ma to present. Science 292: 686–693.

Zachos, J.C., Delaney, M.L., Wara, M.W., Petrizzo, M.R., Bralower, T.J., Bohaty, S., Brill, A. and Premoli-Silva, I. 2003. A transient rise in tropical sea surface temperature during the Paleocene-Eocene thermal maximum. Science 302: 1551–1554.

Zanazzi, A., Kohn, M.J., MacFadden, B.J. and Terry, D.O., Jr. 2007. Large temperature drop across the Eocene–Oligocene transition in central North America. Nature 445: 639–642.

Zeebe, R.E. 2012. History of seawater carbonate chemistry, atmospheric CO_2, and ocean acidification. Annual Review of Earth and Planetary Sciences 40: 141–165.

Ziegler, A.M., Hansen, K.S., Johnson, M.E., Kelly, M.A., Scotese, C.R. and Van der Voo, R. 1977. Silurian continental distributions, paleogeography, climatology and biogeography. Tectonophysics 40: 13–51.

2

South Atlantic Circulation and Variability from a Data Assimilating Model

Elbio D. Palma[1,*] and *Ricardo P. Matano*[2]

Introduction and Background

The ocean is a key component of the climate system due to its large heat capacity and its ability to transport large amounts of heat and freshwater from low to high latitudes through the global ocean circulation. This meridional transport is accomplished through a complex system of three dimensional surface and deep currents known as the global overturning circulation (Talley 2013). The complexity of the global overturning circulation is most evidenced in the South Atlantic Ocean (SA)—the portion of the Atlantic Ocean that extends from the equator to the Subtropical Front (~45°S). The upper circulation of the SA has two distinct circulation regimes: the tropical, which is characterized by a complex system of currents and countercurrents straddling the region between the tropics and the Equator, and the subtropical regime, which extends farther south and is characterized by an anticyclonic gyre and western and eastern boundary currents at the continental boundaries (Stramma and England 1999) (Fig. 2.1a). The northwestward flow of the South Equatorial Current

---

[1] Departamento de Física, Universidad Nacional del Sur e Instituto Argentino de Oceanografia (IADO-CONICET), Bahía Blanca, Argentina.
[2] College of Earth Ocean and Atmospheric Sciences (CEOAS), Oregon State University, Corvallis, Oregon, USA.
 E-mail: rmatano@coas.oce.orst.edu
* Corresponding author: uspalma@criba.edu.ar

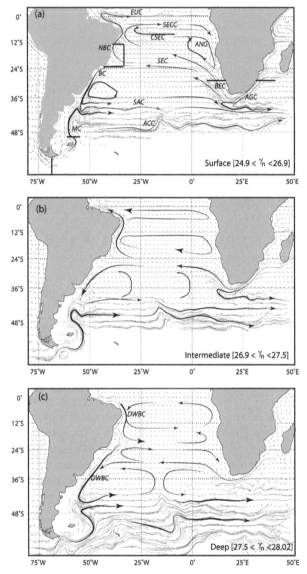

Figure 2.1. Horizontal mean circulation (in neutral density layers) simulated by SODA in the SA. (a) Surface layer, (b) Intermediate layer, (c) Deep layer. Large arrows indicate the main flow direction in each layer. Shown are the Aghulas Current (AGC), the Benguela Current (BEC); the South Equatorial Current (SEC); the North Brazil Current (NBC); the Brazil Current (BC); the Malvinas Current (MC); the South Atlantic Current (SAC); the Central branch of the SEC (CSEC); the South Equatorial Countercurrent (SECC); the Equatorial Undercurrent (EUC); and the Antarctic Circumpolar Current (ACC). ANG indicates the location of the Angola Gyre. DWBC in panel (c) is the Deep Western Boundary Current. The black lines in panel (a) indicates the approximate location where the major currents transports are computed (see Table 2.1).

(SEC) marks the boundary between these regimes. After approaching the coast of Brazil the SEC splits into the equatorward and poleward flows of the North Brazil (NBC) and the Brazil Current (BC). At the eastern side of the basin the upper ocean circulation is dominated by the relatively small Angola anticyclone and, farther south, by the Benguela Current (BC) and the Benguela Upwelling jet (Lutjeharms 1996). The circulation patterns at intermediate levels are similar to those up above but the subtropical gyre is shifted poleward by approximately 10° (Stramma and England 1999, Fig. 2.1b). The deep layers of the South Atlantic are dominated by the poleward flow of the Deep Western Boundary Current (DWBC), which carries salty and oxygen depleted North Atlantic Deep Water (NADW, Stramma and England 1999, Fig. 2.1c). Branches of the deep western boundary current are known to split from the western boundary and spread eastward into the interior of the basin near 8°S and 17°S (Garzoli et al. 2015; Hogg et al. 2005). The pathways of these branches in the interior of the ocean basin are not well known but observations show NADW within the Cape Basin (Beal and Bryden 1999; van Sebille et al. 2012).

Our knowledge of the temporal variability in the SA circulation is more limited than that of its mean circulation. There are several studies of the SA's most energetic regions [(e.g., the Brazil/Malvinas Confluence (BMC), the Agulhas Retroflection (AR), etc.], but there are few studies of the variability of the basin-scale circulation. Venegas et al. (1997; 1998) and Palastanga et al. (2002) investigated the low-frequency variability of the SA's Sea Surface Temperature (SST) from approximately 40 years of climatological data. They identified three low-frequency modes of variability with periods of about 14, 6, and 4 years that, according to their analyses, are associated with variations of sea-level pressure. Sterl and Hazeleger (2003) and Haarsma et al. (2005) added that these modes are generated by anomalous winds through turbulent heat fluxes, Ekman transport, and wind-induced mixing. Witter and Gordon (1999) using 4 years of Sea Surface Height (SSH) from TOPEX-POSEIDON (1993–1996) identified two modes of variability. The first mode was characterized by a sea level maximum in the eastern portion of the basin. The time series associated with this mode indicates that the anticyclonic circulation in the SA's subtropical gyre underwent a transition during 1993–1996, from stronger to weaker than average circulation. They also observed a similar temporal signature in the zonal wind, suggesting a coupling between the oceanic and atmospheric variations. The second mode was associated with interannual variations in the BMC. A more recent analysis suggests an increase of the Sea Surface Salinity (SSS) of the SA's subtropical gyre and a decrease of the salinity of the region around 40°S (Sato and Polito 2008). Such changes imply a slowdown of the southern limb of the subtropical gyre.

The Meridional Overturning Circulation (MOC) is a simplified version of the global overturning circulation that considers the zonally integrated transports across a particular ocean basin or the globe (Rahmstorf 2006). The SA portion of the global MOC (SAMOC hereafter) involves a deep, southward flow of North Atlantic Deep Water (NADW) and a compensating northward flow of warm and salty surface waters, and cooler and fresher Antarctic Intermediate Waters (AAIW) and Antarctic Bottom Waters (AABW). This circulation pattern, with warm surface and intermediate waters flowing towards the equator and cold deep waters towards the pole produces an equatorward heat flux. In fact, the SA is the only ocean basin that extends to high latitudes and transports heat from the poles to the equator. Although this anomalous heat flux was recognized by the middle of the last century (Rintoul 1991; Talley 2003) the sources for the surface and intermediate waters entering into the SA remain largely undetermined. Gordon (1986) proposed the so-called *warm path* hypothesis, which contends that the SA gets most of its surface and intermediate waters through eddies and filaments detached at the retroflection of the Agulhas Current. Rintoul (1991), and later Broecker (1991), proposed the alternative *cold path* hypothesis arguing that AAIW injected through the Drake Passage are converted to surface waters through air-sea interactions and then exported northward. It has been more than two decades since the publication of Gordon and Rintoul's influential articles, yet there is still no consensus on the origins of the SA's northward mass outflow. Some studies support the *cold-path* theory (England et al. 1994; Macdonald 1998; de las Heras and Schlitzer 1999; Nof and Gorder 1999; Sloyan and Rintoul 2001; You 2002), others the *warm-path* theory (Holfort and Siedler 2001; Weijer et al. 2002; Donners and Drijfhout 2004; Souza et al. 2011), and still others argue that both paths are important (Matano and Philander 1993; Macdonald and Wunsch 1996; Poole and Tomczack 1999). The controversy reflects the difficulty of identifying the pathways of the MOC in the SA, which are interwoven with those of the local wind-driven circulation and shrouded by the intense mesoscale variability of the basin.

Observation-based estimates of the SAMOC are largely based on hydrographic sections. Lumpkin and Speer (2007) combined the hydrographic data with an inverse model to estimate a mean overturning transport of ~18 Sv (1 Sv = 10^6 m^3/s) and a mean Meridional Heat Flux (MHT) of ~0.68 PW ± 0.2 PW (1 PW = 10^{15} watts) at ~35°S. Hydrographic snapshots, however, cannot be used to estimate variations of the SAMOC since high frequency variations are likely to alias as artificial trends or low-frequency variations (Meinem et al. 2013). To address this problem during the last decade an international consortium of scientists have strived to make consistent observations of the meridional mass and heat flux along the AX18 line (~35°S) [http://www.aoml.noaa.gov/phod/

SAMOC_international]. Garzoli and Baringer (2007) estimated that during the period 2002–2006 the MHT across AX18 was 0.55 ± 0.11 PW. Dong et al. (2009) used the same data set to evaluate the mass and heat transports across a similar cross-section and for the period 2002–2007. They estimated a time mean mass transport of 17.9 ± 2.2 Sv and a northward heat flux of 0.55 ± 0.14 PW. They observed that the SAMOC variability is not only affected by the variability of the boundary currents at both sides of the basin but also by the variability of the interior flow. Meinem et al. (2013) combined data from mooring arrays deployed along 34.5°S to produce a time series of the SAMOC during the period of March 2009 and December 2010. They estimated a mean transport of 21.3 Sv and very energetic variations of the meridional transport at periods smaller than ~3 months. The observational estimates collected by the SAMOC consortium are in general agreement with those obtained from numerical models. Perez et al. (2011) evaluated two high-resolution models, which produced consistent estimates of the mean mass and heat transports of ~16 Sv and 0.41 PW. The models showed meridional increments of the mass and heat fluxes, which are also consistent with observations (Lumpkin and Speer 2007).

In this chapter we seek to simultaneously characterize the structure and variability of the SA circulation and the SAMOC for the period 1958–2008. To do so in a dynamically consistent framework we use the results of a state-of-the-art data assimilating model. Details of the model configuration and forcing as well as the methods of analysis are given next. After that we present a description and discussion of the mean SA circulation as well as its long-term variability. Finally we summarize the main findings of our analysis.

Data and Methods

To study the mean state and variability of the SA ocean circulation, we employed the results of a state of the art ocean general circulation model, the multiyear Simple Ocean Data Assimilation (SODA). The ocean model is based on the Los Alamos implementation of the Parallel Ocean Program numerics (version 2.0.1; Smith et al. 1992). The model resolution is an average 0.4 (longitude) X 0.25 (latitude) with 40 vertical z-levels (10 m spacing for the surface layers and 250 m for the deeper layers). We used SODA reanalysis data version 2.2.4 where the surface boundary conditions are provided from a new atmospheric reanalysis (Compo et al. 2011) designated as 20CRv2. In addition to the surface wind stress, the atmospheric model provides the solar radiation, specific humidity, cloud cover, 2 m air temperature, precipitation and 10 m wind speed for computing heat and freshwater fluxes. Vertical diffusion of momentum, heat, and salt is based on a nonlocal KPP scheme and horizontal diffusion for subgrid-scale processes is based on a biharmonic mixing. More details of the model implementation and simulation strategies are described in Giese and Ray (2011).

The model is constrained by observed temperature and salinity profiles using a sequential assimilation algorithm. In addtion to *in situ* (ICOADS 2.5; Woodruff et al. 2011) and satellite SST the assimilated data set employs the World Ocean Database 2009 (Boyer et al. 2009); including all available hydrographic profile data, as well as ocean data from moored time series (TAO/Triton), and Argo drifters. Details of the data assimilation algorithm are described in Carton et al. (2000). Model output, such as temperature, salinity, and velocity are averaged over 5 day intervals. These average fields are then mapped onto a uniform global 0.5 (longitude) X 0.5 (latitude) grid using the horizontal grid Spherical Coordinate Remapping and Interpolation Package with second order conservative remapping (Jones 1999). The mapping shifts the locations of the temperature and horizontal velocity grids, which are offset in the original model, to the same set of remapped grid point locations while conserving the mass transport. In our analysis we will employ monthly means of the model output for the period 1958–2008 (from http://soda.tamu.edu/assim/).

To characterize the interannual variability of the oceanic circulation in the SA, we computed EOF modes (Bjornsson and Venegas 1997) from the model SSH, SST fields, as well as the curl of the wind stress (CURLW) used to force the model. Prior to calculation, the model fields were smoothed in space with a 6° loess filter. This value is large enough to damp the dominance of the highly energetic western boundary regions and small enough to allow us to quantify the relative contributions from large- and meso-scale modes. To separate the interannual from the seasonal and intra-annual variability we computed the EOFs from annual averages of the model output. Additionally, and to discriminate the more energetic time scales, we employ standard spectral analysis and wavelet techniques (Lau and Weng 1995).

South Atlantic Circulation and Variability

The mean circulation

Here we discuss the water masses and the time-mean (1958–2008) horizontal and meridional overturning circulation obtained from the model results.

Water masses

The model main water masses in the SA ocean are represented in Fig. 2.2 which shows a meridional cross-section at 25°W of potential temperature and salinity. The upper part of the section shows the warm water sphere

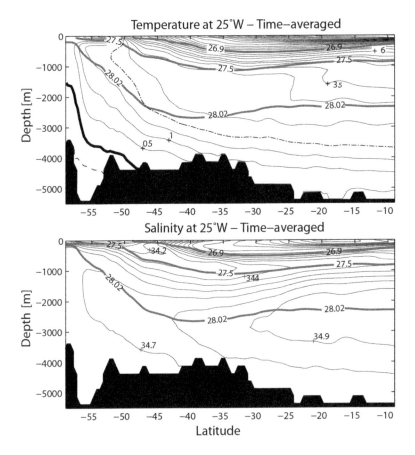

Figure 2.2. Record mean Temperature (a) and Salinity (b) distribution on a meridional section along 25°W extracted from the model. Contour interval is 0.5°C until 4°C and 1°C thereafter for temperature and 0.1 psu for salinity. The black thick line in panel (a) indicates the 0°C contour. The dashed-dotted line indicates the 2°C contour, approximately the upper limit of AABW. The dashed line indicates the –0.5°C contour. Gray thick lines in (a) and (b) indicates the neutral density surfaces that define the surface ($\gamma_n > 26.9$), intermediate ($26.9 < \gamma_n > 27.5$) and deep ($27.5 < \gamma_n > 28.02$) layers used to compute the horizontal circulation depicted in Fig. 2.1.

with a shallow thermocline near the Equator and a deeper one in the subtropical South Atlantic (top panel). From the salinity section (bottom panel) we note the intrusion of Antarctic Intermediate Water (AAIW) from the south into the southern Atlantic while the cold temperature tongue touching the bottom and spreading northward represents Antarctic Bottom Water (AABW). In-between these water masses, we found the southward spreading North Atlantic Deep Water (NADW). In what follows we use neutral density layers (Jackett and McDougall 1997) to define surface ($\gamma_n < 26.9$), intermediate ($26.9 < \gamma_n < 27.5$) and deep layers ($27.5 < \gamma_n < 28.01$) for analyzing the horizontal circulation. Note that the intermediate layer lies between the core of AAIW and NADW as defined by the salinity properties of these water masses (Fig. 2.2, bottom panel).

Mean horizontal (Gyre) circulation

To illustrate the realism of the SODA simulation, we discuss the schematic depth-averaged (in each neutral density layer) horizontal flow field obtained from the model results. The transport magnitude of the main currents, its temporal variability, and the associated acronyms are listed in Table 2.1. The surface (thermocline) circulation in the upper layer, composed mainly by South Atlantic Central Water (SACW), is dominated by a broad anticyclonic subtropical gyre (Fig. 2.1a). At the southern tip of Africa, the Agulhas Current (AGC) intrudes into the SA and retroflects at about 20°E. The northern branch of this current take a NW path and joins a strong Benguela Current (BEC), the eastern boundary current of the gyre. The BEC extends farther downstream into the broad northwestward flow of the South Equatorial Current (SEC). At about 18°S this flow bifurcates into a stronger northward flowing branch—the North Brazil Current (NBC)—and a weaker southward flow—the Brazil Current (BC), which is the western boundary current of the subtropical gyre. The location of the bifurcation point in the model is close to the observations (Stramma et al. 1995). Further south, the BC collides with the northward path of the Malvinas Current (MC) near 38°S, forming the Brazil/Malvinas Confluence (BMC) (Gordon and Greengrove 1986; Olson et al. 1988; Matano 1993). Two narrow and relatively weak recirculation cells are seen eastward of the BC. Similar recirculating cells have been reported by Tsuchiya et al. (1994). From there the main flow proceeds southward until ~45°S where it turns east forming the predominantly zonal path of the South Atlantic Current (SAC) closing in this manner the gyre. The BEC therefore receives

Table 2.1. Main ocean currents in the South Atlantic Ocean (excluding equatorial currents) with the mean transport and standard deviation for the 1958–2008 period extracted from the model results. The cross-sections where the transport was estimated are indicated in Fig. 2.1a. Note that only the ACC transport is computed between land masses.

Acronym	Current name	Mean transport	Stdv
AGC	Aghulas Current	53.2 Sv (30°S)	6.7 Sv
BEC	Benguela Current	4.2 Sv (30°S)	1.7 Sv
SEC	South Equatorial Current	6.6 Sv (33°W)	9.0 Sv
NBC	North Brazil Current	2.0 Sv (11°S)	7.4 Sv
BC	Brazil Current	4.5 Sv (22°S)	4.7 Sv
MC	Malvinas Current	40.0 Sv (51°S)	4.9 Sv
ACC	Antarctic Circumpolar Current	153.0 Sv	5.4 Sv

water from the South Atlantic (via the SAC) and from the Indian Ocean (via the AGC, Lutjeharms 1996). The cross-equatorial flow in the surface layer is concentrated in the western equatorial Atlantic, off the coast of Brazil, although at this level the NBC shows a strong retroflection with its main branch turning east at the Equator joining the strong eastward flow of the Equatorial Undercurrent (EUC). A similar behavior of the NBC at the surface layer has been reported by Stramma and England (1999) based on historical observations. Northeast of the subtropical gyre there is a smaller cyclonic gyre centered approximately at 0°W and 12°S that is the surface signature of the Angola Gyre (Gordon and Bosley 1991). North of 9°S there are additional tropical currents flowing zonally represented by the model: a strong eastward jet, the South Equatorial Countercurrent (SECC, ~4°S) that joins the eastward branch of the Angola Gyre near the western coast of Africa and a weaker westward current related with the central branch of the South Equatorial Undercurrent (CSEU, ~8°S). This system of zonal equatorial currents corresponds well with the observational description presented by Stramma and England (1999). South of the subtropical gyre the circulation is dominated by the eastward flow of the Antarctic Circumpolar Current (ACC).

The mean flow field at the AAIW level is shown in Fig. 2.1b. The main difference with the surface circulation is a poleward shift of the subtropical gyre. The northern portion of the SEC reaches the shelf at about 23°S and turns northward, feeding the NBC and sustaining a strong cross-hemispheric flow. A stronger, westward component, of the SEC reaches the Brazilian shelf and bifurcates near 27°S. The main path of the current turn southward as the western boundary current of the subtropical gyre. The location of these turning points are in agreement with direct current observations and geostrophic transport computations at the AAIW layer (Boebel et al. 1997). A weak (recirculating) contribution of northward flowing intermediate water from the SAC/ACC is visible near 18°W and 0°W. Earlier models have also shown the southward shift of the subtropical gyre with depth and similar recirculations of AAIW near 20°W (Stramma and England 1999; Marchesiello et al. 1998; Garzoli and Matano 2011). Concurrently with the southern shift of the subtropical gyre, the tropical NE cyclonic gyre (the Angola Gyre) shrinks, weakens and moves southward. The system of zonal equatorial currents also shows new features. For example, both the EUC and the SEC change from an eastward path at the upper layer to a westward direction at the intermediate layer. This reversal of direction with depth at the Equator was described by Schott et al. (1995) who designated the westward flow as the Equatorial Intermediate Current. It is also interesting to note that, as described by Gordon et al. (1992), the model shows an important amount of water entering the AAIW layer from the Indian Ocean south of Africa via the AGC.

The circulation in the deep layer is dominated by a strong southward western boundary current along the continental shelf of South America (DWBC), composed in large part by NADW (Fig. 2.1c). A branch of this current separates from the coast near the Vitoria-Trinidade seamount chain at 15°S, where it breaks down into large cyclonic eddies located mainly on the inshore side and zonal jets that flow away towards the interior (i.e., at 15°S and 20°S). A substantial portion of the DWBC detrained at the seamounts reaches the African coast, where it turns northward and then westward near the Equator, closing an anticyclonic gyre. Part of the eastward current at 20°S [the Namib Col Current, Speer et al. 1995] recirculates northward near 0°W or gets trapped by small recirculating cells near the eastern boundary. The Namib Col Current was also reproduced in global high resolution models (Stramma and England 1999; Garzoli and Matano 2011) and sigma coordinate modes of the South Atlantic (Marchesiello et al. 1998). After a cyclonic loop, however, a large part of the DWBC returns to the western boundary and flows southward until 40°S where it meets, at this level, with the northward flowing MC. After colliding and mixing the DWBC leaves the western boundary and flows eastward with the SAC. The eastward flow of NADW at 40°S is in good agreement with the analysis of *in situ* data (Tsuchiya et al. 1994). A big recirculating anticyclonic cell is formed when part of the eastward flow turns northward near 18°W and joins a broad westward flow centered at ~32°S. West of this longitude a similar but stronger, anticyclonic cell that extends to the Aghulas retroflection region is formed. This western cell is filled with smaller and intense recirculating patterns. Both anticyclonic gyres have also been reported from results of high resolution models (Stramma and England 1999; Garzoli and Matano 2011). Close to the western coast of South Africa there is a southward flow of NADW, but this path terminates after joining a large cyclonic cell near Cape Basin and the strong westward flow of the AGC extension.

Meridional overturning circulation

The MOC can be characterized through the meridional overturning streamfunction, which is defined as

$$\Psi(y,z) = \int_{0(y)}^{L(y)} \int_{-H}^{z} v(x,y,z') \, dz' dx$$

Here v is the meridional component of the ocean current and the first integration is conducted zonally across the Global Ocean (MOC) or the SA basin (SAMOC). Figure 2.3 shows such a streamfunction constructed with the SODA model velocity fields both for the Global Ocean (top panel) and for the Atlantic basin (bottom panel). Note that the MOC is not defined for the Atlantic basin south of 37°S. The contours can be interpreted as flowlines, and positive (negative) values indicating clockwise (anticlockwise) circulation. SODA represents most of the familiar gross features of the global MOC that have been reconstructed mainly from data-constrained models.

Roughly the meridional circulation can be divided in three layers. There is a wide clockwise cell covering both hemispheres and centered approximately at 1000 m that carries an average of 15 Sv (Fig. 2.3a). The upper branch of this middle cell is associated with AAIW and the lower

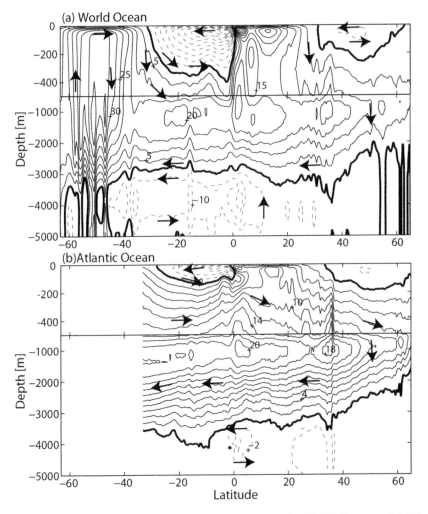

Figure 2.3. Time-mean overturning circulation for (a) the World Ocean and (b) the Atlantic Ocean. Negative values of the stream function are gray dashed lined and indicate counterclockwise overturning. The bold black line indicates the zero contour. The contour interval for the world ocean is 5 Sv, and the interval for Atlantic basin is 2.5 Sv.

branch, extending until ~3000 m with NADW (see Fig. 2.1b). There are several, rather narrow, closed overturning cells inside the main one centered at ~1300 m and located near 35°N, the Equator, 20°S and 50°S. The small cells, particularly the one located near 35°N represents an important recirculation (~10 Sv) of the NADW formed in the North Atlantic. There is also a stronger clockwise recirculating cell between 40°S and 60°S, which is called the Deacon cell (Döös and Webb 1994; Speer et al. 2000) that is sustained by the northward Ekman transport at the latitude of the ACC. Neither of them contribute significantly to the interhemispheric volume transport. On top of this middle layer, two subtropical upper-thermocline wind-driven recirculating cells rotating counterclockwise are visible, with the southern one much more intense than its northern hemisphere counterpart. Below the intermediate cell in the bottom layer, there is a cell of deep (AABW) rotating in the opposite direction, with about 5 Sv flowing into the basin at the bottom, extending until ~10°N, and then retroflecting and leaving it between 3000 m and 3500 m. Recirculation patterns of ~5 Sv are also visible in the bottom cell, particularly near the Equator.

The SAMOC has a similar structure to the MOC, with a large middle inter-hemispheric cell carrying a similar amount of water as the MOC (~15 Sv, Fig. 2.3b). The intensity of SAMOC in the upper and bottom cells, however, is weaker than those of the MOC. This indicates that the inter-hemispheric cell is largely associated with the Atlantic Ocean circulation, while the deep cell (AABW) is mostly constrained to the Pacific and Indian Ocean sectors of the ACC. The upper, Ekman cells are not only less intense but also shallower compared with the MOC, particularly in the SA (<150 m).

The volume transport of the overturning circulation at 24°N in the Atlantic was estimated from hydrographic section data (Roemmich and Wunsch 1985) as 17 Sv. This figure was later updated by Ganachaud and Wunsch (2000) using observations and inverse models to be about 15±2 Sv and by Talley et al. (2003), who estimated 18±5 Sv of NADW formation. More recently Lumpkin and Speer (2007) computed 18.0±2.5 Sv at 24 N and 16.2±3 Sv at 11°S, close to the SODA estimation. Similar structure and intensity for the Global and Atlantic MOC have been computed using results from eddy resolution numerical models (Jane and Martozke 2001) and low resolution models constrained by observations (Kohl et al. 2007). The intensity of the SAMOC obtained from the model at 34°S (~20±3 Sv), however is larger than the one computed from results of two eddy resolving models by Pérez et al. (2011) who reported a maximum of 15.5 ± 3 Sv and the 12.4±2.6 Sv of Lumpkin and Speer (2007). The discrepancy may be related to the relatively small recirculating patterns present in the model results at ~1500 m depth.

EOF analysis

To characterize the variability of the SA circulation we calculated the Empirical Orthogonal Functions (EOFs) of the SSH, SST and CURLW fields. The time-mean fields are substracted from the model data prior to the EOF computation.

SSH

From the initial 4-years of the Topex/Poseidon mission Witter and Gordon (1999) identified two dominant modes of SSH variability in the SA. The first mode suggested a weakening of the subtropical gyre during the 1993–1996 period and the second mode interannual variations in the BMC. Here we extend the preliminary analysis of Witter and Gordon using the 50-years of the SODA experiment. The first three EOFs modes from the model's SSH account for approximately 40% of the total variance (Fig. 2.4). The small difference between the amount of variance accounted by the first and second modes indicates that these modes are not stable and, therefore, that their relative ordering depends on the particulars of data processing (e.g., spatial and temporal smoothing). In fact, the spatial and temporal structure of the different EOF modes depends on the characteristics of the spatial and temporal smoothing used during the data processing. For example, use of monthly (instead of yearly) time series leads to the emergence of a dominant mode of variability with maxima over the BMC and the ARR. The second EOF mode of the monthly data calculation, however, is similar to the first EOF mode of the yearly calculation and the same reordering applies to the 3rd mode. Thus, the EOF analysis is robust in the sense that the changes of the EOFs patterns produced by changes of the smoothing parameters are those expected from the emergence of physical phenomenon with different temporal and spatial scales. In what follows we will describe these modes in the particular order ascribed by our particular choice of parameters but noting that the relative ordering is not statistically significant.

The first EOF is the only one that shows amplitude maxima over the tropical region (<15°S). The second and third modes show SSH maxima in the subpolar region: the second in a longitudinal band similar to that the first mode (30°W–10°W) and the third in the region next to the Drake Passage. At the temporal and spatial scales considered in this analysis the bulk of the SA variability occurs in the subtropical region, between 20°S and 45°S. The first EOF mode shows a dipole structure over the subtropical region, with peaks displaced to the west and east of the ARR and the BMC. The heavy spatial and temporal smoothing used for the present analysis eliminated most of the high-frequency, short wavelength variability associated with eddies and meanders leaving only the large-scale, low-frequency tails associated with

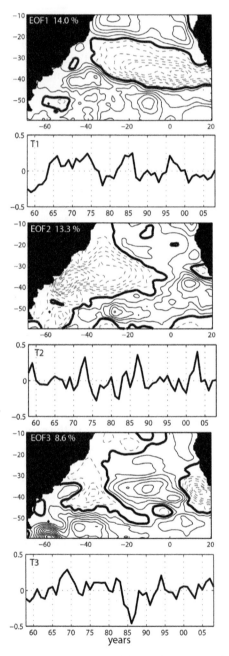

Figure 2.4. Spatial patterns (EOF1, EOF2, and EOF3) and time series of expansion coefficients (T1, T2, T3) of the three EOF modes of SSH. The spatial patterns have a contour interval of 0.1. Negative contours are dashed. Zero line is thicker. The time series are normalized by the standard deviation.

these phenomena. Thus, the SSH minimum of the first mode in the eastern Atlantic is not centered at the Agulhas retroflection itself but it is displaced farther west. Likewise, the maximum in the western region is not centered at the BMC but farther east. The large scale, low-frequency manifestations of the BMC and ARR variability appear in the second and third mode. The second mode has an absolute minimum centered over the BMC that extends to most of the subtropical gyre, while the third mode has a dipole structure near the ARR. Standard spectral analysis of the time series of these three modes (not shown) show spectral peaks at ~12, 8, 6, 5, 3 and 2 years.

The EOF analysis indicates that the SSH variability in the subtropical region can be characterized as a dipole structure with centers in the BMC and the ARR. The region of influence of these centers is roughly limited by the Middle Atlantic Ridge, which runs meridionally through ~20°W. The fact that the times series of the first three EOFs have spectral maxima at ~12 years indicates that this is the dominant large-scale, low frequency mode of SSH variability of this region. The first and second EOFs have secondary spectral peaks at 8 and 6 years and 5 and 3 years. Since the first EOF have an absolute amplitude maximum over the ARR while the second EOF has an absolute maximum over the BMC it seems reasonable to surmise that the 8 and 6 years peaks represent dominant modes of variability in the eastern Atlantic while the 5 and 3 years peak represent dominant modes of variability in the western Atlantic. The amplitude structure of the third EOFs is dominated by a dipole structure over the eastern Atlantic. This dipole has a maximum over the ARR and a minimum farther west. As we shall show the location of this minimum corresponds with a minimum of the wind stress curl thus suggesting that this particular mode of variability represent the rise and fall of the SSH in response to Ekman pumping.

Wind stress curl

The EOFs of CURLW show different spatial and temporal characteristics than those of the SSH (Fig. 2.5). The three EOFs modes show absolutely maxima in the subpolar region, a reflection of the intensity and variability of the winds in that area. The spatial pattern of the first EOF in the subtropical region is characterized by the dominance of a monopole centered at approximately 38°S and 0° to 20°W. The time series of this mode has spectral peaks at 7.6, 5 and 2.8 years. The second EOF mode show a dipole pattern with centers in the western region, between 40°W and 20°W. The time series of this mode has spectral peaks at 8, 4.1 and 3.3 years. The amplitude pattern of the third EOF mode in the subtropics is dominated by meridionally banded structures with a maximum and a minimum spreading westward from the coasts of Namibia and the ARR. The time series of this mode has peaks at 17, 7.2 and 3.8 years. Table 2.1 shows the correlation among the

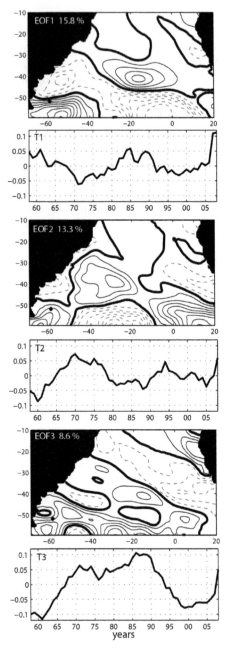

Figure 2.5. Spatial patterns (EOF1, EOF2, and EOF3) and time series of expansion coefficients (T1, T2, T3) of the three EOF modes of the curl of the wind (CURLW). The spatial patterns have a contour interval of 0.1. Negative contours are dashed. Zero line is thicker. The time series are normalized by the standard deviation.

different SSH and wind curl modes. The first SSH mode is the only one that is significantly correlated to the wind and, therefore, the one that is more influenced by Ekman pumping effects. This mode is weakly correlated to the second wind curl mode ($r = 0.24$) and more strongly correlated to the third wind curl mode ($r = 0.6$). A comparison between the spatial amplitudes of all these modes indicates that the wind effect is more dominant in the southeastern Atlantic, where the South Atlantic High is located, and in the middle of the gyre. There the elongated maximum of the wind stress curl centered near 38°S and between 0° to 20°W seems to correspond with the minima of the first EOF of the SSHs emanating from the ARR and the maxima in the third EOFs. There are no significant peaks of wind curl variability at the decadal time scales, which indicates that the 12 year peak shown in the three leading EOFs of the SSH represent an internal mode of variability. The lack of significant correlation between the SSH and wind curl for the second and third EOF modes indicates that these patterns of SSH variability do not represent Ekman pumping effects but variations of the geostrophic flow.

SST

The first mode of SST variability of the SODA record is characterized by an amplitude maximum over the central and western portion of the subtropical gyre and a horse-shoe shaped structure of opposing sign surrounding it (Fig. 2.6). The second mode is a meridionally slanted dipole structure with peaks in the eastern tropical and the western subtropical regions. This mode has already been identified in previous analyses, which indicates that this is the most robust mode of SST variability in this region (Venegas et al. 1998; Palastanga et al. 2002; Sterl and Hazeleger 2003; Haarsma et al. 2005). This dipole is dominated by interdecadal variations and its development has been associated to a coupled mode of air/sea interaction (Venegas et al. 1998; Sterl and Hazeleger 2003). The third mode of SODA SST variability is a monopole centerd over the subtropical gyre. This mode resembles the first mode describe by Venegas et al. (1998) and Palastanga et al. (2002). Its time series has spectral peaks at 15 and 6 years. SODA has skill in representing the SST variability associated with El Niño. The correlation between El Niño 2.0 and 3.4 indexes computed from satellite SSTs and the model SST is very high ($r = 0.89$). The time series of the second EOF mode is the most correlated to El Niño ($r = 0.35$).

Table 2.2 show that there are strong correlations between SSH and SST modes, thus suggesting that an important portion of the SST variability of the SA is not driven by air/sea interactions, as previously postulated, e.g., Venegas et al. (1998), Haarsma et al. (2005), but by advective effects associated with the geostrophic circulation. The time series of the first

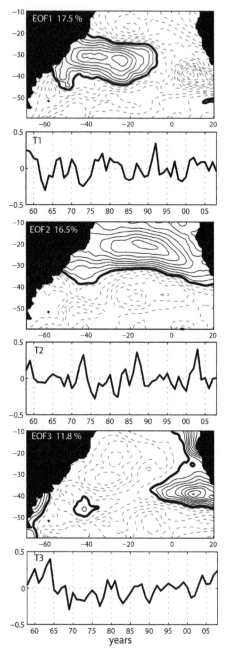

Figure 2.6. Spatial patterns (EOF1, EOF2, and EOF3) and time series of expansion coefficients (T1, T2, T3) of the three EOF modes of the SST. The spatial patterns have a contour interval of 0.1. Negative contours are dashed. Zero line is thicker. The time series are normalized by the standard deviation.

Table 2.2. Correlation coefficients between the three first EOF modes of SSH and the three first EOF modes of CURLW and SST. Numbers in brackets are the 95% significance levels calculated for each correlation. Highly significant correlations are in bold face.

	E1 (CURLW)	E2 (CURLW)	E3 (CURLW)
E1 (SSH)	0.04 [0.28]	**−0.25 [0.24]**	**−0.46 [0.31]**
E2 (SSH)	0.03 [0.26]	0.13 [0.25]	−0.04 [0.30]
E3 (SSH)	0.00 [0.29]	0.20 [0.24]	−0.20 [0.28]
	E1 (SST)	E2 (SST)	E3 (SST)
E1 (SSH)	−0.30 [0.36]	−0.05 [0.36]	**0.48 [0.38]**
E2 (SSH)	**0.65 [0.30]**	−0.06 [0.32]	0.18 [0.34]
E3 (SSH)	−0.10 [0.36]	**−0.40 [0.30]**	−0.16 [0.39]

SST mode, for example, is strongly correlated with the corresponding time series of the second SSH mode ($r = 0.65$). The strong similarity of the amplitude patterns of these two EOFs indicates that most of the variability of this particular SST mode is driven by advective effects. The correlation between the second SST mode and the third SSH mode is smaller ($r = 0.4$) and the similitude of the amplitude patterns is largely restricted to the southeastern portion of the subtropical gyre where the amplitude maxima of the SSH mode roughly corresponds with the amplitude minimum of the SST. There is no obvious correspondence between SST and SSH patterns north of ~30°S, which indicates a smaller contribution of advective effects to the development of this particular mode. The third SST mode is more strongly correlated to the first SSH mode ($r = 0.48$). The amplitude patterns of these two modes shows correspondence in the regions surrounding the BMC and the ARR, thus suggesting that this particular mode is driven by recirculation effects associated with these energetic boundary regions.

Meridional overturing circulation and heat transport

Dong et al. (2009) examined the variability of the SAMOC along the AX18 line (~35°S) using a trans-basin XBT high-density line. They estimated that, during the 2002–2007 period, the mean overturning transport was 17.9 ± 2.2 Sv and the mean MHT was 0.55 ± 0.14 PW. More recently, Meinem et al. (2013) combined data from several mooring arrays deployed along 34.5°S to produce a time series of the SAMOC during the period of March 2009 to December 2010 and estimated a mean overturning transport of 21.3 Sv. The observational estimates of the SAMOC strength are consistent with those calculated from model simulations. Perez et al. (2011), for example, estimated a mean overturning transport of 15.6 ± 3 Sv and a mean MHT of 0.5 ± 0.15 PW at 34.5°S. All these estimates are in general agreement with the values calculated from SODA, which, predicts an overturning mean value of 20 ± 3 Sv and a mean MHT of 0.8 PW ± 0.3 PW at 33.5°S (Fig. 2.7a). The MOC and the MHT at this latitude are strongly correlated ($r = 0.9$) therefore in what follows we will focus our discussion on the MHT.

The MHT variability in the SODA simulation is characterized by a relatively strong seasonal variation of 0.2 PW with a maximum during the austral winter and a minimum during the austral summer (Fig. 2.7b, see insert). Similar variations were also reported by Dong et al. (2014) from two climate models. The MHT also has substantial interannual variations, which are characterized by an increase of amplitude during the last few decades. To characterize the inter-annual variability we substracted the seasonal cycle from the de-trended time series of the MHT and computed a 13 month running mean (Fig. 2.7b). A wavelet analysis of this time series shows that the MHT variability is concentrated in 5 and 2.5 years peaks during the 1970–1980 and 1990–1996 periods (Fig. 2.7c). It also shows a transition of the MHT around 1982 when the maximum energy shifts from 5 to 2.5 years. Note also that in the 14 year period 1976–1990 the MHT is higher than the mean, while in 1990–2004 period it is lower (Fig. 2.7b, thick gray line).

The MHT also show important meridional variations. Lumpkin and Speer (2007), for example, show that according to inversions of hydrographic data the MHT in the SA has an increase from 0.62 ± 0.15 PW at 32°S to 0.74 ± 0.36 PW at 11°S. A similar meridional increment of the MHT was also reported in high resolution models (Pérez et al. 2011). A Hövmoller diagram characterizes the meridional and temporal variations of the MHT in the SODA model (Fig. 2.8a). Lower latitudes have a larger amplitude of MHT variability but there is no change of phase in the latitudinal range considered herein; that is, the MHT variations are highly coherent from the tropics to the middle latitudes. The interannual variations of the MHT are larger than the decadal variations. A spectral analysis (not shown) indicates a strong 5 year peak, which is close to the ENSO periodicity. The correspondence between MHT and ENSO is more clearly shown in the comparison of the Hövmoller diagram with the time evolution of El Niño 3.4 Index (Fig. 2.8b). The MHT anomaly appears stronger during particularly intense El Niño years (i.e., 1965) and weaker during La Niña years (i.e., 2000). The latitudinal dependence of the MHT averaged over the full period (1958–2008) and at particularly strong Niño/Niña years is shown in Fig. 2.8c. The 50 year-mean MHT shows a 0.12 PW increment from 33°S to 15°S, very close to the observed figure reported by Lumpkin and Speer (2007). During El Niño years of 1965 and 1983 the latitudinal increase rise to ~0.4 PW in the same latitude range. During La Niña years (1990; 2000), the model shows a weaker meridional gradient.

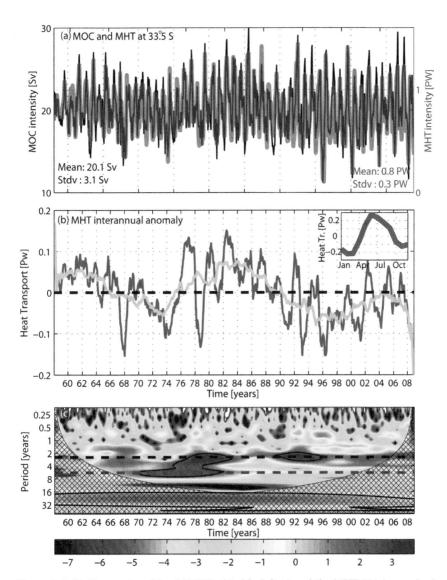

Figure 2.7. (a) Time series of the SAMOC (thin black line) and the MHT (thick gray line) at 33.5°S. (b) Detrended and de-seasonaled time series of MHT smoothed with a 13-month running mean (thick black line) and smoothed with a 5-year running mean (gray line). The insert shows the seasonal mean time-series of the MHT. (c) wavelet spectra of the time series of the MHT. The black contours show regions with a confidence level higher than 95%. Hatched regions indicate the cone of influence. The black and blue horizontal dashed lines indicate the 2.5 and 5-year period. The time series for the wavelet analyses have been normalized thus the color bar does not have units.

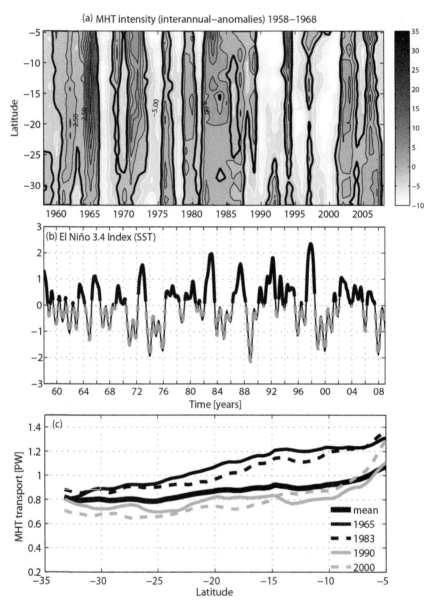

Figure 2.8. (a) Hövmoller diagram of MHT anomalies in the SA Ocean (units in TW = 10^{-3} PW). Black lines indicate positive anomalies and white lines negative ones. The thick black line is the zero contour. (b) Time series of El Niño 3.4 Index. Black lines are related with Niño years, gray dashed lines with La Niña years. (c) Latitudinal variability of the MHT in the SA for selected years.

Conclusion

We characterized the SA Ocean circulation patterns produced by the data-assimilating SODA model. The modeled patterns are in reasonable agreement with observations and other model results.

EOF analysis of different model variables reveal the existence of strong interdecadal oscillations. The leading modes of SSH anomalies are dominated by variations with periods of ~12 years. The bulk of the variability occurs in the subtropical region. The spatial distribution of the first mode can be characterized as a dipole structure with centers in the BMC and the ARR roughly divided by the Middle Atlantic Ridge (~20°W). There are secondary spectral peaks at 8 and 6 years and 5 and 3 years in the first and second EOF. Evaluation of the spatial structure of these modes suggest that the 8 and 6 years peaks represent dominant modes of variability in the eastern Atlantic while the 5 and 3 years peak represent dominant modes of variability in the western Atlantic.

Examination of the first three EOF modes of CURLW indicates that the wind effect is more dominant in the southeastern Atlantic and in the middle of the gyre. In particular, the maximum of the third EOF mode of CURLW seems to correspond with centers of minima (first mode) and maxima (third mode) of the SSH as emanating from the ARR, suggesting that these particular modes are associated with the ocean response to Ekman pumping. However, there are no significant peaks of CURLW variability at the decadal time scales, implying that the 12 year peak shown in the three leading EOFs of the SSH represent an internal mode of variability.

The first mode of SST variability is characterized by an amplitude maximum over the central and western portion of the subtropical gyre and a horse-shoe shaped structure of the opposing sign surrounding it. The second mode is a meridionally slanted dipole structure with peaks in the eastern tropical and the western subtropical region already identified in previous analyses, which indicates that this is the most robust mode of SST variability in this region (Venegas et al. 1998; Palastanga et al. 2002; Sterl and Hazeleger 2003; Haarsma et al. 2005). Its time series has spectral peaks at 15 and 6 years. The third mode of SST variability is a monopole centered over the subtropical gyre. This mode resembles the first mode describe by Venegas et al. (1998) and Palastanga et al. (2002). There are strong correlations between SSH and SST modes, thus suggesting that an important portion of the SST variability of the SA is not driven by air/sea interactions, as previously postulated, e.g., Venegas et al. (1998), Haarsma et al. (2005), but by advective effects associated with the geostrophic circulation.

The SAMOC and MHT predicted by the model results are also consistent with the wide range of previous estimates derived from observations and numerical models. The mean value computed from SODA at 33°.5S are close to the recently estimated values using data from mooring arrays (Meinem et al. 2013) but is larger than the values estimated from two high-resolution numerical models for the period 1986–1998 (Perez et al. 2011). The model SAMOC and MHT anomalies showed a large latitudinal coherence (i.e., positive or negative anomalies) and strong interannual variability that peak at a 5 year period, close to ENSO periodicity. The anomaly appears stronger during particularly intense El Niño years (i.e., 1965) and weaker during La Niña years (i.e., 2000). The 50 year-mean MHT shows a 0.12 PW increment from 33°S to 15°S, close to the observed figure reported by Lumpkin and Speer (2007). During El Niño years of 1965 and 1983 the latitudinal increase rise to ~0.4 PW in the same latitude range. During La Niña years (1990; 2000), the model shows a weaker meridional gradient.

Acknowledgments

E. Palma acknowledge the financial support of Agencia Nacional de Promoción Científica y Tecnológica (ANCYPT-Grant PICT12-0467), Universidad Nacional del Sur (Grant 24/F053), and Ministerio de Ciencia, Tecnología e Innovación Productiva (MINCyT/CONAE Grant 001). R. Matano acknowledge the financial support of NASA through Grants NNX08AR40G and NNX12AF67G, NOAA through Grant NA13OAR4310132 and the National Science Foundation through Grant OCE-0928348.

References

Beal, L.M. and Bryden, H.L. 1999. The velocity and vorticity structure of the Agulhas Current at 32°S. J. Geophys. Res. (Oceans) 104, C3: 5151–5176.
Bjornsson, H. and Venegas, S.A. 1997. A manual for EOF and SVD analyses of climate data, McGill University, CCGCR Report No. 97-1, Montréal, Québec: 1–52.
Boebel, O., Schmid, C. and Zenk, W. 1997. Flow and recirculation of Antarctic Intermediate Water across the Rio Grande Rise. J. Geophys. Res. (Oceans) 102: 20967–20986.
Boyer, T.P., Antonov, J.I., Baranova, O.K., Garcia, H.E., Johnson, D.R., Locarnini, R.A., Mishonov, A.V., O'Brien, T.D., Seidov, D., Smolyar, I.V. and Zweng, M.M. 2009. World Ocean Database. 2009. NOAA Atlas NESDIS, vol. 66, edited by S. Levitus, DVDs, U.S. Govt. Print. Off., Washington, D.C.: 1–256.
Broecker, W.S. 1991. The great Conveyor Belt. Oceanography 4(2): 79–89.
Carton, J.A., Chepurin, G.A., Cao, X. and Giese, B.S. 2000. A Simple Ocean Data Assimilation analysis of the global upper ocean 1950–1995, Part I: Methodology. J. Phys. Oceanogr. 30: 294–309.
Compo, G.P., Whitaker, J.S., Sardeshmukh, P.D., Matsui, N., Allan, R.J., Yin, X., Gleason, Jr., B.E., Vose, R.S., Rutledge, G., Bessemoulin, P., Brönnimann, S., Brunet, M., Crouthamel, R.I., Grant, A.N., Groisman, P.Y., Jones, P.D., Kruk, M.C., Kruger, A.C., Marshall, G.J., Maugeri, M., Mok, H.Y., Nordli, Ø., Ross, T.F., Trigo, R.M., Wang, X.L., Woodruff, S.D. and Worleyu, S.J. 2011. The twentieth century reanalysis project. Q. J. R. Meteorol. Soc. A. 137: 1–28.
de las Heras, M.M. and Schlitzer, R. 1999. On the importance of intermediate water flows for the global ocean overturning. J. Geophys. Res. (Oceans) 104: 15515–15536.

Donners, J. and Drijfhout, S. 2004. The Lagrangian view of South Atlantic interocean exchange in a global ocean model compared with inverse model results. J. Phys. Oceanogr. 34: 1019–1035.

Dong, S., Garzoli, S.L., Baringer, M.O., Meinen, C.S. and Goni, G.J. 2009. Interannual variations in the Atlantic meridional overturning circulation and its relationship with the net northward heat transport in the South Atlantic. Geophys. Res. Lett. 36: L20606: 1–5.

Dong, S., Baringer, M.O., Goni, G.J., Meinen, C.S. and Garzoli, S.L. 2014. Seasonal variations in the South Atlantic Meridional Overturning Circulation from observations and numerical models. Geophys. Res. Lett. 41(13): 4611–4618.

Döös, K. and Webb, D.J. 1994. The Deacon Cell and the other meridional cells of the Southern Ocean. J. Phys. Oceanogr. 24(2): 429–442.

England, M.H., Garcon, V.C. and Minster, J.-F. 1994. Chlorofluorocarbon uptake in a World Ocean model, 1. Sensitivity to the surface gas forcing. J. Geophys. Res. (Oceans) 99: 25215–25233.

Ganachaud, A. and Wunsch, C. 2000. The oceanic meridional overturning circulation, mixing, bottom water formation and heat transport. Nature 408: 453–457.

Garzoli, S.L. and Baringer, M.O. 2007. Meridional heat transport determined with expendable bathythermographs. Part II: South Atlantic transport. Deep Sea Res. Part I 54(8): 1402–1420.

Garzoli, S.L. and Matano, R.P. 2011. The South Atlantic and the Atlantic meridional overturning circulation. Deep Sea Res. Part II, 58(17-18): 1837–1847.

Garzoli, S.L., Baringer, M.O., Dong, S., Perez, R.C. and Yao, Q. 2013. South Atlantic meridional fluxes. Deep Sea Res. Part I, 71: 21–32.

Garzoli, S.L., Dong, S., Fine, R., Meinen, C.S., Perez, R.C., Schmid, C. and Yao, Q. 2015. The fate of the Deep Western Boundary Current in the South Atlantic. Deep Sea Research Part I, Oceanographic Research Papers 103: 125–136.

Giese, B.S. and Ray, S. 2011. El Niño variability in simple ocean data assimilation (SODA), 1871–2008. J. Geophys. Res. (Oceans) 116, C02024: 1–17.

Gordon, A.L. 1986. Interocean exchange of thermocline water. J. Geophys. Res. (Oceans) 91: 5037–5046.

Gordon, A.L. and Greengrove, C.L. 1986. Geostrophic circulation of the Brazil–Falkland confluence. Deep-Sea Res. 33: 573–585.

Gordon, A.L. and Bosle, K.T. 1991. Cyclonic gyre in the tropical South Atlantic. Vol. 38. Deep Sea Res. Suppl I: S323–S343.

Gordon, A.L., Weiss, R.F., Smethie, W.M., Jr. and Warner, M.J. 1992. Thermocline and Intermediate Water Communication Between the South Atlantic and Indian Oceans. J. Geophys. Res. (Oceans) 97: 7223–7240.

Haarsma, R.J., Campos, E.J., Hazeleger, W., Sevrijns, C., Piola, A.R. and Molteni, F. 2005. Dominant modes of variability in the South Atlantic: a study with a hierarchy of ocean-atmosphere models. J. Climate 18: 1719–1735.

Hogg, N.G. and Thurnherr, A.M. 2005. A zonal pathway for NADW in the South Atlantic. Journal of Oceanography 61(3): 493–507.

Holfort, J. and Siedler, G. 2001. The meridional overturning oceanic transports of heat and nutrients in the South Atlantic. J. Phys. Oceanogr. 31: 5–29.

Jackett, D.R. and McDougall, T.J. 1997. A neutral density variable for the world's oceans. J. Phys. Oceanogr. 27: 237–263.

Jayne, S.R. and Marotzke, J. 2001. The dynamics of ocean heat transport variability. Rev. Geophys. 39: 385–411.

Jones, P.W. 1999. First- and second-order conservative remapping schemes for grids in spherical coordinates. Mon. Weather Rev. 127: 2204–2210.

Kohl, A., Stammer, D. and Cornuelle, B. 2007. Interannual to Decadal Changes in the ECCO Global Synthesis. J. Phys. Oceanogr. 37: 313–337.

Lau, K.-M. and Weng, H. 1995. Climate signal detection using wavelet transform: How to make a time series sing. Bull. Amer. Meterol. Soc. 12: 2391–2402.

Lumpkin, R. and Speer, K. 2007. Global ocean meridional overturning. J. Phys. Oceanogr. 37(10): 2550–2562.

Lutjeharms, J.R.E. 1996. The exchange of water between the South Indian and South Atlantic Oceans. *In*: Wefer, G., Berger, W.H., Siedler, G. and Webb, D. (eds.). The South Atlantic: Present and Past Circulation. Springer-Verlag, Berlin-Heidelberg, Germany: 122–162.

Macdonald, A.M. and Wunsch, C. 1996. An estimate of global ocean circulation and heat fluxes. Nature 382: 436–439.

Macdonald, A.M. 1998. The global ocean circulation: a hydrographic estimate and regional analysis. Prog. Oceanogr. 41: 281–382.

Marchesiello, P., Barnier, B. and de Miranda, A.P. 1998. A sigma-coordinate primitive equation model for studying the circulation in the South Atlantic. Part II: Meridional transports and seasonal variability. Deep Sea Res. 45: 573–608.

Matano, R.P. 1993. On the separation of the Brazil Current from the coast. J. Phys. Oceanogr. 23: 79–90.

Matano, R.P. and S.G.H. Philander. 1993. Heat and mass balances of the South Atlantic Ocean calculated from a numerical model. J. Geophys. Res. (Oceans) 98: 977–984.

Meinen, C.S., Speich, S., Perez, R.C., Dong, S., Piola, A.R., Garzoli, S.L., Baringer, M.O., Gladyshev, S. and Campos, E.J.D. 2013. Temporal variability of the Meridional Overturning Circulation at 34.5°S: Results from two pilot boundary arrays in the South Atlantic. J. Geophys. Res. (Oceans) 118: 6461–6478.

Nof, D. and Van Gorder, S. 1999. A different perspective on the export of water from the South Atlantic. J. Phys. Oceanogr. 29: 2285–2302.

Olson, D., Podestá, G., Evans, R. and Brown, O. 1988. Temporal variations in the separation of the Brazil and Malvinas Currents. Deep-Sea Res. 35: 1971–1990.

Palastanga, V., Vera, C.S. and Piola, A.R. 2002. On the leading modes of sea surface temperature variability in the South Atlantic Ocean. Clivar Exchanges 25: 1–4.

Perez, R.C., Garzoli, S.L., Meinen, C.S. and Matano, R.P. 2011. Geostrophic velocity measurement techniques for the meridional overturning circulation and meridional heat transport in the South Atlantic. J. Atmos. Oceanic Technol. 28: 1504–1521.

Poole, R. and Tomczak, M. 1999. Optimum multiparameter analysis of the water mass structure in the Atlantic Ocean thermocline. Deep-Sea Res. I, 46: 1895–1921.

Rahmstorf, S. 2006. Thermohaline ocean circulation. *In*: Elias, S.A. (ed.). Encyclopedia of Quaternary Sciences. Elsevier, Amsterdam: 10 pp.

Rintoul, S.R. 1991. South Atlantic interbasin exchange. J. Geophys. Res. (Oceans) 96: 2675–2692.

Roemmich, D. and Wunsch, C. 1985. Two transatlantic sections: meridional circulation and heat flux in the subtropical North Atlantic. Deep-Sea Res. Part a-Oceanographic Research Papers 32: 619–664.

Sato, O.T. and Polito, P.S. 2008. Influence of salinity on the interannual heat storage trends in the Atlantic estimated from altimeters and Pilot Research Moored Array in the tropical Atlantic data. J. Geophys. Res. (Oceans) 113, C02008: 1–11.

Schott, F.A., Tramma, L. and Fischer, J. 1995. The warm water inflow into the western tropical Atlantic boundary regime, spring, 1994. J. Geoph. Res. (Oceans) 100: 24745–24760.

Sloyan, B.M. and Rintoul, S.R. 2001. Circulation, renewal, and modification of Antarctic Mode and Intermediate Water. J. Phys. Oceanog. 31: 1005–1030.

Smith, R.D., Dukowicz, J.K. and Malone, R.C. 1992. Parallel ocean general circulation modeling. Physica D 60: 38–61.

Souza, J.M.A.C., de Boyer Montegut, C., Cabanes, C. and Klein, P. 2011. Estimation of the Agulhas ring impacts on meridional heat fluxes and transport using ARGO floats and satellite data. Geophysical Research Letters 28, L21602: 1:5.

Speer, K.G., Siedler, G. and Talley, L. 1995. The Namib Col Current. Deep-Sea Res. 42: 1933–1950.

Speer, K., Guilyardi, E. and Madec, G. 2000. Southern Ocean transformation in a coupled model with and without eddy mass fluxes. Tellus 52: 554–565.

Sterl, A. and Hazeleger, W. 2003. Coupled variability and air-sea interaction in the South Atlantic Ocean. Climate Dynamics 21: 559–571.

Stramma, L., Fischer, J. and Reppin, J. 1995. The North Brazil Undercurrent. Deep Sea Res. Part I, 42: 773–795.

Stramma, L. and England, M. 1999. On the water masses and mean circulation of the South Atlantic Ocean. J. Geophys. Res. (Oceans) 104: 20863–20883.

Talley, L.D. 2003. Shallow, intermediate, and deep overturning components of the global heat budget. J. Phys. Oceanogr. 33: 530–560.

Talley, L.D. 2013. Closure of the global overturning circulation through the Indian, Pacific, and Southern Oceans: Schematics and transports. Oceanography 26(1): 80–97.

Tsuchiya, M., Talley, L.D. and McCartney, M.S. 1994. Water mass distributions in the western South Atlantic; A section from South georgia Island (54S) northward across the equator. J. Mar. Res. 52: 55–81.

van Sebille, E., Johns, W.E. and Beal, L.M. 2012. Does the vorticity flux from Agulhas rings control the zonal pathway of NADW across the South Atlantic? J. Geophys. Res. (Oceans) 117(C05037): 1–12.

Venegas, S.A., Mysak, L.A. and Straub, D.N. 1997. Atmosphere-ocean coupled variability in the South Atlantic. J. Climate 10: 2904–2920.

Venegas, S.A., Mysak, L. and Straub, D. 1998. An interdecadal climate cycle in the South Atlantic and its links to other basins. J. Geophys. Res. (Oceans) 103: 24723–24736.

Weijer, W., de Ruijter, W.P.M., Sterl, A. and Drijfhout, S.S. 2002. Response of the Atlantic overturning circulation to South Atlantic sources of buoyancy. Global Planet. Change 34: 292–311.

Witter, D.L. and Gordon, A.L. 1999. Interannual variability of South Atlantic circulation from 4 years of TOPEX/POSEIDON satellite altimeter observations. J. Geoph. Res. (Oceans) 104: 20927–20948.

Woodruff, S.D., Worley, S.J., Lubker, S.J., Ji, Z., Freeman, J.E., Berry, D.I., Brohan, P., Kent, E.C., Reynolds, R.W., Smith, S.R. and Wilkinson, C. 2011. ICOADS Release 2.5: Extensions and enhancements to the surface marine meteorological archive. Int. J. Climatol. 31: 951–967.

You, Y. 2002. Quantitative estimate of Antarctic Intermediate Water contributions from Drake Passage and the southwest Indian Ocean to the South Atlantic. J. Geophys. Res. (Oceans) 107, (C43031): 1–20.

3

The Issue of Fossil Fuels at the Ocean

Emissions to the Sea and Contribution to Global CO$_2$

A.V. Botello,* G. Ponce-Velez, L.A. Soto and S.F. Villanueva

Introduction

The vast ocean surface of the globe is known not only for its thermo-regulator capability but also for its significant capacity for absorbing anthropogenic CO$_2$ (Brewer 2009). Since the Industrial Revolution, the burning of fossil fuels has historically represented the primary source of energy to sustain economic productivity. However, they also have become one of the main anthropogenic stressors in the world oceans approximately since the mid 20th century. Its gas emissions into the atmosphere have been closely correlated with the increasing global CO$_2$ and linked to shifts in climate perturbations.

Presently, there is a consensus in recognizing the synergetic effect between climatic and anthropogenic factors. This effect is of such complexity that is hard to discern its consequences upon ecosystem processes (Mollman et al. 2012). Indeed, the attempt to assess changes in the stability and resilience functions of coastal or oceanic ecosystems represent challenging issues that remain unsolved. Future scenarios concerning fossil fuels input in the ocean are difficult to establish. Temporal and spatial ecosystem variability contributes to increasing the uncertainty in our interpretation

Universidad Nacional Autónoma de México, Instituto de Ciencias del Mar y Limnología. Circuito Exterior, Ciudad Universitaria s/n México, D.F. 04510.
* Corresponding author: gatoponcho2015@gmail.com

of anthropogenic disturbances. Global climate change is no longer a controversial issue. Large scale phenomena like El Niño or la Niña can create excessive periods of moisture or prolonged dryness episodes. Its effects provoke shifts in the evaporation-precipitation rates, increases watershed, CO_2 and sea-level rise (Karnauskas et al. 2015). These in conjunction cause loss of biodiversity, promote the life-cycle of invasive species and acidification processes. The black-tide produced by accidental oil spills can reach coastal regions when the severe climatic phenomena disrupt the normal evaporation-precipitation balance.

The Intergovernmental Panel on Climate Change (IPCC) has strongly emphasized the need to assess the risk and vulnerability of coastal systems given the predicted global atmospheric warming. This can cause changes in the circulation pattern of major ocean currents, its surface thermal regime, and in the average sea level. These factors have been closely associated with the frequency and intensity of recent hydro-meteorological hazards (storms and hurricanes), whose consequences have produced: severe floods in low-lands, coastal erosion, and alarming habitat alterations in mangroves, coral reefs and coastal lagoons systems. Another disturbing environmental scenario caused by the increasing trend of atmospheric CO_2 is the acidification of oceanic waters. Upsetting the delicate balance of atmosphere-sea water CO_2 system provokes irreversible change in the carbonate system that is the main pH regulating mechanism in the ocean. Increasing levels of acidification of the oceans represents a grave threat to all the benthic organisms that rely on the calcification process (fixation of $CaCO_3$) for the construction of their exoskeletons.

The relevance of analyzing the implications of fossil fuels emission into the ocean arises from the fact that oil provides 40 to 43% of all energy used by the world; approximately, oil accounts for nearly 40% of global warming emissions (IPCC 2011). Excessive build-up of CO_2 in the atmosphere and the concomitant global warming poses a threat to the ocean's health and human society.

The IPCC and the International Energy Agency have emphatically expressed their concern about the excessive use of coal in industrial operations, a practice that unfortunately, is expected to continue for several decades. Both agencies coincided that such a trend makes it much harder to center our attention in promoting cleaner fossil fuel technologies. (www. carbontracker.org/wp-content/uploads/2014/09/Unburnable-Carbon-Full-rev2-1.pdf)

This chapter intends to update the current state of knowledge on the following issues:

- Historical trends in exploration/exploitation of fossil fuels and inputs into the ocean.

- Holistic interpretation on the environmental consequences caused by accidental oil spills in marine ecosystems (loss or detrimental effects upon ecosystem functions, such as stability and resilience capacities).
- Social and economic implications of oil emissions in the ocean: as overexploitation of marine living resources; loss of aesthetic and recreational ecosystem properties.

Main World Oil Producers

According to the most recent census conducted by the U.S. Energy Information Administration in 2014 (U.S.E.I.A.), the world's top five oil producers are: the U.S. (13.9 thousand barrels/day) followed by Saudi Arabia (11.69 thousand barrels/day), Russia (10.89 thousand barrels/day), China (4.6 thousand barrels/day), and Canada (4.49 thousand barrels/day). These countries are not only important oil producers but they also represent the largest global energy consumers. Their economy relies heavily on the consumption of fossil fuels. In spite of the international efforts to reduce the fossil fuel emissions into the atmosphere, the prospect of reducing significantly human-caused CO_2 may take several decades. Presumably, the ocean has been credited for sequestering 28% of anthropogenic CO_2 since 1750 (Sabine et al. 2004). These authors used the global data set resulting from the World Ocean Climate Experiment (WOCE) to estimate a cumulative oceanic sink from 1800 to 1994 of ~433 ± 70 $GtCO_2$, or around half the fossil fuel emissions over this time period. Therefore, the ocean has constituted a net sink for anthropogenic CO_2 over the past 200 years.

In the IPCC (2013) Report, it is acknowledged that the atmospheric CO_2, methane, and nitrous oxide have increased to unprecedented levels for the past 800,000 years. It is estimated that the ocean has absorbed nearly 30% of anthropogenic CO_2, thus promoting acidification.

The new Carbon Tracker Initiative report seems to indicate we cannot trust market forces to achieve these climate goals—as the discrepancy between 'allowed atmospheric emissions space' and proven carbon reserves is simply too big—and a large part of these reserves are already owned by energy companies (Fig. 3.1). (www.bitsofscience.org; www.ft.com)

According to the report the top 100 coal companies and the top 100 oil companies presently own 745 gigatonnes CO_2 worth of fossil fuel reserves. Burning this stock would raise atmospheric CO_2 concentrations higher than the 450 limit. The entirety of proven reserves owned by private and public companies and governments is equivalent to 2,795 gigatonnes of CO_2, almost five times as much as the 450 Scenario allows us to consume.

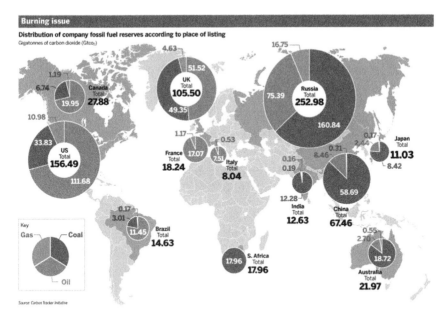

Figure 3.1. Global distribution of fossil fuel reserves.

In the period between 2000 and 2011 the world has not shown much understanding of carbon scarcity. Fossil fuel consumption led to the release of 321 gigatonnes of CO_2 over the last decade, 36% of the total allowed emissions space of 886 Gt between 2000 and 2050. (www.bitsofscience.org)

Data published by British Petroleum (BP), of the top 10 countries producing oil and natural gas.

	Country	2013	2014	v.i. (%)	%
1	United States	10,069	11,644	15.6	13.1
2	Saudi Arabia	11,393	11,505	1	13
3	Russia	10,777	10,838	0.6	12.2
4	Canada	3,977	4,292	7.9	4.8
5	China	4,216	4,246	0.7	4.8
6	Arab Emirates	3,648	3,712	1.8	4.2
7	Iran	3,525	3,614	2.5	4.1
8	Iraq	3,141	3,285	4.6	3.7
9	Kuwait	3,135	3,123	-0.4	3.5
10	Mexico	2,875	2,784	-3.2	3.1
	Mundial total	86,579	88,673	2.4	66.6

North America, Central and South America oil production in the last five years (2010–2014) bbl/day.

	2010	2011	2012	2013	2014
North America	16,115.90	16,685.40	17,915.30	19,330.90	21,167.80
Canada	3,441.70	3,596.90	3,855.90	4,073.10	4,383.30
Mexico	2,978.60	2,960.00	2,940.70	2,915.10	2,811.90
United States	9,695.60	10,128.50	11,118.70	12,342.80	13,972.60
Central & South America	7,881.70	8,058.10	8,002.20	8,126.40	8,407.50

Recent Emissions Trends

In 2012, global CO_2 emissions were 31.7 $GtCO_2$; this represents a 1.2% year-on-year increase in emissions, about half the average annual growth rate since 2000, and four percentage points less than in 2010, year of initial recovery after the financial crisis. Emissions in countries that are not included in Annex I of the United Nations Framework Convention on Climate Change as amended on 11 December 1997, continued to increase (3.8%), albeit at a lower rate than in 2011 while emissions in Annex I countries of this Convention, decreased by 1.5%. In absolute terms, global CO_2 emissions increased by 0.4 $GtCO_2$ in 2012, driven primarily by increased emissions from coal and oil in non-Annex I countries (OECD-IEA 2014) (Fig. 3.2).

Emissions from fuel

Although coal represented 29% of the world global total primary energy supply (TPES) in 2012, it accounted for 44% of the global CO_2 emissions due to its heavy carbon content per unit of energy released, and to the fact that 18% of the TPES derives from carbon-neutral fuels (Fig. 3.3). In contrast to gas, coal is nearly twice as emission-intensive on average.

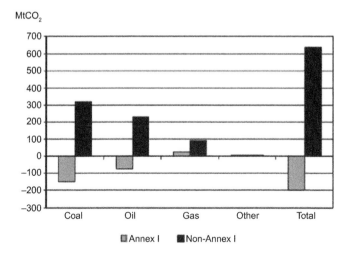

Figure 3.2. Change in CO_2 emissions (2011–2012). In 2012, emissions from coal and oil increased in non-Annex I countries and decreased in Annex I countries (OECD-IEA 2014).

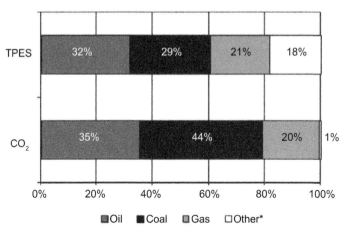

*Other includes nuclear, hydro, geothermal, solar, tide, wind, biofuels and waste.

Figure 3.3. World primary energy supply and CO_2 emissions: share by fuels in 2012. Globally, coal combustion generates the largest share of CO_2 emissions, although oil still is the largest energy source (OECD-IEA 2014).

Nearly 90% of the CO_2 emissions originate from fossil-fuel combustion and, therefore, are determined by the following three main factors:

- Energy demand or the level of energy-intensive activity; in particular, related to power generation, basic materials industry and road transport;
- Changes in energy efficiency;
- Shifts in the fuel mix, such as: from carbon-intensive coal to low-carbon gas, or from fossil fuels to nuclear or renewable energy.

In general, important drivers of specific fossil-fuel consumption are the fuel price, and relative price differences between coal, oil products, and natural gas. Indeed, energy policies also are aimed to manage fossil-fuel use (Fig. 3.3A).

These resources have evolved significantly during the last decade, following 10 years of rather stable relative contributions among fuels. In 2002 in fact, oil still held the largest share of emissions (41%), three percentage points above of coal (Fig. 3.4).

In 2012, CO_2 emissions from the combustion of coal increased by 1.3% to 13.9 $GtCO_2$. Currently, coal supports much of the growing energy demand of those emerging economies (such as China and India) where energy-intensive industrial production is growing rapidly and large coal reserves exist with limited reserves from other energy sources.

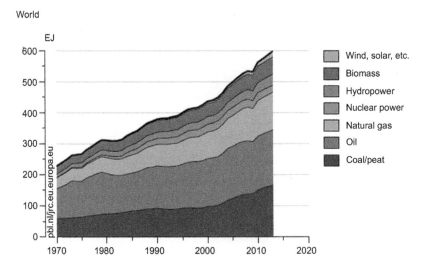

Figure 3.3A. Total primary energy supplies by type.

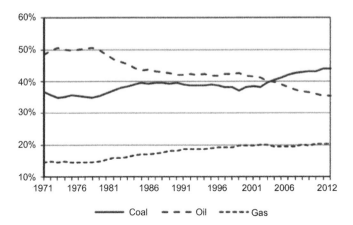

Figure 3.4. Fuels shares in global CO_2 emissions. The fossil fuel mix changed significantly in the last 10 years, with coal replacing oil as the largest source of CO_2 emissions (OECD-IEA 2014).

Emissions by region

Non-Annex I countries, collectively, represented 55% of global CO_2 emissions in 2012. At the regional level, annual growth rates varied greatly: emissions rate growth in China (3.1%) was lower than in previous years. However, emissions increased strongly in Africa (5.6%), Asia excluding China (4.9%) and in the Middle East (4.5%). Emissions in Latin America (4.1%) and Annex II Asia Oceania (2.5%) exhibited a more moderate rate, while emissions decreased in Annex II North America (–3.7%), Annex II Europe (–0.5%) and Annex I EIT (–0.8%) (Fig. 3.5).

Global CO_2 emissions reached a new high of 35.3 billion tones (Gt) CO_2 (Fig. 3.5A) in 2013, which is an increase of 0.7 Gt or 2.0% compared to the previous year. This moderate increase is similar to the actual recorded in 2012 of 0.6 Gt or 1.7%. After an average annual increase of CO_2 emissions of 1.1 Gt or 3.8% per year since 2003—when excluding the effect of the credit crunch recession years 2008 and 2009—the annual increases in 2012 and 2013 in global CO_2 emissions are about half of the increases in the preceding decade, albeit higher than the average annual increase in the 1995–2002 period. The average annual increase over the 1995–2002 period after the recession in the former Soviet Union was about 1.2% or 0.4 $GtCO_2$ per year. The increase in emissions over the 2012–2013 period is much smaller than expected, given that in 2012 and 2013 the global economy grew by 3.4 and 3.1%, respectively, which is slightly less than the average annual 3.9% growth rate of GDP since 2003 (again excluding the 2008–2009 years) (IMF 2014). Within the total increase in global emissions in 2013, there are remarkable differences between countries.

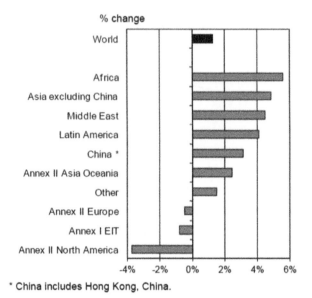

Figure 3.5. Change in CO_2 emissions by region (2011–2012). Emissions in Annex II North America fell in 2012; emissions in all non-Annex I regions escalated, and Africa showed the largest relative increase (OECD-IEA 2014).

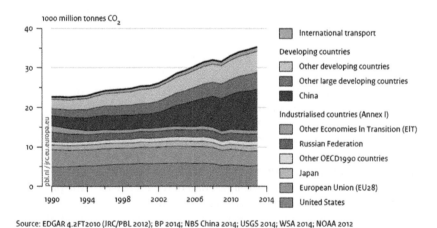

Source: EDGAR 4.2FT2010 (JRC/PBL 2012); BP 2014; NBS China 2014; USGS 2014; WSA 2014; NOAA 2012

Figure 3.5A. Global CO_2 emissions per region from fossil-fuel use and cement production.

Regional differences in contributions to global emissions conceal even larger differences among individual countries. Nearly two-thirds of global emissions for 2012 originated from just 10 countries, with the shares of China (26%) and the United States (16%) far surpassing those of all others. These two countries alone produced 13.3 $GtCO_2$. The top-10 emitting countries include five Annex I countries and five non-Annex I countries (Fig. 3.6).

As different regions and countries have contrasting economic and social structures, the picture would significantly change when moving from absolute emissions to indicators such as emissions per capita or per GDP.

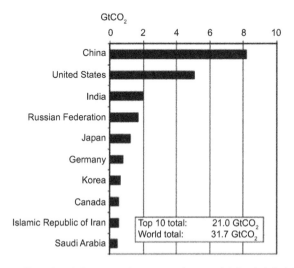

Figure 3.6. Top 10 emitting countries account for two-thirds of global CO_2 emissions in 2012 (OECD-IEA 2014).

Emissions by sector

Two sectors produced nearly two-thirds of global CO_2 emissions in 2012: electricity and heat generation, by far the largest, accounted for 42% while transport accounted for 23% (Fig. 3.7).

Generation of electricity and heat worldwide relies heavily on coal, the most carbon-intensive fossil fuel. Countries such as Australia, China, India, Poland and South Africa produce over two-thirds of their electricity and heat through coal combustion.

Between 2011 and 2012, CO_2 emissions from electricity and heat increased by 1.8%, faster than total emissions. While the share of oil in electricity and heat emissions has declined steadily since 1990, the input of gas increased slightly, and the contribution share of coal increased significantly, from 65% in 1990 to 72% in 2012 (Fig. 3.8). Carbon intensity developments for this sector will strongly depend on the fuel mix used to generate electricity, including the share of non-emitting sources, such as renewable and nuclear, as well as on the potential penetration of CCS technologies.

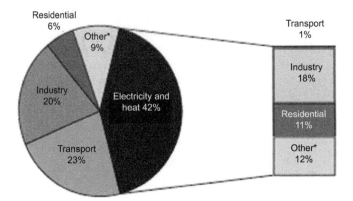

Figure 3.7. World CO$_2$ emissions by sector. Two sectors combined, generation of electricity and heat and transport represented nearly two-thirds of global emissions in 2012 (OECD-IEA 2014).

* Other includes commercial/public services, agriculture/forestry, fishing, energy industries other than electricity and heat generation, and other emissions not specified elsewhere.

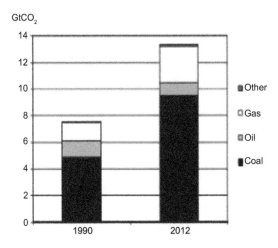

Figure 3.8. CO$_2$ emissions from electricity and heat almost doubled between 1990 and 2012, driven by the large increase of generation from coal (OECD-IEA 2014).

As for transport, the fast emissions growth was driven by emissions from the road sector, which increased by 64% since 1990 and accounted for about three quarters of transport emissions in 2012 (Fig. 3.9). It is interesting to note that despite efforts to limit emissions from international transport, emissions from marine and aviation bunkers, 66 and 80% higher in 2012 than in 1990, respectively, expanded even faster than those from road.

Non-OECD Americas includes Argentina, Bolivia, Brazil, Colombia, Costa Rica, Cuba, Dominican Republic, Ecuador, El Salvador, Guatemala,

GtCO$_2$

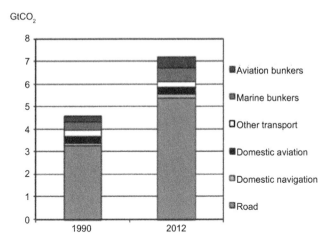

Figure 3.9. Global CO$_2$ emissions from transport (OECD-IEA 2014).

Haiti, Honduras, Jamaica, Netherlands Antilles, Nicaragua, Panama, Paraguay, Peru, Trinidad and Tobago, Uruguay, Venezuela and Other Non-OECD Americas. Other Non-OECD Americas includes Antigua and Barbuda, Aruba, Bahamas, Barbados, Belize, Bermuda, British Virgin Islands, Cayman Islands, Dominica, Falkland Islands (Malvinas), French Guyana, Grenada, Guadeloupe, Guyana, Martinique, Montserrat, Puerto Rico (for natural gas and electricity), St. Kitts and Nevis, Saint Lucia, St. Pierre and Miquelon, St. Vincent and the Grenadines, Suriname, and Turks and Caicos Islands.

Annex I Parties includes Australia, Austria, Belarus, Belgium, Bulgaria, Canada, Croatia, the Czech Republic, Denmark, Estonia, Finland, France, Germany, Greece, Hungary, Iceland, Ireland, Italy, Japan, Latvia, Liechtenstein (not available in this publication), Lithuania, Luxembourg, Malta, Monaco (included with France), the Netherlands, New Zealand, Norway, Poland, Portugal, Romania, Russian Federation, the Slovak Republic, Slovenia, Spain, Sweden, Switzerland, Turkey, Ukraine, the United Kingdom and the United States.

The countries that are listed above are included in Annex I of the United Nations Framework Convention on Climate Change as amended on 11 December 1997 by the 12th Plenary meeting of the Third Conference of the Parties in Decision 4/CP.3. This includes countries that were members of the OECD at the time of the signing of the Convention, the EEC, and 14 countries in Central and Eastern Europe and the Former Soviet Union that were undergoing the process of transition to market economies. At its 15th session, the Conference of the Parties decided to amend Annex I to the Convention to include Malta (Decision 3/CP.15). The amendment entered into force on 26 October 2010.

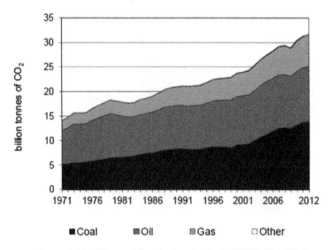

Figure 3.10. CO_2 world emissions by fuel (OECD-IEA 2014).

Wider Caribbean Region

Twenty-seven countries plus a number of territories and dependencies share the waters in the Wider Caribbean Region. There is a striking diversity of cultural and historic background, natural resource endowments, political organization and degrees of socio-economic development.

Presently, the Wider Caribbean Region (WCR) is potentially one of the largest oil producing areas in the world. Major production sites include Louisiana and Texas, USA, the Campeche Bay, Mexico; Lake Maracaibo, Venezuela; and the Gulf of Paria, Trinidad, all of which are classified as production-accident high-risk zones.

The WCR includes several independent and the dependent Island States in the Caribbean Sea, as well as mainland territories of South and Central America. This region covers an area of 4.3–106 km², with an estimated 40% of its human population residing within 2 km of the coast [United Nations Environment Program/Caribbean Environment Program (UNEP/CEP) 2001] (Fig. 3.11). (www.researchgate.net/publication/5941069)

The Wider Caribbean area is defined from the following waypoints:

1. Cost of Florida at 30°N
2. 30°N 77' 30"W
3. 20°N 59'W
4. 7°N20' 50"W
5. Cost of Guyana

Approximately 70% of the Caribbean population lives in coastal cities, towns and villages, a consequence of: the abundance of relatively easy to

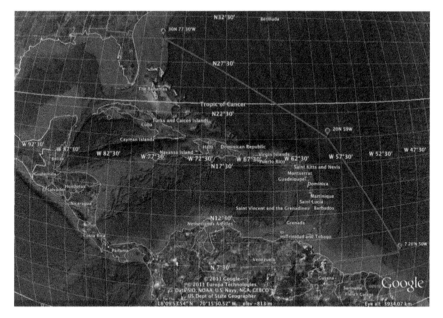

Figure 3.11. Wider Caribbean Area Map. (www.ancomaritime.com/page0/files/Wider_Caribbean_Region.html)

Source: UNEP ROLAC Atlas of the World, Comprehensive Edition, The Times 1994.

navigate and, therefore, very accessible natural harbors; the export oriented economy; the importance of artesian fisheries; and the tourism industry's coastal focus. More than half the population lives within 1.5 km of the coast and international airports, roads and capital cities are commonly situated along the coast (Mimura and others 2007). (www.pnuma.org/deat1/pdf/Climate_Change_in_the_Caribbean_Final_LOW20oct.pdf)

Main sources of petroleum pollution in the Wider Caribbean are: exploration and exploitation, production, storage and transportation, urban and municipal discharges, refining and chemical wastes, normal loading operations, rupture of oleo ducts and accidental spills.

Currently, more than 7 million of barrels are being transported daily in the waters of Caribbean, thus generating intense tanker traffic. It has been estimated that oil discharges from tank washings within the Wider Caribbean could be as high as 7 million barrels/year.

Recent estimates indicate that a total of 6.2 million tons of crude oil is released into the world's oceans from various sources. Close to 2.2 million tons of oil pollutants are attributed to marine transportation. According to Gundlach (1977), 28% of all petroleum derivates released in the oceans end up in coastal zones.

Table 3.1. CO_2 emissions from the consumption of petroleum (Million metric tons).

	2009	2010	2011	2012	2013
North America	2857.059	2850.58	2840.27	2794.69	2821.931
Bermuda	0.71151	0.71202	0.5922	0.61419	0.6381
Canada	271.7684	283.8449	288.9988	290.5136	293.7499
Greenland	0.64787	1.00099	0.6156	0.60494	0.61252
Mexico	263.8425	265.9298	259.1067	262.878	258.1161
Saint Pierre and Miquelon	0.08899	0.09226	0.09166	0.15105	0.15429
United States	2320	2299	2290.865	2239.928	2268.66
Central & South America	841.0166	914.8066	918.1542	944.9014	957.5212
Antarctica	0.24476	0.24481	0.14209	0.09946	0.10088
Antigua and Barbuda	0.66632	0.72231	0.58643	0.58643	0.61135
Argentina	79.61184	98.18628	98.28425	99.34855	102.1794
Aruba	1.03984	1.05483	0.89644	0.87641	0.89196
Bahamas, The	3.60306	3.60685	3.64329	3.83606	4.08049
Barbados	1.47093	1.37267	1.40361	1.27234	1.24454
Belize	0.46869	0.46976	0.47263	0.67518	0.67248
Bolivia	7.21639	7.98172	9.23582	9.62237	9.76599
Brazil	329.9378	362.3507	375.9334	390.2961	403.319
Cayman Islands	0.55311	0.55701	0.53744	0.47297	0.48805
Chile	48.32144	45.58633	41.65663	44.43942	44.7624
Colombia	35.83905	38.05036	41.30329	42.66859	42.9493
Costa Rica	6.60839	6.75625	6.37361	7.01428	7.19288
Cuba	26.74043	25.80145	25.73314	23.90187	23.89966
Dominica	0.12966	0.12999	0.14095	0.13206	0.13531
Dominican Republic	15.01701	15.10939	16.54733	16.91229	17.02307
Ecuador	28.40337	33.12088	32.92674	34.80375	35.78612
El Salvador	5.93246	5.93334	6.34964	6.37496	6.52037
Falkland Islands (Islas Malvinas)	0.04764	0.04766	0.04557	0.04557	0.04861

Table 3.1 contd. ...

...Table 3.1 contd.

	2009	2010	2011	2012	2013
French Guiana	0.82375	0.98756	1.07359	1.0401	1.05365
Grenada	0.27073	0.27035	0.27243	0.43129	0.48902
Guadeloupe	2.12595	2.12921	2.18799	2.32746	2.38928
Guatemala	11.03806	9.61074	11.41701	11.50135	11.4641
Guyana	1.64218	1.63656	1.67985	1.66138	1.66153
Haiti	2.24946	2.12937	2.08969	2.09431	2.0882
Honduras	6.80563	7.11114	7.47578	10.03583	10.00197
Jamaica	9.16112	7.9755	9.48075	12.451	12.53197
Martinique	2.67338	2.67523	2.79553	2.84116	2.91822
Montserrat	0.08324	0.08346	0.08801	0.08801	0.09241
Netherlands Antilles	12.20526	12.42646	12.15432	11.8431	11.77163
Nicaragua	4.28562	4.25815	4.61268	5.07572	5.23846
Panama	15.68038	18.58749	16.94065	15.98283	16.20549
Paraguay	4.0938	4.45232	3.99208	3.8664	3.8766
Peru	25.85795	28.0781	25.73766	24.28152	24.23582
Puerto Rico	23.96874	22.64778	21.74677	20.60569	16.06031
Saint Kitts and Nevis	0.27004	0.25632	0.27299	0.25068	0.26389
Saint Lucia	0.41273	0.43317	0.41898	0.41567	0.42457
Saint Vincent/ Grenadines	0.19548	0.20149	0.20064	0.26886	0.27823
Suriname	2.05052	2.71822	2.18749	2.26822	2.35156
Trinidad and Tobago	5.42513	5.8943	5.90033	6.44412	6.88569
Turks and Caicos Islands	0.16529	0.19585	0.16061	0.15936	0.15901
Uruguay	8.97565	8.01692	8.20867	7.46765	7.33408
Venezuela	93.37972	110.32	100.9761	105.5513	106.5159
Virgin Islands, U.S.	15.21083	14.45133	13.75365	12.40965	9.38291
Virgin Islands, British	0.11372	0.17694	0.11756	0.16008	0.17487

The main sources of fossil hydrocarbons in oceans and coastal zones are:

a. anthropogenic hydrocarbons produced by various human activities
b. biogenic hydrocarbons produced naturally by marine organisms
c. hydrocarbons from natural seeps.

In the open sea, concentrations of hydrocarbons are generally low, and their origin is not always easily determined. In contrast, in inshore waters (bays, estuaries and coastal areas), hydrocarbons may reach high concentrations, due to the direct effects of oil pollution resulting from spills, petrochemical plants and refineries by-products, normal loading operations and troposphere transport.

At present, there is growing evidence that pollution of coastal waters is accelerating in the Wider Caribbean region in view of the rapid industrialization and urban expansion that jeopardize the existing municipal infrastructure capacity.

Based on available information on contaminants input in coastal waters of the region, sewage represents the main source of public health and ecological risk.

Nonetheless, oil is the frequent pollutant detected in the region, and its impact in coastal ecosystems and recreational areas are well documented for both the Gulf of Mexico and the Caribbean Sea areas. (www.cep.unep.org/issues/lbsp.html)

Offshore oil and gas exploitation can become sources of pollution, either in the form of accidental oil spills or from the release of 'produced' from the oil-bearing strata with the oil and the gas at the time of production. The produced water is discharged into the marine environment together with waste drilling chemicals and mud, and may contain substances that exert high oxygen demand, together with toxic poly-aromatic hydrocarbons (PAHs), benzene, ethylbenzene, xylene and heavy metals, such as lead, copper, nickel and mercury. Accidental oil spills from offshore operations are often caused by pipeline breakage, well blowouts, platform fires overflows and equipment malfunctioning. In addition to the accidental oil spills, there is also a significant amount of natural seepage of petroleum hydrocarbons from submarine oil deposits, which contributes to marine pollution. Unlike the previously described sources of oil pollution, natural oil seepages are very difficult to estimate.

Much of the information on oil pollution levels in coastal and marine waters of the Wider Caribbean Region comes from the UNEP-IOC/IOCARIBE CARIPOL (Caribbean Oil Pollution Database) Program initiated in 1979. The data gathered by CARIPOL indicated that the concentration of dissolved/dispersed petroleum hydrocarbons (DDPHs) are generally low in offshore waters, while relatively high levels are found in enclosed coastal areas. Oil refineries and petrochemical plants were also seen as the major sources of coastal oil pollution within the region. NOAA Status and

Trends Programme has been gathering information about the accumulation of petroleum hydrocarbons, particularly toxic compounds, such as PAHs, in sediments and marine organisms along the U.S. Gulf Coast. The CARIPOL Programme has also obtained similar information along the Mexican Gulf coast and the coastal areas of the Caribbean region (Table 3.2). (http://www.cep.unep.org/issues/lbsp.html)

Table 3.2. Major oil spills occurred in the region since 1962 to date.

Year	Spill area	Million liters
1962	ARGEA PRIMA, Guámica, Puerto Rico	11
1967	Leaking pipe, Louisiana, USA	25
1968	WITWATER, Panama	3 diesel oil
1970	oversea platform	10
1971	SANTA AUGUSTA, St. Croix, USVI	13
1973	ZOECOLOCOTRONIS, Cabo Rojo, Puerto Rico	5
1975	GARBIS, Cayos de la Florida USA	24–5
1976	Corpus Christi, TX	1
1977	Unidentified boat. Bahía de Guayanilla, Puerto Rico	2
1978	Howard Star, Tampa, Florida USA	15–20
1979	BURHAH AGATE, Texas, USA	5–41
1979	ATLANTIC EMPRESS. Trinidad y Tobago	158
1979–80	IXTOC-I	528–1,626
1984	Alvenus, Louisiana USA	25
1985	RANGER, Texas, USA	24–52
1986	REFINERY Las Minas, Panamá	8
1991	VISTA BELLA Barge. St. Kitts y Nevis	2
1994	Berman, San Juan Puerto Rico	375
1997	NISOS AMORGOS	3.2
2000	Unknown Ship. La Habana Cuba	0.4
2010	Pozo Macondo. *Deepwater Horizon*. Coast Louisiana. Gulf of Mexico	780,000

In: Beltrán, J., Villasol, A., Botello, A.V. and Palacios, F. 2005.

Energy

Most CO_2 emissions in the region result from fossil fuel use and the region relies heavily on imported oil as its energy source and transport fuel (Table 3.3). Some countries, for example Barbados and Trinidad and Tobago, also draw on their own supplies of oil, bitumen and natural gas (Government

Table 3.3. The Caribbean GHG emissions (excluding land use, land-use change and forestry) (Gg).

Party	CO_2	CH_4	N_2O
Antigua and Barbuda	288.30	4.67	0.01
Bahamas	1 866.20	1.00	1.00
Barbados	1 913.38	85.12	0.16
Belize	598.07	265.04	0.55
Cuba	23 508.14	445.85	16.94
Dominica	76.53	2.98	0.04
Dominican Republic	18 416.75	230.33	9.75
Grenada	135.00	70.02	0.00
Guyana	1 445.80	42.15	1.21
Haiti	156.77	126.19	7.50
Jamaica	8 561.00	58.17	343.36
Saint Kitts and Nevis	70.89	2.83	12
Saint Lucia	268.59	28.37	0.07
Saint Vincent and the Grenadines	95.07	2.97	0.72
Suriname	4 150	41	12
Trinidad and Tobago	14 987.00	55.54	0.76

Source: UNFCCC 2005c, and the focal point for the Dominican Republic.

of Barbados 2001). The use of charcoal and other forest derived fuel products is not widespread in countries in the region, although charcoal is commonly produced and used in Haiti (Government of Haiti 2001). (www.pnuma. org/deat1/pdf/Climate_Change_in_the_Caribbean_Final_LOW20oct.pdf)

Sources

1. Marine transport

The Wider Caribbean Region is potentially one of the largest oil producing areas in the world. About 7 million barrels of crude oil are transported through the area every day generating intense tanker traffic (Fig. 3.12).

Tanker movements through the Caribbean constitute an intricate network of routes to and from producers, trans shipment point, refineries and consumers. The U.S. is the major producer and importer of crude oil and petroleum products in the region, and this fact alone establishes most of the tanker routes.

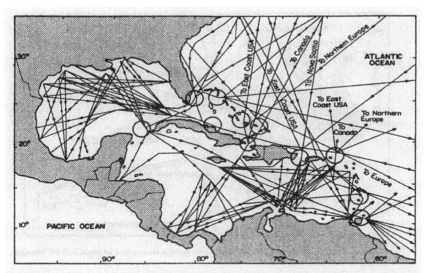

Main trade lanes for petroleum hydrocarbons in the Wider Caribbean and Gulf of Mexico (Reinberg 1984).

Figure 3.12. Main trade lanes for petroleum hydrocarbons in the Wider Caribbean and Gulf of Mexico (Reinberg 1984).

2. Spills

Estimated on the annual inputs rates of oil into the marine environment, at a global scale, oil spills have been reduced during the last years from 2.13×10^6 tons/year to 1.47×10^6 tons/year (Farrington 1985). Nevertheless, both estimates indicate that chronic oil spillage is more substantial than the more noticeable input from tanker accidents.

Accidental or intentional spills of crude or refined petroleum products in the course of tanker operations constitute another significant fraction of the oil reaching the marine environment.

The main risk areas for oil spills from ships are at the passages where the density of traffic is particularly high. The busiest passages are through the Yucatan Channel, the Bahamas Channel and the Florida Strait (ITOPF 2003). According to GESAMP 2007, there are an estimated 250 oil spills in the Gulf of Mexico and the Caribbean Sea annually. (www.cep.unep. org/publications-and-resources/marine-and-coastal-issues-links/oils-hydrocarbons)

Nevertheless, massive oil spills are dramatic and highly visible, receiving considerable attention in the media due to the severe ecological damage caused by the spilled oil. In the Wider Caribbean Region, the Ixtoc-1 oil spill in the shallow waters of the SW Gulf of Mexico is, perhaps the best-known case (Jernelov and Linden 1981; Soto et al. 2014) having produced one of the largest man-made spill prior to the war-related spills in the region of Kuwait in 1991. In the course of the Ixtoc-1 blowout, approximately 0.5 million tons of light petroleum were released to the marine environment over the course of 9 months before the well was finally capped (Table 3.4).

Following the Ixtoc-1 accident, Mexico implemented a long-term survey program in order to evaluate the impacts of oil activities in marine waters and coastal ecosystems in the Gulf of Mexico (1990–2005).

Table 3.4. Jernelov and Linden (1981) based on data from the Ixtoc-1 case analogies with other spills, made an estimation of the fate of the crude oil spilled as follows.

	Percent	Metric tons
Burned at well site	1	5,000
Mechanically removed at well site	5	23,000
Evaporated into the atmosphere	50	238,000
Degraded biologically and chemically	12	58,000
Landed on Mexican beaches	6	29,000
Landed on Texas beaches	1	4,000
Sank to the bottom	25	120,000

Other Spills

A summary of data on oil spills and accidents are included in Tables 3.5 and 3.6, based on information compiled by Corredor (1991).

Exploration and Exploitation

The main non-living resources which are exploited in the Wider Caribbean are oil and gas. The petroleum industry alone generates 70% of Venezuela's national income, and is critical to the economies of Trinidad-Tobago, Mexico, and the United States Gulf coast. In terms of oil reserves, the Wider Caribbean contains more than 5% of the world resources. Offshore oil exploitation is particularly important in the Gulf of Mexico's continental shelf where more than 1000 platforms are currently in operation. Oil exploitation in Lake of Maracaibo, Venezuela; is also important in this respect.

The International Maritime Organization (IMO) estimates that 6.7% of the total offshore oil production, at a global level, is lost through spills as a consequence of pipeline accidents, well blowouts, platform fires, overflows and malfunctions.

Table 3.5. Routes of oil traffic and major oil spill accidents in the Caribbean.
(Source: ITOPF 1996).

No.	Year	Source and spill zone	Millions of liters (type of oil)
1	1971	Saint Augusta, St. Croix, U.S. Virgin Islands	13 (crude)
2	1973	Zoe Colocotronis, Cabo Rojo, Puerto Rico	5 (crude)
3	1975	Garbis, Florida Keys, United States	24–25 (crude)
4	1976	Ruptured pipeline in Corpus Christi, Puerto Rico	1 (crude)
5	1977	Unidentified ship, Guayanilla Bay, Puerto Rico	2 (crude)
6	1978	Howard Star, Tampa, Florida, United States	15–20% crude, 80% bunker
7	1979	Burhah Agate, Texas, United States	5–41
8	1979	Atlantic Empress, off the coast of Trinidad and Tobago	158
9	1979–1980	IXTOC-I marine platform explosion, Campeche, Mexico	528–1626 (crude)
10	1984	Alvenus, Louisiana, United States	25
11	1985	Ranger, marine platform explosion, Texas, United States	24–52
12	1986	Las Minas Refinery, Panama	8 (crude)
13	1991	Vista Bella Barge, off Saint Kitts and Nevis	2 (bunker C)
14	1994	Berman, San Juan, Puerto Rico	375 (gasoil No. 6)
15	1997	Nisos Amorgos, tanker, Gulf of Venezuela	3.2

Table 3.6. Larger oil spills in the GIWA region Caribbean Islands since 1973.

Year	Location	Oil spill (liters)
1973	Zoe Colocotronis, Cabo Rojo, Puerto Rico	5 million crude oil
1975	Garbis, Florida Keys US	24.5 million crude oil
1976	Pipe break, Corpus Christi, TX	1 million crude oil
1977	Unknown Ship, Guayanilla Bay, Puerto Rico	2 million Venezuelan crude oil
1978	Howard Star, Tampa, Florida, US	0.4 million crude oil and 2.5 million bunker oil
1982	Tanker Princess Anne Marie, Pinar del Rio Cuba	5.3 million crude oil, 0.7 million fuel oil
1991	Vista Bella Barge, off side St. Kitts & Nevis	2 million bunker oil
1994	Morris J. Berman, San Juan Puerto Rico	3.7 million fuel oil
1998	Unknown ship, Havana, Cuba	0.06 million diesel oil
2000	Unknown ship, Havana Cuba	0.4 million crude

Source: IOCARIBE 1997; CIMAB 2000

In addition to potential accidents, offshore oil exploration has other impacts on the adjacent coastal region. The deployment of oil rigs and pipelines creates exclusion zones for fishing vessels and shipping. Moreover, the elimination and disposal of an oil platform that have reached the end of their usefulness need also to be considered.

As indicated earlier in this chapter, Mexico is the world's 10th largest oil producer (2.91 million barrels per day (Mb/d) or 3.5% of world total), after Saudi Arabia, Russia, the US, China, Canada, Iran, UAE, Kuwait and Iraq. Its production reached a peak in 2004 but has since been in decline. Between 2004 and 2012, it fell by 24% as production from its main offshore oil field, Cantarell, deteriorated.

Mexico has relied on the US market for more than 75% of its oil exports, which account for 10% of total US imports. However, as the US is expected to become the world's largest oil producer by the end of this decade thanks to its shale revolution, Mexico needs a new exporting strategy. The US drilled 137 deepwater and ultra-deepwater wells in the Gulf of Mexico in 2012 compared with six wells by Mexico (Fig. 3.13).

In the case of Mexico, there have been changes in the oil reserves estimates between 1998 and 2005, such changes are significant and reflect a lack of investment in exploration thus causing a reduction in the relative value of fossil fuel Resources/Production (R/P). It should be noted that if

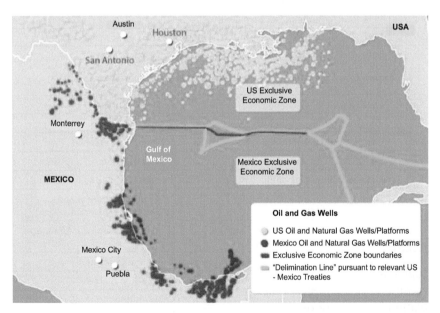

Figure 3.13. Actual sites for oil production in the Gulf of Mexico. www.worldreview.info/content/mexico-courts-foreign-investment-historic-energy-reformhttp://www.worldreview.info/content/mexico-courts-foreign-investment-historic-energy-reform.

only the existing oil reserves (11.83 billion barrels) are considered, the R/P is 9.7 years. This figure seems an underestimation, but in reality it is not.

It is wise to invest a lot of funds in exploration to increase this value without limit, because when there are high numbers of proven reserves, resources are unproductive. The important thing is to maintain the value of R/P in a suitable range over time, adding enough new reserves to replace oil extraction conducted in the year. The big private oil companies have an R/P of between 10 and 15 years (Fig. 3.14). (www.energiaadebate.com/Articulos/marzo2008/Bazanmarzo2008.htmhttp://www.energiaadebate.com/Articulos/marzo2008/Bazanmarzo2008.htm; http://www.energiaadebate.com/Articulos/marzo2008/Bazanmarzo2008.htm; http://www.energiaadebate.com/Articulos/marzo2008/Bazanmarzo2008.htm)

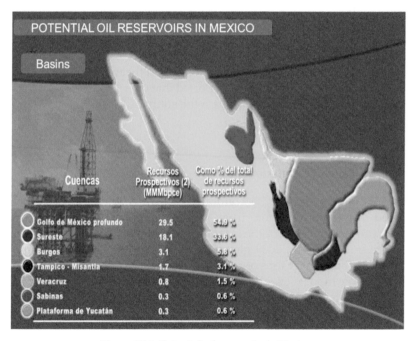

Figure 3.14. Potential oil reservoirs in Mexico.

Accidents

With regard to pollution caused by accidents, the most important pollutant resulting from shipping operations is oil. The results published by the US National Academy of Sciences (1985) indicated that, of the million tons of oil entering the sea from all sources, aproximately 1.5 million tons was caused by marine transportation. Of this, 400,000 tons/year is the result of maritime accidents. In the Wider Caribbean about 7 million barrels of crude oil are transported throughout the region every day, which generates intensive tanker traffic.

Seeps

Literature reviews provide evidence that naturally occurring oil and gas seeps existed in the Gulf of Mexico and the Caribbean Sea for thousands of years. There are many references in the historical literature of 16th and 17th-century Spanish explorers caulking their ships with tar found on the beaches of south Texas and Louisiana.

In 1933, Price reported a seep on the north end of St. Joseph's Island off Texas coast. Levorsen (1954) listed natural offshore seeps in the Gulf of Mexico off Yucatan and in the Caribbean in Vonsetes Bay, Barbados.

Many seeps occur in northern Mexico oil fields in the so-called Golden Lane trend south of Tampico. De Goyler (1932) stated that over 6000 seeps were found in Cerro Azul 400 km south of Brownsville, Texas. Similarly, gas seepages are a relatively common phenomena in the northwestern Gulf of Mexico. Watkins and Worzel (1978) reported that over 19000 seeps probably exist in a small area about 6000 km² on the south Texas shelf.

Discussion

Global fossil-fuel CO_2 emissions-trends

Since 1751 approximately 356 billion metric tons of carbon have been released into the atmosphere from the consumption of fossil fuels and cement production. Half of these fossil fuel CO_2 emissions have occurred since the mid 1980s. In 2009 global fossil-fuel carbon emission estimate was 8378 million metric tons of carbon, represents a slight decline of 0.35% from all—time high (8769 million metric tons of carbon) in 2008. The slight decline is due to the Global Financial Crisis which began in mid-2008 and had obvious economic and energy use consequences particularly in North America and Europe.

Globally, liquid and solid fuels accounted for 76% of the emissions from fossil-fuel burning and cement production in 2009. Combustion of gas fuels accounted for 17.9% (1568 million metric tons of carbon) of the total emissions from fossil fuels in 2009 and reflected a gradually increase global utilization of natural gas. Emissions from center production (413 million metric tons of carbon in 2009) have more than doubled since the mid-1990s and now represent 4.7% of global CO_2 releases from fossil-fuel burning and cement production. Gas flaring which accounted for roughly

2% of global emissions during the 1970s now accounts for less than 1% of global fossil-fuel releases.

What are the potential emissions from global fossil fuel reserves? The Potsdam Climate Institute also calculated the total potential emissions from burning the world's proven fossil fuel reserves (coal, oil and gas). This is based on reserve figures reported at a country level and UNFCCC emissions factors for the relevant fossil fuel types. Oil was split into conventional and unconventional types, whilst coal was split into three different bands to reflect the range of carbon intensity. The total CO_2 potential of the earth's proven reserves comes to 2795 $GtCO_2$; 65% of this is from coal, with oil providing 22% and gas 13%. This means that governments are currently indicating their countries contain reserves equivalent to nearly five times the carbon budget for the next 40 years. Consequently, only one-fifth of the reserves could be burnt unabated by 2050 if we are to reduce the likelihood of exceeding 2°C warming to 20% (Fig. 3.15). (www.carbontracker.org/wp-content/uploads/2014/09/Unburnable-Carbon-Full-rev2-1.pdf; http://www.carbontracker.org/wp-content/uploads/2014/09/Unburnable-Carbon-Full-rev2-1.pdf)

Coal, like all other sources of energy, has a number of environmental impacts, from both coal mining and coal use.

Coal mining raises a number of environmental challenges, including soil erosion, dust, noise and water pollution, and impacts on local biodiversity. Steps are taken in modern coal mining operations to minimize these impacts.

Continuous improvements in technology have dramatically reduced or eliminated many of the environmental impacts traditionally associated with the use of coal in the vital electricity generation and steelmaking industries. Viable, highly effective technologies have been developed to tackle the release of pollutants—such as oxides of sulfur (SOx) and

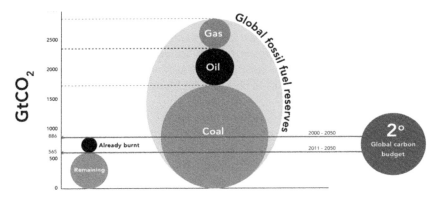

Figure 3.15. Comparison of the global 2°C carbon budget with fossil fuel reserves CO_2 emissions potential.

nitrogen (NOx)—and particulate and trace elements, such as mercury. More recently, greenhouse gas (GHG) emissions, including carbon dioxide (CO_2) and methane (CH_4) have become a concern because of their link to climate change.

There is now growing recognition that technology developments have to be part of the solution to climate change. This is particularly true for coal because its use is growing in so many emerging economies, including the largest and fastest growing countries such as China and India.

Carbon capture use and geological storage (CCUS) technology are the only currently available technology that allows very deep reductions to be made in CO_2 emissions to the atmosphere from fossil fuels at the scale needed.

Failure to widely deploy CCUS will seriously hamper international efforts to address climate change. The Intergovernmental Panel on Climate Change (IPCC)—the pre-eminent body on climate science—has identified CCUS as a critical technology to stabilize atmospheric greenhouse gas concentrations in an economically efficient manner. The IPCC found that CCUS could contribute up to 55% of the cumulative mitigation effort by 2100 while reducing the costs of stabilization to society by 30% or more.

CCUS will be needed across a number of sectors that need to tackle CO_2 emissions, including fossil fuel power stations (coal, gas and oil), steel, aluminum, cement, and chemicals. (www.worldcoal.org/coal-the-environment/http://www.worldcoal.org/coal-the-environment/; http://www.worldcoal.org/coal-the-environment/)

North America, Central and South America oil production in the last five years (2010–2014) bbl/day

The anthropogenic carbon emissions caused by the burning of fossil fuels, cement production and deforestation are having a major impact on the world's largest ecosystem—the oceans. Atmospheric carbon dioxide is the highest it has been for at least the last 15 Ma (Tripati et al. 2009; LaRiviere et al. 2012; Bijma et al. 2013) and probably longer (300 Ma) (Hönisch et al. 2012). The effect is both a warming of the atmosphere and of the oceans (Rayner et al. 2003; IPCC 2007a; Belkin 2009; Sherman et al. 2009; Reid and Beaugrand 2012).

Another direct impact of raised atmospheric CO_2 is ocean acidification, through its entry into marine surface waters and its chemical reaction with the water (Caldeira and Wickett 2003; Caldeira 2007; Cao and Caldeira 2008). The physical and chemical impacts of CO_2 emissions are not limited to the direct effects of warming and a lowering of ocean pH. There are processes associated with warming and acidification, many of which are summarized in this chapter. These factors can when combined, amplify their potential impact on ocean's life. For example, global warming will increase surface ocean stratification, which in turn will affect the surface-water light regime and nutrient input from deeper layers. This will impact primary production (Rost et al. 2008). Oxygen transport to the deep sea by down welling water masses will be weakened by freshening because of increased melt water input from Greenland and high-Arctic glaciers, altering patterns of ocean mixing, slowing down the conveyor belt and leading to progressive depletion of the ocean's oxygen inventory. The potential effects of these factors are further exacerbated by other anthropogenic stresses, such as pollution, eutrophication and overfishing, which have destabilized some ecosystems and significantly reduced many species' populations, thus limiting the potential for adaptation. The geological record suggests that the current acidification is potentially unparalleled in at least the last 300 million years of Earth history, and raises the possibility that we are entering an unknown territory of marine ecosystem change (Hönisch et al. 2012). This chapter summarizes the observed impacts of the last century, and the predicted impacts of a continued elevation of CO_2 on the marine environment.

Although the human-induced pressures of overexploitation and habitat destruction are the main causes of recently observed extinctions (Dulvy et al. 2009), climate change is increasingly adding to this. Changes in ocean temperatures, chemistry, and currents mean that many organisms will find themselves in unsuitable environments, potentially testing their ability to survive. Adaptation is one mean of accommodating environmental change; migration is another. However, global warming asks for a pole ward migration whereas ocean acidification would require an equator ward migration as colder waters acidify faster.

Given that consumption of petroleum represents 35% of world primary energy use, with natural gas and coal making up 24 and 28%, respectively, it is an understatement to say that the world economy depends critically on these non-renewable resources. The default position for some governmental agencies, such as the US Energy Information Administration appears to be that the fossil-fuel resource base is large enough that economic considerations are of primary importance.

Increasing demand for energy comes from worldwide economic growth and development. Global Total Primary Energy Supply (TPES) more than doubled between 1971 and 2012, mainly relying on fossil fuels (Fig. 3.16).

Despite the growth of non-fossil energy (such as nuclear and hydropower), considered as non-emitting, the share of fossil fuels in the world energy supply is relatively unchanged over the past 41 years. In 2012, fossil sources accounted for 82% of the global TPES.

Growing world energy demand from fossil fuels plays a key role in the upward trend in CO_2 emissions (Fig. 3.17). Since the Industrial Revolution,

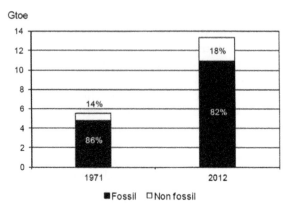

* World primary energy supply includes international bunkers.

Key point: Fossil fuels still account for most – over 80% – of the world energy supply.

Figure 3.16. The world primary energy supply (includes international bunkers). Fossil fuels still account for most—over 80%—of the world energy supply (OECD-IEA 2014).

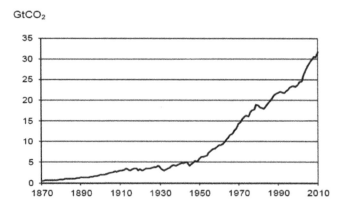

Source: Carbon Dioxide Information Analysis Center, Oak Ridge National Laboratory, US Department of Energy, Oak Ridge, Tenn., United States.

Key point: Since 1870, CO_2 emissions from fuel combustion have risen exponentially.

Figure 3.17. Trend in CO_2 emissions from fossil fuel combustion. Since 1870, CO_2 emissions from fuel combustion have risen exponentially.

annual CO_2 emissions from fuel combustion dramatically increased from near zero to almost 32 $GtCO_2$ in 2012. Next a brief overview of recent trends in energy-related CO_2 emissions, as well as in some of the socio-economic drivers of emissions are provided.

Conclusions

The review of three issues related to fossil fuel emissions into the ocean—trends in exploration/exploitation and inputs in the ocean, accidental oil spills, and social and economic implications-revealed that the development of the world society continues to rely heavily on the combustion of fossil fuels as the primary source of energy. Fuel emissions into the atmosphere account roughly for 40% of the global warming provoked by the excessive CO_2 build-up. Consequently, ocean's health is threatened by the resulting temperature increase and acidification of oceanic waters. For the past 200 years, world oceans have acted as a natural sink for CO_2. In absolute terms, global CO_2 emissions increased by 0.4 $GtCO_2$ in 2012, driven primarily by increased emissions from coal and oil. In 2002, coal combustion was responsible for the largest CO_2 atmospheric emission (41%). However, oil consumption surpassed this percentage by three points. Unfortunately, the prospect of reducing the injection of greenhouse gases into the atmosphere may take several decades.

In the period 2011–2012, significant increases in CO_2 emissions were regionally recorded in Africa (5.6%), Asia (4.9), and Latin America (4.1%), while in North America (–3.7%) and Europe (–0.5%) decreasing trends were noted. Nonetheless, since different regions and countries have distinct economic and social structures, the scenario would significantly change when moving from total emissions to indicators such as emissions per capita or GDP. Nearly two-thirds of global emissions for 2012 originated from just ten countries, with the shares of China (26%) and the United States (16%) far surpassing those of all others.

Recent estimates indicate that a total of 6.2 million tons of crude oil is released into the world's oceans from various sources. For instance, in the Caribbean waters, more than 7 million of barrels are being transported daily. The Wider Caribbean Region (WCR) is one of the largest oil-producing areas in the world. Major production sites include Louisiana and Texas, USA, the Campeche Bay, Mexico; Lake Maracaibo, Venezuela; and the Gulf of Paria, Trinidad, all of which are classified as production-accident high-risk zones. In the open sea, concentrations of hydrocarbons are low, and their origin is not always readily determined. In contrast, in inshore waters (bays, estuaries, and coastal areas), hydrocarbons may reach high concentrations, due to the direct effects of spills, petrochemical plants and refineries by-products, normal loading operations and troposphere transport. At present, there is growing evidence that pollution of coastal waters is accelerating in the WCR given the rapid industrialization and urban expansion. The data gathered by Caribbean Environment Program (CARIPOL) indicated that the concentration of dissolved/dispersed petroleum hydrocarbons (DDPHs) are normally low in offshore waters, while relatively high levels are found in enclosed coastal areas.

Accidental or intentional spills of crude or refined petroleum products in the course of tanker operations constitute a significant fraction of the oil reaching the marine ecosystem. According to GESAMP 2007, there are an estimated 250 oil spills in the Gulf of Mexico and the Caribbean Sea annually. Massive oil spills are dramatic and highly visible, receiving considerable attention in the media due to the severe ecological damage caused by the spilled oil. In the Wider Caribbean Region, the Ixtoc-1 oil spill in the shallow waters of the SW Gulf of Mexico, and the Deepwater Horizon in the northern Gulf, are perhaps the best-known cases. Data on oil spills and accidents are here updated.

Continuous improvements in technology have dramatically reduced or eliminated many of the environmental impacts traditionally associated with the use of coal. Viable, highly efficient technologies have been developed to cope with the release of pollutants—such as oxides of sulfur (SOx) and nitrogen (NOx)—and trace elements, such as mercury. There is now mounting recognition that technology developments have to be part of the solution to climate change. This applies in particular to coal because its escalating use in many emerging economies, including the largest and fastest growing countries such as China and India. Carbon capture use and geological storage (CCUS) technology are the only currently available technology that allows very deep reductions to be made in CO_2 emissions to the atmosphere from fossil fuels. The physical and chemical impacts of CO_2 emissions are not limited to the direct effects of warming and a lowering of ocean pH, many of which are summarized in this chapter. These factors can, when combined, amplify their potential impact on ocean's life.

References

Belkin, I.M. 2009. Rapid warming of large marine ecosystems. Prog. Oceanogr. 81: 207–213.
Beltrán, J., Villasol, A., Botello, A.V. y Palacios, F. 2005. Condición actual del ambiente marino-costero de la región del Gran Caribe. pp. 1–24. *In*: Botello, A.V., Rendón-von Osten, J., Gold-Bouchot, G. y Agraz-Hernández, C. (eds.). Golfo de México Contaminación e Impacto Ambiental: Diagnóstico y Tendencias, 2da Edición. Univ. Autón. de Campeche, Univ. Nal. Autón. de México, Instituto Nacional de Ecología. 696 p.
Bijma, J., Pörtner, H.O., Yesson, Ch. and Rogers, A.D. 2013. Climate change and the oceans—What does the future hold?. Marine Pollution Bulletin 74 (2013): 495–505.

CIMAB. 2000. Dictamen técnico de muestras colectadas en el derrame de petróleo en la costa de ciudad de La Habana. Dictamen Técnico. La Habana, 9 p.

Corredor, J. 1991. State of pollution by oil and marine Debris in the Wider Caribbean Region. IOC-IO-CARIBE/UNEP Workshop Cartagena Colombia. June 1991 (Unpublished manuscript).

DeGoyler, E. 1932. Oil associated with igneous rocks in Mexico. Am. Assoc. Pet. Geol. Bull. 16: 779–808.

Dulvy, N., Pinnegar, J. and Reynolds, J. 2009. Holocene extinctions in the sea. *In*: Turvey, S. (ed.). Holocene Extinctions. Oxford University Press, Oxford, UK.

Farrington, J.W. 1985. Oil pollution: A decade of monitoring and research. Oceanus 28: 2–12.

GESAMP. 2007. Estimates of Oil Entering the Marine Environment From Sea-based Activities. GESAMP (IMO/FAO/UNESCO-IOC/UNIDO/WMO/IAEA/UN/UNEP Joint Group of Experts on the Scientific Aspects of Marine Environmental Protection). 2007. Estimates of oil entering the marine environment from sea-based activities. Rep. Stud. GESAMP No. 75, 96 pp.

Hönisch, B., Ridgwell, A. and Schmidt, D.N. 2012. The geological record of ocean acidification. Science 335: 1058. http://dx.doi.org/10.1126/science.1208277.

IOCARIBE. 1997. Regional Marine Pollution Emergency. Information and training center Wider Caribbean. COI/REMPEITC/Caribbean 43 p.

ITOPF. 1996. An assessment of the risk of oil spills and the state of preparedness in 13 International tanker owners pollution federation limited, 28 p.

IPCC. 2007a. Climate Change 2007: The Physical Science Basis. Solomon, S., Qin, D., Manning, M., Chen, Z., Marquis, M., Averyt, K., Tignor, M.M.B. and Miller, H.L. (eds.). Working Group 1 Contribution to the Fourth Assessment Report of the Intergovernmental Panel on Climate Change (IPCC). Technical Summary and Chapter 10 (Global Climate Projections).

Jernelov, A. and Linden, O. 1981. IXTOC-I: A case study of the world's largest oil spill. Ambio: 299–306.

LaRiviere, J.P., Ravelo, A.C., Crimmins, A., Dekens, P.S., Ford, H.L., Lyle, M. and Wara, M.W. 2012. Late Miocene decoupling of oceanic warmth and atmospheric carbon dioxide forcing. Nature 486: 97–100.

Levorsen, A.I. 1954. Geology of petroleum. Freeman, San Francisco, Calif, 703 pp.

OECD/IEA. 2014. CO_2 emissions from fuel combustion. France, 136 pp.

Rayner, N.A., Parker, D.E., Horton, E.B., Folland, C.K., Alexander, L.V., Rowell, D.P., Kent, E.C. and Kaplan, A. 2003. Global analyses of sea surface temperature, sea ice, and night marine air temperature since the late nineteenth century. J. Geophys. Res. 108(D14): 4407. doi: http://dx.doi.org/10.1029/2002JD002670.

Reid, P.C. and Beaugrand, G. 2012. Global synchrony of an accelerating rise in sea surface temperature. J. Mar. Biol. Assoc. UK 92: 1435–1450.

Reinberg, L. Jr. 1984. Waterborne trade of petroleum and petroleum products in the Wider Caribbean Region. Final Reort N° CG-W-10-84 US Department of transportation and US Coast Guard. Washington D.C., 106 p.

Rost, B., Zondervan, I. and Wolf-Gladrow, D. 2008. Sensitivity of phytoplankton to future changes in ocean carbonate chemistry: current knowledge, contradictions and research directions. Mar. Ecol. Prog. Ser. 373: 227–237.

Sherman, K., Belkin, I.M., Friedland, K.D. O'Reilly, J. and Hyde, K. 2009. Accelerated warming and emergent trends in fisheries biomass yields of the world's large marine ecosystems. Ambio 38: 215–224.

Tripati, A., Roberts, C. and Eagle, R. 2009. Coupling of CO_2 and ice sheet stability over major climate transitions of the last 20 million years. Science 326: 1394–1397.

Soto, L.A., Botello, A.V., Licea-Durán, S., Lizárraga-Partida, M.L. and Yañez-Arancibia, A. 2014. The environmental legacy of the Ixtoc-I oil spill in Campeche Sound, southwestern Gulf of Mexico. Front. Mar. Sci., 07 November 2014.

http://dx.doi.org/10.3389/fmars.2014.00057.

(UNEP/CEP) United Nations Environment Program/Caribbean Environment Program. The Coastal Zone, 2001.

UNFCCC. 2005c. UNFCCC, 2005c. Sixth Compilation and synthesis of initial national communications from Parties not included in Annex I to the Convention, FCCC/SBI/2005/18/Add.2. Climate Change Secretariat, Bonn, Germany.

US National Academy of Sciences. 1985. Petroleum in the Marine Environment. National Academy of Science, Workshop on inputs, fates and the effects of petroleum in the marine environment. Airlie Mouse. Virginia. May 21–25, 1973. 107 p.

www.ancomaritime.com/page0/files/Wider_Caribbean_Region.html

www.bitsofscience.org

www.carbontracker.org/wp-content/uploads/2014/09/Unburnable-Carbon-Full-rev2-1.pdf

www.cep.unep.org/issues/lbsp.html

www.cep.unep.org/publications-and-resources/marine-and-coastal-issues-links/oils-hydrocarbons

www.energiaadebate.com/Articulos/marzo2008/Bazanmarzo2008.htmhttp://www.energiaadebate.com/Articulos/marzo2008/Bazanmarzo2008.htm

www.ft.com

www.pnuma.org/deat1/pdf/Climate_Change_in_the_Caribbean_Final_LOW20oct.pdf

www.researchgate.net/publication/5941069

www.worldcoal.org/coal-the-environment/http://www.worldcoal.org/coal-the-environment/

www.worldreview.info/content/mexico-courts-foreign-investment-historic-energy-reformhttp://www.worldreview.info/content/mexico-courts-foreign-investment-historic-energy-reform

4

Continent Derived Metal Pollution Through Time

Challenges of the Global Ocean

Luiz Drude de Lacerda[1,*] and *Jorge Eduardo Marcovecchio*[2]

Introduction

The global demand for materials and energy grows apace since the middle of last century and no stopping seems feasible in the next decades. Despite increasingly strict controls of emissions from anthropogenic sources, the demand growth results in an increment of the release into the environment of by-products able to contaminate natural ecosystems. On the other hand, the increasing social and economic pressures towards sustainable development will require the recovery of degraded areas, the more efficient use of natural resources and effective monitoring of environmental conditions.

Whereas most contaminant sources are located inland, their emissions, through the atmosphere and watersheds, and human-built structures such as pipelines and outfalls, will eventually reach the ocean. A key step in the transfer of materials to the ocean is continent–ocean interface, which serves a crucial role in the functioning of the Earth System and contributes to the maintenance of a significant portion of the global population and numerous production chains. The magnitude and rate of change in human

[1] Instituto de Ciências do Mar (LABOMAR), Universidade Federal do Ceará, Av. Abolição 3207, Fortaleza, CE, 60.165-081, Brasil.
[2] Instituto Argentino de Oceanografía (IADO–CONICET/UNS), Florida 7000, Edificio E-1, 8000 Bahía Blanca, Argentina.
  E-mail: jorgemar@iado-conicet.gob.ar
[*] Corresponding author: ldrude@pq.cnpq.br

activities that modify hydrological systems and the transport of material into the ocean have intensified over the last 150 years. They have produced significant changes in the structure and functioning of oceanic ecosystems. Thus, there is an urgent need to formulate public policies that permit the mitigation of impacts and the adaptation of populations affected by this scenario and that contribute to the welfare and conservation of natural capital in this important region, as well as hampering the export of pollution to the oceans and increasing environmental pressures upon their resources.

The interaction between the continent and the ocean harbors processes that originate in the watersheds of tributary rivers, induced by water masses dynamics on the continental shelf and in the ocean, and by atmospheric processes, which ultimately define the magnitude and characteristics of the flux of materials, energy and organisms across the interfaces (Fig. 4.1). Currently, this interface operationally encompasses a stretch of 100 km of shoreline to the continental slope, which corresponds to less than 25% of the planet's surface. It houses nearly 50% of the human population and 75% of megacities with more than 10 million inhabitants and produces approximately 90% of global fisheries. This interface is the focus of industrial and transportation development and a significant source of mineral resources, including petroleum and natural gas. It is a primary tourist destination and a reservoir of biodiversity and ecosystems, which the functioning of the planet depends on. Changes in anthropogenic origin generally reduce the availability of resources and services that impede sustainable development and cause significant depreciation of natural capital. The usual disciplinary fragmentation of the managers responsible for public policies diminishes our ability to understand interdisciplinary

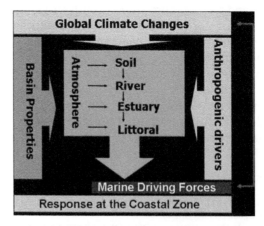

Figure 4.1. Conceptual model of the processes that act on the continent-ocean interface. This figure highlights the interdependence between the different environmental compartments inter-influenced via the transfer of material and energy shaped by global and regional changes in the continent and ocean.

processes that act on the continent-ocean interface and to understand and mitigate the impacts of regional and global changes. This fragmentation also limits our ability to direct management strategies, which hinders the decision-making process. This obstacle prompts the design of unreliable scenarios of adaptation to environmental changes.

The ecological and biogeochemical sensitivity of natural ecosystems to pollutants depends on the characteristics of each particular region, including the human activities installed there. Therefore, it is necessary to delineate consistent indicators of the support capacity of natural ecosystems, to provide reliable scenarios for the implementation of future activities and their management. This is especially important in marine ecosystems where monitoring of all aspects, both biological and abiotic, eventually affected by pollution is virtually impossible. The definition of these indicators requires a deep understanding of the biogeochemical processes that control the flow and cycling of substances in this new scenario of global and regional changes induced by human activity, i.e., the biogeochemistry of the Anthropocene.

Natural ecosystems have an intrinsic capacity to immobilize contaminants (supporting capacity) which is dependent on their fundamental ecological properties. However, all the planet's ecosystems are in varied degree of alteration due to anthropogenic actions. Therefore, their support capacity is in different stages of utilization. The ability of a given ecosystem to immobilize a given contaminant is therefore also a function of the degree of alteration or degradation of the system. Relatively healthy systems are able to immobilize a particular contaminant load allowing small remobilization. Once the system is approaching its maximum capacity of immobilization, small additional contaminant loads can result in great remobilization of the same (Stigliani 1995; Lacerda 2003). We can assume that presently, most marine ecosystems are already very close to reach their maximum support capacity, in such a way that small changes induced by global changes can result in significant impacts at the regional and local level.

Changing the Nature of Anthropogenic Sources of Pollutants

J.O. Nriagu (Nriagu 1993; 1996) coined in the late 20th century the term 'The Silent Epidemic' to characterize the contamination by metals originated from anthropogenic activity, that have been the target of increasingly stringent environmental legislation during the last two decades of the 20th century. As a result, much of the point sources of metals in industrialized countries were greatly reduced. However, the legacy of the indiscriminate use of metals over the past 150 years remains potentially available for remobilization following changes in land use (Lacerda 2007). Today, much of the problems of environmental contamination originate in the emission from diffuse sources difficult to control and even to quantify (Pirrone et al. 1998; Pacyna et al. 1999). Figure 4.2 illustrates the main sources of Hg to the atmosphere. Diffuse sources such as fossil fuel burning and artisanal gold mining are the main current source of Hg to the atmosphere, contributing with over 62% of the total. These sources are usually more difficult to control and even quantify.

On the other hand, significant changes have occurred in not only the sources, but also their global distribution. Figure 4.3 shows the comparison between the current emissions of Hg to the global atmosphere. It is clearly a redirection of emissions from industrialized countries to developing or even underdeveloped nations, where less stringent legislation results

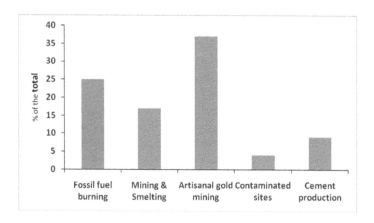

Figure 4.2. Global Hg emissions to the atmosphere from anthropogenic sources (UNEP 2013).

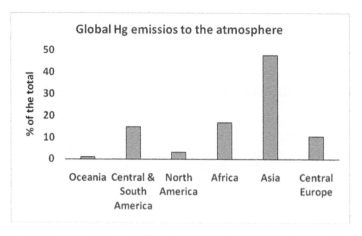

Figure 4.3. Relative contribution of different geographical areas to the overall Hg input to the atmosphere.

in higher emissions of Hg to the environment. These emissions occur in environments whose ecological processes are unknown or only poorly comprehended and where the man-nature interaction still corresponds to the livelihood of a large part of the population, in particular food security. Similar situation occurs with many other contaminants, especially metals and POPs (Persistent Organic Pollutants).

The growing significance of diffuse sources also results in that even in less urbanized and industrialized sectors, such as the northeast coast of Brazil for example; runoff estimates show that the sum of contributions of different anthropogenic sources for most coastal basins, even excluding industrialization, is significantly larger than the contribution of natural sources (Lacerda et al. 2008; 2009; 2011). Figure 4.4 shows the relative contribution from natural, including the physical and chemical denudation of soils and the atmospheric deposition, and different anthropogenic sources of nutrients (N, P) and environmentally significant trace metals (Cu, Zn and Hg) to this region of the Brazilian coast formerly tough as a touristic destination where human signatures are still not evident. Thus, the inclusion of anthropogenic sources in biogeochemical models that involve these substances becomes mandatory; however, no systematic inventories of these emissions are inexistent for most global seas.

Another significant aspect of the geographical changes in the location of the major sources of contaminants is the magnitude and rapidity of the regional changes in most developing countries and, unfortunately, in many preserved areas of the planet. These changes may cause events of 'resurgence' of contaminants whose sources have already disappeared or decreased significantly and, therefore, will have their mobilization mechanisms controlled by biogeochemical processes inherent of each ecosystem rather than the intensity and size of the sources themselves (Lacerda and Bastos 2012).

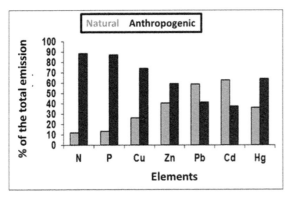

Figure 4.4. Relative contributions of natural and anthropogenic sources for 19 estuaries of the northeast coast of Brazil (CE, RN and PB), obtained from data in Lacerda et al. (2008; 2009; 2011).

The Scenario of Regional Changes

Changes of land use and water resources have enormous potential for changing the dynamics of contaminants control processes through the environment. For example, the conversion of forest areas in pasture and agriculture in the Amazon after the disappearance of the artisanal gold mining, the main source of Hg during the decades of 1980 and 1990, still retain high concentrations of Hg in fish and humans (Bastos et al. 2006), due to remobilization of Hg deposited in soils during that period. At the continent-ocean interface where most land-use changes presently takes place, the situation is much more intense and the process much faster. The diversion of waters from different basins to provide freshwater for a growing population has enormous impacts on the transfer of pollutants from continental watersheds to the ocean. At the Paraíba do Sul River basin to the basin of the Bay of Sepetiba, in Southeastern Brazil, water diversion resulted in an increase in sedimentation rates of the Bay (Fig. 4.5). In addition, an extra contribution of about 30% resulted in the total flux of Hg and 20% of Zn and Pb for this Bay from outside its original watershed (Molisani et al. 2006).

On the other hand, the effects of damming and watershed diversion in the rivers in the semi-arid northeastern coast of Brazil have the opposite effect (Fig. 4.6). Water withdrawn in artificial dams results in the decrease in the flow of freshwater to the ocean and the extensive erosion of the coastline, in many sectors occupied by mangroves. These ecosystems have a high capacity of accumulation of metals. Erosion of mangrove substrate results in mobilization of sediments enriched in metals and their release and availability to the biota. Simultaneously the changes of soil uses that may result in changes in metals fluxes to the ocean, river diversion also induces changes in the sources of metals themselves. Changes in agricultural

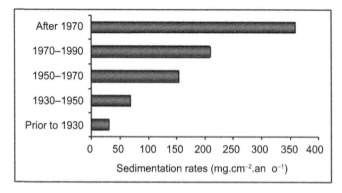

Figure 4.5. Sedimentation rates in Sepetiba Bay, southeastern Brazil, during the 20th century (Lacerda et al. 2002).

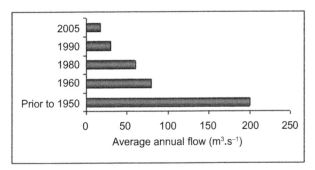

Figure 4.6. River flow to the Atlantic Ocean from the Jaguaribe River, northeastern Brazil during the past 50 years (Marins et al. 2002).

practices and the introduction of new practices can also result in increased flows of metals to the environment. In several estuaries of the Brazilian Northeast, the introduction of irrigated fruit growing, allowed by damming and river diversion, has increased the load of metals from this source due to the increase use of fertilizers and chemicals, typical of intensive agriculture, which have varying concentrations of metals. Similarly, the expansion of aquaculture has brought new challenges to the understanding of this recent source contribution to the total flow of metals into estuaries (Lacerda et al. 2006; 2011). The transport of these metals for marine ecosystems, however, will depend on transfer process of material (e.g., soil) that is not directly proportional to the magnitude of the source, making it difficult to their modeling.

The extent to which anthropogenic interference vectors overlap and even alter the dynamics of contaminants, which hinder the understanding of natural fluctuations and the system response to global changes, is another important aspect in the understanding of the interaction between the continent and the ocean in the scenario of the Anthropocene. The knowledge of the contribution of small and medium-sized rivers, and even some large rivers recently altered by anthropogenic activities, for global entries to the oceans is scarce or poorly updated, due to lack of studies and monitoring of local and regional authorities more consistent (Maybeck and Vörösmarty 2006). Thus, the order of magnitude of the global transport of continental materials into the ocean is still a controversial topic. Global estimates result from currently accepted models and extrapolations (Meybeck 2005). However, newer models and greater database already provide variations of these estimates three times larger than those traditionally accepted. Possibly, the differences found are due to optimizations of mathematical models and the highest number of data for the reference bases used to increased anthropogenic loads, but also due to global climate change. These observations emphasize the need for local and regional studies to better estimate continental flows for the oceans, particularly in the South Atlantic Ocean. Moreover, as these estimates turn out to affect estimates of concentrations of chemical species in the ocean, when chemical balances are involved, and/or where there is the presence of gaseous phase, larger uncertainty occurs.

In some sectors of the Brazilian coast, there is sufficient knowledge to confirm the interdependence between drainage basins and the ocean. The drainage basin of the Paraná River, for example, interferes with coastal and oceanic processes in the North, up to latitude 22° (Möller et al. 2008), with a huge export of sediments which influences strongly the geochemistry of continental shelf of the coastal and shelf areas of southeast and southern Brazil (Burone et al. 2013). Recently, studies on the rate of sedimentation and geochemistry of continental shelf to the north of Rio de Janeiro, show that land uses change even in medium-sized basins, such as the Rio Paraíba do Sul, are able to change the quantity and quality of the sediments deposited on the continental shelf (Wanderley et al. 2013). This change affects the transport of substances beyond the continental slope, gauged by chemical species distribution and geochemical fractions and elementary relations indicators of impacts vectors (Dittmar et al. 2012).

In a similar way, studies developed in waters from the South Atlantic show comparable results. Piola et al. (2005) reported that the discharge of La Plata River ($\sim$22,000 $m^3.s^{-1}$) produce a strong impact and numerous modifications on South Western Atlantic shelf waters, including the transport of a huge volume of continental materials towards the ocean. In addition, and according to Ronco et al. (2008) human activities are the main cause of serious pollution of surface waters, sediments and soils within La Plata basin due to both point and non-point industrial, agricultural and urban sources, and these emissions will finally accumulate in the continental shelf. Moreover, this transference processes could be significantly increased due to periodical extreme variations within the precipitation regime from the Paraná – La Plata hydrological basin (Boulanger et al. 2005; Depetris 2007). The distribution and concentration of nutrients (e.g., nitrate, silicate) and dissolved oxygen clearly indicate the combined effect of the Malvinas current, Brazil current and La Plata basin outflow (Braga et al. 2008) which function as a fertilization process supporting the seasonal primary production and fishery productivity within this region.

These kind of studies has also been developed in other coastal systems along the Atlantic littoral of Argentina. Thus, the transport of continental materials to the ocean through Bahía Blanca estuary has been (and still is being) assessed (e.g., Marcovecchio et al. 2008), including different impacts and effects linked to nutrients (Negrin et al. 2012; Spetter et al. 2015), trace metals (Botté et al. 2010; Fernández Severini et al. 2013; Simonetti et al. 2013) or POPs (Arias et al. 2009; 2010; 2011; Oliva et al. *in press*) within different abiotic or biological compartments of the estuary. Finally, both Gaiero et al. (2003) and Depetris et al. (2005) have also reported results similar to those previously pointed out for numerous rivers from Patagonia, in the southern Argentina, representing a significant source of continental materials for the Southern Atlantic Ocean.

The understanding of this transport is crucial, for example, for the characterization of the environmental effects of offshore exploitation of oil and natural gas and dredging (Lacerda and Marins 2006; Lacerda et al. 2013a; Aguiar et al. 2013).

Unfortunately, the existing knowledge on the role of watersheds in the continent–ocean interaction is still very restricted. Preliminary data suggest a strong biogeochemical control exercised by plume of different rivers along the northeast and southeastern coast of Brazil on the dynamics of nutrients, Suspended Solids (SS) and Greenhouse Gases (GHG) in the water column, and even in the sediment. The hinterland of the semi arid Brazilian Northeast witnesses' fast economic transformations mainly associated with industrial decentralization and may result in indirect impacts in more distant coastal basins. For example, the Salgado River basin, a tributary of the river Jaguaribe, in the eastern coast of Ceará State (Fig. 4.7), has an emergent industrial pole, associating leather, textile and

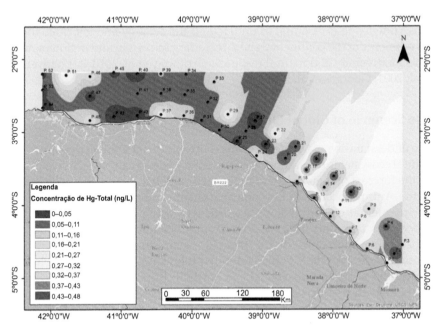

Figure 4.7. Mercury distribution in coastal waters of the semi arid northeast as a proxy of regional changes in the region's watersheds. The clear pattern of higher concentrations to the east are a result of such changes.

metallurgy industries. The industrial sector in the region has grown 16% in the period 2006–2012.

In addition, the intensive use of chemical fertilizers, of pesticides on crops, the increasing loads of domestic effluents discharged by cities, waste from livestock and deforestation of large areas, results in an increasing emission of metals to the adjacent coastal areas. Figure 4.7 presents Hg distribution in surface waters on the continental shelf adjacent of this region. Mercury is a very reliable proxy of overall anthropogenic emissions the easily visualized pattern of higher concentrations to the east of the continental shelf relative to the western sector strongly suggest the impacts of changing land use in the continual watersheds.

Even considering the magnitude of the natural fluxes uploaded by rivers, evidence based on the distribution of ^{210}Pb activity in sediments of the continental shelf sediment profiles; strongly suggest the anthropogenic influence in the composition of the sediments brought to the sea. Nery and Bonoto (2011), for example, associated changes in ^{210}Pb activity to mining, along the northern coast of Brazil. Wanderley et al. (2013) associated these changes to agricultural inputs and related changes in sedimentation rates and sediment composition to changes in sugar cane agricultural practices in the adjacent coastal plains. However, sudden fluctuations in the chemistry of interstitial waters, reported in these studies, suggest that other drivers influence the distribution of continental materials on the continental shelf. This knowledge is essential to understand the ecology and geochemistry of sediments of the continental shelf, including the origin and maintenance of its biodiversity and natural resources (Anthony et al. 2010; 2012; Joyeux et al. 2001; Georges et al. 2007).

The Scenario of Global Climate Change

The understanding of the mechanisms controlling contaminant dynamics in natural ecosystems throughout the Quaternary period can provide subsidies on the forecast of their behavior in the face of global change characteristics of the Anthropocene. To implement measures aiming the management environmental contamination, in particular of the marine environment, it becomes necessary to have a deep understanding of biogeochemical processes responsible for the transfer, accumulation and cycling of contaminants over time. Figure 4.8 shows the variation of the atmospheric deposition of Hg in a coastal lake in northern Brazil. The analysis of this temporal pattern shows great sensitivity of this metal to the climate changes that occurred along the Quaternary. It is possible, for example, to clearly depict an increase of Hg deposition in more recent periods related to warmer conditions verified in the last millennium. Such variations relate to known changes in the global weather conditions (Lacerda et al. 1999).

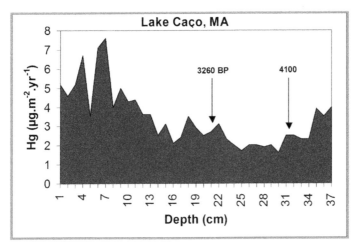

Figure 4.8. Variation of the atmospheric deposition of Hg in Lake Caço, Maranhão state northeastern Brazil, over the past 5,000 years (adapted from Lacerda et al. 1999).

Stupar et al. (2014) studied the deposition and accumulation of Hg in sediments from the Suquía River and the lagoons of Plata and Mar Chiquita, all of them in the province of Cordoba (in the central Argentina). The concentrations of Hg appeared to be stable along the last century, including in sites close to the big cities located within the studied area (e.g., Córdoba city, ~3.3 million inhabitants), with mean values close to 27 µg Hg.kg^{-1} sed.; most of this mercury was linked to organic matter and oxo(hydr)oxides of Fe and Mn, and associated to extremely fine grain sediments (silt and clay). However, a significant higher peak of Hg (~131 µg Hg.kg^{-1}) was determined in the core portion corresponding to 1985–1992, and which can be presumably related to Hudson volcano eruption that occurred at that time. This kind of process also affects the continent–ocean transfer considering the studied systems mentioned (Suquía River, Plata lagoon and Mar Chiquita lagoon) outflows to the Paraná River, which finally discharges on La Plata River and consequently into the South Atlantic Ocean.

It is a well-known fact that land use is one of the determinant sources of metals to associated aquatic systems (Kang et al. 2010; Wuana and Okieimen 2011). Inappropriate land use has been discussed as a factor that can affect human health over many years, and land-use change directly reflects changing human activities in recent decades (Patz et al. 2004; Pielke 2005). Land-use change can lead to vegetation damage, soil, water and land degradation, and local and global climate change, either directly or indirectly. Land use can also affect the fate of contaminants in soil. Keeping in mind that aquatic environments are the final sink of continental materials drainage, human health risks may increase as a result of the bioaccumulation of toxic substances through the food web, and land use determines the start of the food web (Zhao et al. 2012). Unfortunately, rapid and significant changes in land use are expected as a response to global climate change.

Climate model scenarios provide the best available information for assessing future impacts of climate change on the water quality and ecology of surface water bodies (Kundzewicz et al. 2007; Bates et al. 2008). Climate change is expected to have far-reaching consequences for river regimes, flow velocity, hydraulic characteristics, water levels, inundation patterns, residence times, changes in wetted areas and habitat availability, and connectivity across habitats (Brown et al. 2007). More intense rainfall and flooding could result in increased loads of suspended solids (Lane et al. 2007), sediment yields (Wilby et al. 1997), *Escherichia coli* and contaminant metal fluxes (Hartland et al. 2012; Visser et al. 2012) associated with soil erosion and fine sediment transport from the land (Leemans and Kleidon 2002). Consequently, changes within this direction have a direct effect on the transport of materials (trace metals among them) from the continental systems to the associated marine ones.

Climate change will have a powerful effect on the environmental fate and behavior of chemical toxicants through variation on physical, chemical and biological drivers of partitioning between the atmosphere, water, soil/sediment and biota (Moe et al. 2013; Hooper et al. 2013). These processes include air-surface exchange, wet and dry deposition and reaction rates, such as photolysis, biodegradation, oxidation and methylation. Both temperature and precipitation, as altered by climate change, are expected to have the largest influence on the partitioning of chemical toxicants (Nizzetto et al. 2010; Wöhrnschimmel et al. 2013). Other processes, such as snow and ice melt, biota lipid dynamics and organic carbon cycling, will be potentially altered by climate change producing significant increases in fugacity and contaminant concentrations (MacDonald et al. 2005).

The impacts of climate change on trace metal contamination have been fully discussed for marine ecosystems (Schiedek et al. 2007). The potential increased risk of flooding due to climate change has implications for the inundation of contaminated land causing an increased risk of contaminants being remobilized in floodwater and of contaminated sediment and water reaching the freshwater and marine environment. Trace metal contamination often accumulates in the topsoil and the contaminant leaching is therefore controlled by the location of the water table. As a result, high concentrations of metals in surface water are often found during periods with high groundwater levels and high discharge rates (Rozemeijer and Broers 2007). In addition, total concentrations of heavy metals with high adsorption capacities to suspended solids also increase, due to increased resuspension of contaminated suspended sediment under high river discharge rates

(Mulholland et al. 1997; De Weert et al. 2010). The process of transference of continental materials to the ocean is quite complex and could include numerous steps, which could enlarge its complexity as well as the length of the corresponding path, as suggested by Fig. 4.9.

This kind of phenomenon has driven the generation of numerous research programs linked to continent–sea transfer processes all over the world since starting 21st century. Within this framework the occurrence of studies on different environmental scenarios was widely reported in many countries. So, the extensive review by Phillipart et al. (2011) includes the analysis of numerous ecosystems from Europe observed from the transfer of materials viewpoint, even in different conditions, such as land uses and climate change and local anthropogenic activities, offering numerous scientific papers as a result of these researches. Moreover, in Latin America several similar programs have been developed within which that from Brazil has proven to be one of the most robust (Lacerda et al. 2001; 2002). Within this context numerous scientific publications have been presented, including the mentioned integral analysis for different coastal ecosystems from Brazil. Molisani et al. (2006) have studied the water discharge and sediment load from an anthropogenically-modified catchment into the Sepetiba Bay, in the Rio de Janeiro state; also Molisani et al. (2007) have assessed the Hg transport from the Paraiba do Sul River watershed into Sepetiba Bay. Similarly, Lacerda et al. (2008) have studied both the natural and anthropogenic inorganic nutrients (N and P) input to estuaries within

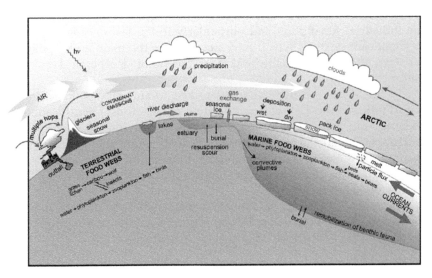

Figure 4.9. Delivery of contaminants emitted from northern industrial regions to the Arctic where they may be concentrated in biota or removed through degradation and burial (after Macdonald et al. 2005).

Ceará State coast, in NE Brazil; in a similar way, Lacerda et al. (2013b) have reported the pluriannual watershed discharges of Hg in the Jaguaribe River estuary (tropical semi-arid), in the northeastern Brazil and showed the already significant impact of regional land-use changes, mostly due to damming, and global climate changes resulting in a further decrease of river flux due to decreasing precipitation and increasing Hg bioavailability.

In the last years (since 2010) this Brazilian research program has been extended to several areas of Argentina, and the first results have appeared within this scientific framework. Delucchi et al. (2011) evaluated the impact of organotin compounds on coastal environments from Patagonia, in the southern Argentina. Fernández Severini et al. (2013) reported heavy metal contents within suspended particulate matter and the transference to zooplankton in Bahía Blanca estuary, while La Colla et al. (2015) working in the same environment have informed of the load of both dissolved and particulate metals within that anthropogenically stressed system. Finally, and following the same trend, Marcovecchio and Lacerda (2014) presented a comparative study on land-originated pollutant discharge into semi-arid estuarine environments from Brazil (the Jaguaribe River estuary) and Argentina (the Bahía Blanca estuary), which represents the first step in the integration of knowledge between both research groups.

Concluding Remarks

The imbalance of heavy metals dynamics and circulation within marine environments has been fully modified during the last centuries, after the Industrial Revolution, but particularly in the last decade. It is due to significant changes which have occurred in the main anthropogenic activities affecting the quality and magnitude of biogeochemical cycles in natural systems.

In this sense, between changes linked to anthropogenic activities which produce these effects, modification of soils uses, intensification of chemical and petrochemical industrial production, intensive agriculture and husbandry, spread out of intensive aquaculture, enlargement of harbors and port activities, should be highlighted among others. In addition, the events linked to climate change should also be considered within this framework analysis.

Undoubtedly the contribution of inland activities significantly impacts the dynamics of metals in coastal system, and in the marine environment, as has been explicitly detailed in previous paragraphs.

References

Aguiar, J.E., Lacerda, L.D., Miguens, F.C. and Marins, R.V. 2013. The geostatistics of the metal concentrations in sediments from the eastern Brazilian continental shelf in areas of gas and oil production. Journal of the South American Earth Sciences 51: 91–104.

Anthony, E.J., Gardel, A., Gratiot, N., Proisy, C., Allison, M.A., Mead, A., Dolique, F. and Fromard, F. 2010. The Amazon-influenced muddy coast of South America: A review of mud-bank–shoreline interactions. Earth-Science Reviews 103: 99–121.

Anthony, E.J., Gardel, A., Proisy, C., Fromard, F., Gensac, D., Peron, C., Walcker, R. and Lesourd, S. 2012. The role of fluvial sediment supply and river-mouth hydrology in the dynamics of the muddy, Amazon-dominated Amapá–Guianas coast, South America: A three-point research agenda. Journal of South American Earth Sciences 44: 18–24.

Arias, A.H., Spetter, C.V., Freije, R.H. and Marcovecchio, J.E. 2009. Polycyclic aromatic hydrocarbons (PAHs) distribution in water column, native mussels (*Brachydontes* sp. and *Tagelus* sp.) and fish (*Odontesthes* sp.) from an industrialized South American estuary. Estuarine, Coastal & Shelf Science 85: 67–81.

Arias, A.H., Vazquez-Botello, A., Tombesi, N.B., Ponce-Vélez, G., Freije, R.H. and Marcovecchio, J.E. 2010. Presence, distribution and origins of Polycyclic Aromatic Hydrocarbons (PAHs) in sediments from Bahía Blanca estuary, Argentina. Environmental Monitoring & Assessment 160(1-4): 301–314.

Arias, A.H., Pereyra, M.T. and Marcovecchio, J.E. 2011. Multi-year monitoring of estuarine sediments as ultimate sink for DDT, HCHs and other organochlorinated pesticides in Argentina. Environmental Monitoring & Assessment 172(1-4): 17–32.

Bastos, W.R., Gomes, J.P.O., Oliveira, R.C., Almeida, R., Nascimento, E.L., Bernardi, J.V.E., Lacerda, L.D., Silveira, E.G. and Pfeiffer, W.C. 2006. Mercury in the environment and riverside population in the Madeira River Basin, Amazon, Brazil. The Science of the Total Environment 368: 344–351.

Bates, B.C., Kundzewicz, Z.W., Wu, S. and Palutikof, J.P. (eds). 2008. Climate Change and Water. Technical Paper VI of the Intergovernmental Panel on Climate Change, IPCC Secretariat, Geneva, Switzerland.

Botté, S.E., Freije, R.H. and Marcovecchio, J.E. 2010. Distribution of several heavy metals in tidal flats sediments within Bahía Blanca Estuary (Argentina). Water, Air & Soil Pollution 210: 371–388.

Boulanger, J.-P., Leloup, J., Penalba, O., Rusticucci, M., Lafon, F. and Vargas, W. 2005. Observed precipitation in the Paraná-Plata hydrological basin: long-term trends, extreme conditions and ENSO teleconnections. Climate Dynamics 24: 393–413.

Braga, E.S., Chiozzini, V.C., Berbel, G.B.B., Maluf, J.C.C., Aguiar, V.M.C., Charo, M., Molina, D., Romero, S.I. and Eichler, B.B. 2008. Nutrient distributions over the Southwestern South Atlantic continental shelf from Mar del Plata (Argentina) to Itajaí (Brazil): Winter–summer aspects. Continental Shelf Research 28: 1649–1661.

Brown, L.E., Hannah, D.M. and Milner, A.M. 2007. Vulnerability of alpine stream biodiversity to shrinking glaciers and snowpacks. Global Change Biology 13(5): 958–966.

Burone, L., Ortega, L., Franco-Fraguas, P., Mahiques, M., Garcia-Rodriguez, F., Venturini, N., Marin, Y., Brugnoli, E., Nagai, R., Muniz, P., Bicego, M., Figueira, R. and Salaroli, A. 2013. A multiproxy study between the Rio de la Plata and the adjacent South-western Atlantic inner shelf to assess the sediment footprint of river vs. marine influence. Continental Shelf Research 55: 141–154.

Delucchi, F., Narvarte, M.A., Amin, O.A., Tombesi, N.B., Freije, R.H. and Marcovecchio, J.E. 2011. Organotin compounds in sediments of three coastal environments from the Patagonian shore, Argentina. International Journal of Environment and Waste Management 8(1/2): 3–17.

Depetris, P.J., Gaiero, D.M., Probst, J.L., Hartmann, J. and Kempe, S. 2005. Biogeochemical output and typology of rivers draining patagonia's Atlantic seaboard. J. Coast. Res. 21(4): 835–844.

Depetris, P.J. 2007. The Paraná River under extreme flooding: a hydrological and hydro-geochemical insight. Interciencia 32(10): 656–662.

De Weert, J., Streminska, M., Hua, D., Grotenhuis, T., Langenhoff, A. and Rijnaarts, H. 2010. Nonylphenol mass transfer from field-aged sediments and subsequent biodegradation in reactors mimicking different river conditions. Journal of Soils and Sediments 10(1): 77–88.

Dittmar, T., Rezende, C.E., Manecki, M., Niggemann, J., Ovalle, A.R.C. and Bernardes, M.C. 2012. Continuous flux of dissolved black carbon from a vanished tropical forest biome. Nature Geoscience 5: 6218–622.

Fernández Severini, M.D., Hoffmeyer, M.S. and Marcovecchio, J.E. 2013. Heavy metals concentrations in zooplankton and suspended particulate matter in a Southwestern Atlantic temperate estuary (Argentina). Environmental Monitoring & Assessment 185(2): 1495–1513.

Gaiero, D.M., Probst, J.-L., Depetris, P.J., Bidart, S.M. and Leleyter, L. 2003. Iron and other transition metals in Patagonian riverborne and windborne materials: geochemical control and transport to the southern South Atlantic Ocean. Geoch. Cosmoch. Acta 67(19): 3603–3623.

Georges, J.Y., Fossette, S., Billes, A., Ferraroli, S., Fretey, J., Gremillet, D., Le Maho, Y., Myers, A.E., Tanaka, H. and Hays, G.C. 2007. Meta-analysis of movements in Atlantic leatherback turtles during the nesting season: conservation implications. Marine Ecology Progress Series 338: 225–232.

Hartland, A., Fairchild, I.J., Lead, J.R., Borsato, A., Baker, A., Frisia, S. and Baalousha, M. 2012. From soil to cave: Transport of trace metals by natural organic matter in karst dripwaters. Chemical Geology 304-305: 68–82.

Hooper, M.J., Ankley, G.T., Kristol, D.A., Maryoung, L.A., Noyes, P.D. and Pinkerton, K.E. 2013. Interactions between chemical and climate stressors: a role for mechanistic toxicology in assessing climate change risks. Environmental Toxicology & Chemistry 32(1): 32–48.

Joyeux, J.C., Floeter, S.R., Ferreira, C.E.L. and Gasparini, J.L. 2001. Biogeography of tropical reef fishes: the South Atlantic puzzle. Journal of Biogeography 28: 831–841.

Kang, J.-H., Lee, S.W., Cho, K.H., Ki, S.J., Cha, S.M. and Kim, J.H. 2010. Linking land-use type and stream water quality using spatial data of fecal indicator bacteria and heavy metals in the Yeongsan river basin. Water Research 44: 4143–4157.

Kundzewicz, Z.W., Mata, L.J., Arnell, N.W., Döll, P., Kabat, P., Jiménez, B., Miller, K.A., Oki, T., Sen, Z. and Shiklomanov, I.A. 2007. Freshwater resources and their management. pp. 173–210. In: Parry, M.L., Canziani, O.F., Palutikof, J.P., van der Linden, P.J. and Hanson, C.E. (eds.). Climate Change 2007: Impacts, Adaptation and Vulnerability. Contribution of Working Group II to the Fourth Assessment Report of the Intergovernmental Panel on Climate Change. Cambridge University Press, Cambridge, UK.

Lacerda, L.D., Ribeiro, M.G., Cordeiro, R.C., Siffedine, A. and Turcq, B. 1999. Mercury atmospheric deposition during the past 30,000 years in Brazil. Ciência & Cultura Journal of the Brazilian Society for the Advancement of Science 51: 363–371.

Lacerda, L.D., Marins, R.V., Barcellos, C. and Knoppers, B.A. 2001. River basin activities, impact and management of anthropogenic trace metal and sediment fluxes to Sepetiba Bay, Southeastern Brazil. pp. 207–218. In: von Bodungen, B. and Turner, R.K. (eds.). Science and Integrated Coastal Management. Dahlem University Press.

Lacerda, L.D., Marins, R.V. and Barcellos, C. 2002. Anthropogenic fluxes of sediments and trace metals of environmental significance to Sepetiba Bay, SE Brazil. pp. 99–104. In: Lacerda, L.D., Kremer, H.H., Kjerfve, B., Salomons, W., Marshall-Crossland, J.I. and Crossland, J.C. (eds.). South American Basins: LOICZ Global Change Assessment and Synthesis of River Catchment – Coastal Sea Interaction and Human Dimensions. LOICZ Reports & Studies No.21.

Lacerda, L.D. 2003. Updating global mercury emissions from small-scale gold mining and assessing its environmental impacts. Environmental Geology 43: 308–314.

Lacerda, L.D. and Marins, R.V. 2006. Geoquímica de sedimentos e o monitoramento de metais na plataforma continental Nordeste Oriental do Brasil. Geochimica Brasiliensis 20: 123–135.

Lacerda, L.D. 2007. Biogeoquímica de contaminantes no Antropoceno. Oecologia Brasiliensis 11(2): 139–144.

Lacerda, L.D., Molisani, M.M., Sena, D. and Maia, L.P. 2008. Estimating the importance of natural and anthropogenic sources on N and P emission to estuaries along the Ceará State Coast NE Brazil. Environmental Monitoring & Assessment 141: 149–164.

Lacerda, L.D., Santos, J.A. and Lopes, D.V. 2009. Fate of copper in intensive shrimp farms: bioaccumulation and deposition in pond sediments. Brazilian Journal of Biology 69: 851–858.

Lacerda, L.D., Soares, T.M., Costa, B.G.B. and Godoy, M.D.P. 2011. Mercury Emission Factors from Intensive Shrimp Aquaculture and its relative importance to the Jaguaribe River Estuary, NE Brazil. Bulletin of Environmental Contamination and Toxicology 87: 657–661.

Lacerda, L.D., Bastos, W.R. and Almeida, M.D. 2012. The impacts of land use changes in the mercury flux in the Madeira River, Western Amazon. Anais da Academia Brasileira de Ciências 84: 69–78.

Lacerda, L.D., Campos, R.C. and Santelli, R.E. 2013a. Metals in water, sediments and biota of an offshore oil exploration area in the Potiguar Basin, Northeastern Brazil. Environmental Monitoring and Assessment 185: 4427–4447.

Lacerda, L.D., Dias, F.J.S., Marins, R.V., Soares, T.M., Godoy, J.M.O. and Godoy, M.L.D.P. 2013b. Pluriannual watershed discharges of Hg into a tropical semi-arid estuary of the Jaguaribe River, NE Brazil. Journal of the Brazilian Chemical Society 24(11): 1719–1731.

La Colla, N.S., Negrin, V.L., Marcovecchio, J.E. and Botté, S.E. 2015. Dissolved and particulate metals dynamics in a human impacted estuary from the SW Atlantic. Estuarine, Coastal and Shelf Science 166: 45–55.

Lane, S.N., Reid, S.C., Tayefi, V., Yu, D. and Hardy, R.J. 2007. Interactions between sediment delivery, channel change, climate change and flood risk in a temperate upland environment. Earth Surface Processes and Landforms 32: 429–446.

Leemans, R. and Kleidon, A. 2002. Regional and global assessment of the dimensions of desertification. pp. 215–232. In: Reynolds, J.F. and Stafford-Smith, D.M. (eds.). Global Desertification. Do Humans Cause Deserts? Dahlem University Press, Berlin, Germany.

Macdonald, R.W., Harner, T. and Fyfe, J. 2005. Recent climate change in the Arctic and its impact on contaminant pathways and interpretation of temporal trend data. Science of the Total Environment 342: 5–86.

Marcovecchio, J.E., Botté, S.E., Delucchi, F., Arias, A.H., Fernández Severini, M.D., De Marco, S.G., Tombesi, N.B., Andrade, S.J., Ferrer, L.D. and Freije, R.H. 2008. Pollution processes in Bahía Blanca estuarine environment. In: Neves, R., Baretta, J. and Mateus, M. (eds.). Perspectives on Integrated Coastal Zone Management in South America. Part B: From Shallow Water to the Deep Fjord: The Study Sites. IST Scientific Publishers, Lisbon (Portugal), Chapter 28: 303–316. (ISBN: 978-972-8469-74-0).

Marcovecchio, J.E. and Lacerda, L.D. 2014. Procesos de contaminación en ambientes costeros semi-áridos: comparación entre ambientes templados (estuario de Bahía Blanca, Argentina) y tropicales (estuario de Jaguaribe, Brazil). In: III Reunión Argentina de Geoquímica de Superficie –III RAGSU–, Mar del Plata (Argentina), 2–5 Diciembre de 2014.

Marins, R.V., Freire, G.S.S., Maia, L.P., Lima, J.P.R. and Lacerda, L.D. 2002. Impacts of land-based activities on the Ceará coast, NE Brazil. pp. 92–98. In: Lacerda, L.D., Kremer, H.H., Kjerfve, B., Salomons, W., Marshall-Crossland, J.I. and Crossland, J.C. (eds.). South American Basins: LOICZ Global Change Assessment and Synthesis of River Catchment – Coastal Sea Interaction and Human Dimensions. LOICZ Reports & Studies No. 21.

Meybeck, M. 2005. Riverine quality at the Anthropocene: Propositions for a global space and time analysis, illustrated by the Seine River. Aquatic Sciences 64: 376–393.

Meybeck, M. and Vörösmarty, C. 2006. Fluvial filtering of land-to-ocean fluxes: from natural Holocene variations to Anthropocene. Comptes Rendus Geoscience 337: 107–123.

Moe, S.J., De Schamphelaere, K., Clements, W.H., Sorensen, M.T., Van der Brink, P.J. and Liess, M. 2013. Combined and interactive effects of global climate change and toxicants on populations and communities. Environmental Toxicology & Chemistry 32(1): 49–61.

Molisani, M.M., Kjerfve, B., Silva, A.P. and Lacerda, L.D. 2006. Water discharge and sediment load to Sepetiba Bay from an anthropogenically-altered drainage basin, SE Brazil. Journal of Hydrology 331: 425–433.

Molisani, M.M., Kjerfve, B., Barreto, R. and Lacerda, L.D. 2007. Land–sea mercury transport through a modified watershed, SE Brazil. Water Research 41: 1929–1938.

Moller, O.O., Piola, A.R., Freitas, A.C. and Campos, E.D. 2008. The effects of river discharge and seasonal winds on the shelf off southeastern South America. Continental Shelf Research 28: 1607–1624.

Mulholland, P.J., Best, G.R., Coutant, C.C., Hornberger, G.M., Meyer, J.L., Robinson, P.J., Stenberg, J.R., Turner, R.E., Vera-Herrera, F. and Wetzel, R.G. 1997. Effects of climate change on freshwater ecosystems of the southeastern United States and the Gulf Coast of Mexico. Hydrological Processes 11(8): 949–970.

Negrin, V.L., González Trilla, G.L., Kandus, P. and Marcovecchio, J.E. 2012. Decomposition and nutrients dynamics in a *Spartina alterniflora* marsh of the Bahía Blanca estuary, Argentina. Brazilian Journal of Oceanography 60(2): 259–263.

Nery, J.R.C. and Bonotto, D.M. 2011. 210Pb and composition data of near-surface sediments and interstitial waters evidencing anthropogenic inputs in Amazon River mouth, Macapá, Brazil. Journal of Environmental Radioactivity 102: 348–362.

Nizzetto, L., MacLeod, M., Cabrerizo, A., Dachs, J., Di Guardo, A., Ghirardello, D., Jarvis, A., Lindroth, A., Ludwig, B., Monteith, D., Perlinger, J.A., Scheringer, M., Schwendenmann, L. and Semple, K.T. 2010. Present, and Future Controls on Levels of Persistent Organic Pollutants in the Global Environment. Environmental Science & Technology 44: 6526–6531.

Nriagu, J.O. 1993. A legacy of mercury pollution. Nature 363: 589.

Nriagu, J.O. 1996. A history of global metal pollution. Science 272: 223–224.

Oliva, A.L., Quintas, P.Y., La Colla, N.S., Arias, A.H. and Marcovecchio, J.E. 2015. Distribution, sources and potential ecotoxicological risk of Polycyclic Aromatic Hydrocarbons in surface sediments from Bahía Blanca estuary, Argentina. Archives of Environmental Contamination and Toxicology (in press). doi:10.1007/s00244-015-0169-0.

Pacyna, J.M., Scholtz, T. and Pirrone, N. 1999. Global emissions of anthropogenic mercury to the atmosphere. *In*: Villas Boas, R.C. and Lacerda, L.D. (eds.). Mercury as a global pollutant. V International Conference, Rio de Janeiro, Centro de Tecnologia Mineral (CETEM-CNPq), Rio de Janeiro. Abstracts.

Pirrone, N., Allegrini, I., Keeler, G.J., Nriagu, J.O., Rossman, R. and Robbins, J.A. 1998. Historical atmospheric mercury emissions and depositions in North America compared to mercury accumulation in sedimentary records. Atmos Environ. 32: 929–940.

Patz, J.A., Daszak, P., Tabor, G.M., Aguirre, A.A., Pearl, M., Epstein, J., Wolfe, N.D., Kilpatrick, A.M., Foufopoulos, J., Molyneux, D., Bradley, D.J. and Members of the Working Group on Land Use Change and Disease Emergence. 2004. Unhealthy landscapes: policy recommendations on land use change and infectious disease emergence. Environmental Health Perspectives 112: 1092–1099.

Philippart, C.J.M., Anadón, R., Danovaro, R., Dippner, J.W., Drinkwater, K.F., Hawkins, S.J., Oguz, T., O'Sullivan, G. and Reid, P.C. 2011. Impacts of climate change on European marine ecosystems: Observations, expectations and indicators. Journal of Experimental Marine Biology and Ecology 400: 52–69.

Pielke, R.A. 2005. Land use and climate change. Science 310: 1625–1626.

Piola, A.R., Matano, R.P., Palma, E.D., Möller, O.O. Jr. and Campos, E.J.D. 2005. The influence of the Plata River discharge on the western South Atlantic shelf. Geophysical Research Letters 32: L01603 (4 pp).

Ronco, A., Peluso, L., Jurado, M., Rossini, G.B. and Salibian, A. 2008. Screening of sediment pollution in tributaries from the Southwestern coast of the Río de La Plata estuary. Latin American Journal of Sedimentology and Basin Analysis 15(1): 67–75.

Rozemeijer, J.C. and Broers, H.P. 2007. The groundwater contribution to surface water contamination in a region with intensive agricultural land use (Noord-Brabant, The Netherlands). Environmental Pollution 148(3): 695–702.

Schiedek, D., Sundelin, B., Readman, J.W. and Macdonald, R.W. 2007. Interactions between climate change and contaminants. Marine Pollution Bulletin 54(12): 1845–1856.

Simonetti, P., Botté, S.E., Fiori, S.M. and Marcovecchio, J.E. 2013. Burrowing crab (*Neohelice granulata*) as a potential bioindicator of heavy metals in the Bahía Blanca estuary, Argentina. Archives of Environmental Contamination & Toxicology 64: 110–118.

Spetter, C.V., Buzzi, N.S., Fernández, E.M., Cuadrado, D.G. and Marcovecchio, J.E. 2015. Assessment of the physicochemical conditions sediments of a polluted tidal flat colonized by microbial mats in Bahía Blanca Estuary (Argentina). Marine Pollution Bulletin 91: 491–505.

Stigliani, W.M. 1995. Global perspective and risk assessment. pp. 331–343. *In*: Salomons, W., and Stigliani, W.M. (eds.). Biogeodynamics of Pollutants in Soils and Sediments. Springer, Berlin Heidelberg New York.

Stupar, Y.V., García, M.G., Schmidt, S., Schäfer, J., Huneau, F., Le Coustumer, P., Piovano, E. and Blanc, G. 2014. Registro histórico de mercurio em la cuenca del río Suquía, Córdoba, Argentina. pp. 303–324. *In*: Marcovecchio, J.E., Botté, S.E. and Freije, R.H. (eds.). Procesos Geoquímicos Superficiales em Iberoamérica. Soc. Iberoamer. Fís. Quím. Ambiental (SIFyQA), Salamanca (España). ISBN: 978-84-937437-6-5.

UNEP. 2013. Global Mercury Assessment. United Nations Environmental Programme, Paris.

Wanderley, C.V.A., Godoy, J.M., Godoy, M.L.D.P., Rezende, C.E., Lacerda, L.D., Moreira, I. and Carvalho, Z.L. 2013. Evaluating sedimentation rates in the estuary and shelf region of the Paraiba do Sul River, Brazil. Journal of the Brazilian Chemical Society http://dx.doi.org/10.5935/0103-5053.20130268.

Wilby, R.L., Dalgleish, H.Y. and Foster, I.D.L. 1997. The impact of weather patterns on contemporary and historic catchment sediment yields. Earth Surface Processes & Landforms 22: 353–363.

Wöhrnschimmel, H., MacLeod, M. and Hungerbuhler, K. 2013. Emissions, fate and transport of persistent organic pollutants to the arctic in a changing global climate. Environmental Science & Technology 47: 2323–2330.

Wuana, R.A. and Okieimen, F.E. 2011. Heavy Metals in Contaminated Soils: A Review of Sources, Chemistry, Risks and Best Available Strategies for Remediation. ISRN Ecology, Vol. 2011, Art. ID 402647, 20 pp. doi:10.5402/2011/402647.

Zhao, H., Xia, B., Fan, C., Zhao, P. and Shen, S. 2012. Human health risk from soil heavy metal contamination under different land uses near Dabaoshan Mine, Southern China. Science of the Total Environment 417-418: 45–54.

5

Emerging Pollutants in the Global Change Scenario

Bernd Markert,[1,*] *Stefan Fränzle,*[2] *Simone Wünschmann*[1] and *Peter Menke-Glückert*[†3]

Introduction: Emerging Pollutants—General Aspects and Features

It is important to note and understand that 'emerging pollutants' need not be 'novel' to the environment and biosphere, that is, they need not be anthropogenic; the property which renders them 'emerging pollutants' simply is the fact that their possible role in environmental pollution was discovered more or less recently. We shall see this is partly too mainly due to a profound change in the manner we look upon interactions among 'environment', local biota and some chemical introduced to it.

Emerging pollutants, can likewise be biogenic compounds which become significant pollutants only because either patterns of use of certain parts of environment or abundances of emitting organisms did change, and anthropogenic compounds which may or may not be introduced and applied as biocides, pharmaceuticals (including antibiotics and orally administered anticonceptive agents [the 'pill']) or just be selected for reasons of their peculiar chemical stability and corresponding solvent, flame retardant, etc. properties (Schüürmann and Markert 1998; Markert et al. 2003; Fraenzle et al. 2005; 2012; Markert et al. 2015).

[1] Environmental Institute of Scientific Networks, Fliederweg 17, D-49733 Haren, Germany.
[2] University of Dresden, International Graduate School Zittau, Department of Biological and Environmental Sciences, Research Group of Environmental Chemistry, Markt 23, D-02763 Zittau, Germany.
[3] Augustastraße 51, D-53173 Bonn, Germany.
* Corresponding author: Markert@EISN-Institute.de

There stability translates into persistent behavior while hormones[1] will cause an 'amplified' change of biochemical processes already at very low ambient levels (for 17α-ethynylestradiol [for brevity, EE2], levels of ≈1 pmol/l [≈0.3 ng/l,[2] Oehlmann and Schulte-Oehlmann 2003 may compromise reproduction in many kinds of aquatic and semi aquatic vertebrates), thereby hitting entire trophic chains and the chemical state of surface waters. EE2 is an unpolar compound, capable of undergoing enrichment by biomagnifications along the trophic chain. Top-level predators (piscivorous fishes, birds of prey like osprey, white-bald eagle or red kite [*Milvus milvus*], crocodiles) may be removed from such a lake altogether by either deliberate fishing or hunting or (if unwanted) by simply keeping them from reproduction (e.g., by hormone activity producing intersex/imposex states, or by endocrinic reduction of Ca transfer into bird eggshells by agents such as DDT, making the eggs break during breeding or already during delivery), causing the entire biogeochemistry and nutrient feedback loops to change down the entire trophic ladder. In such cases/systems, effects of eutrophication will be much more difficult to cope with (either was done on the level of piscivorous fishes [pike, pike-perch] in the famous Ontario Experimental Lakes Area), as predicted in ecological stoichiometry approaches.

For such compounds acting as hormones, the size of their effects—to be understood in terms of ecological stoichiometry (consumer-driven nutrient recycling)—by far exceeds that which would be 'commonly' expected from the amounts of the compounds used. In addition, some of these xenohormones (mainly, xenoestrogens) were used in quite substantial amounts and in application scenarios eventually releasing most of it to the environment (nonylphenyl-based detergents, phthalate ester plastic softeners, tributyltin salts).

To give another example and mode of action of 'indirect' hazardous effects, eutrophication of both fresh waters and shallow ocean regions brought about blooms of cyanobacteria which then cause massive emissions of microcystines (hepatotoxic cyclic heptapeptides) which pose a challenge in tap water production and a direct hazard to swimmers who sooner or later would ingest some of this water. With other, 'really' anthropogenic

---

[1] Note that many compounds act as hormones just to a few species or taxa, e.g., the role of 2-methyl cyclohexanes and –cyclohexenes as both pheromones and agents controlling and enabling pupification of certain [lignophagous] beetle [-larvae]!

[2] Such levels are maintained by average use of anticonceptive pills in Germany (total consumption some 50 kg/a) or the Netherlands (14 kg/a) if EE2 has an ambient lifetime of a few months; as it may be much more persistent, nobody really knows whether the aquatic and groundwater levels measured now were released recently or can be traced back to the much higher doses of EE2 in anticonceptive pills used in the later 1960s and 1970s.

compounds the risks associated with them would be detected only after[3] their large-scale introduction as solvents or for other purposes, like with CFCs (which as such are non-toxic but produce higher sun-burn UV levels at surface by destroying parts of the stratospheric ozone) or polychlorinated biphenyls. With the advent of ecotoxicology (in 1969), our perception of how some chemical—whether of bio-, anthropogenic or geogenic (volcanoes!)

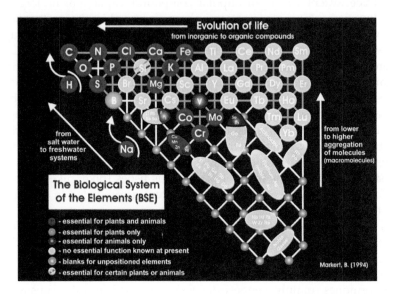

Figure 5.1. The Biological System of the Elements (BSE) for terrestrial plants (glycophytes) (Markert 1994). The diagram shows relationships among the elements together with their corresponding essential functions (*colors*), extent of biochemical functions and the corresponding capacity to form macromolecules by condensation reactions (*vertical arrow at right side of diagram*). Whereas in 'pure' geochemistry oxophilic metals produce the most complicated condensation products, i.e., clay minerals, there was a shift towards non-metal-based structures during chemical and biological evolution which afforded polymeric structures based on the latter (C, N, O) (*horizontal arrow to the left*). The diagonal arrow refers to changes of concentration from ocean- to freshwater. There is substantial decrease of concentrations in some elements (Mg, Sr, Cl, Br) from ocean to freshwater requiring them to be enriched by biomasses if their biochemical use is to be continued. Such kinds of enrichment can only be accomplished by means of certain biochemical features which involve properties and/or components of the corresponding biogenic materials—many of which are specific for one species at least in their particular combination.

---

[3] Even though nowadays the legal framework for permission of circulating certain chemicals in the EU (REACH) or keeping on doing so tries to anticipate such effects from QSAR studies on various groups of chemicals, the very term of emerging pollutants is telling, telling us that there are severe limits to this kind of anticipation. Moreover, such studies are executed only for substances meant to be circulated, sold or emitted in substantial amounts (>> 1 ton/year in the EU). For comparison, the anticonceptive hormone EE2 which arguably is a standard and commonplace broadly applied chemical now is used at a level of some 200 kg/a all over the EU!

origins or taken from any combination of these principal sources—would act in and interact with both environment and biota did alter fundamentally.

In other cases we already know that some technical innovation, such as the introduction of the three-way catalytic exhaust gas converter, brings about novel emissions into the environment—that is, particles which escape from the ceramic matrix of the converter, containing nanoparticulate Pd, Rh and Pt besides of Ce, Zr in the matrix—but cannot yet estimate what will be either the pathway of such possible pollutants in biota and environment nor whether, what kind of or how large a damaging impact would be. Or whether there are predictable damages which can be attributed to given sources. With respect to inorganic chemistry and to have a more multi-element approach to physiological and toxicological functions of elements a so called Biological System of Elements (Fig. 5.1) has been established about two decades ago (Markert 1994).

Meanwhile further research has shown that there are exceptions of the role of certain elements to specific animals/plants. Therefore the original BSE was extended into a modified BSE (Fig. 5.2; Markert et al. 2015) where Cr, Ba, Rb, Sr and Li take probably following role:

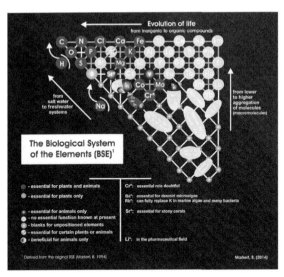

Figure 5.2. Extended Biological System of the Elements after Markert (1994) showing exceptions of the role of certain elements to specific animals/plants (Markert et al. 2015).[4]

---

[4] Related findings of trace element research are given here for instance: Adriano 1992; Baker 2008; De Bievre 2006; De Bievre et al. 2011; Diatta et al. 2015; Freitas et al. 2007; Frontasyeva et al. 2004; 2007; Hooda 2010; Knox et al. 2001; Lacerda and Salomon 2012; Loppi and Bonini 2000; Marcovecchio and Ferrer 2005; Marcovecchio et al. 2006; Maret 2016; Paoli et al. 2013; Prasad et al. 2010; 2013; Rinklebe and DuLaing 2011; Steinnes et al. 2011; Tabors and Lapina 2012; Van Nevel et al. 1998; Wolterbeek et al. 2010; Wünschmann et al. 2008.

- The essential role of chromium is doubtful for plants and animals;
- Barium is essential for desmid microalgae;
- Rubidium can fully replace K in marine algae and many bacteria;
- Strontium is essential for stony corals;
- Lithium has a beneficial effect to animals (humans) in the pharmaceutical field.

In yet other cases we do not understand origins of environmental analytes which are known to be e.g., phytotoxic: trichloroacetic acid is produced from H-CFCs (more precisely, corresponding C_2- or C_3-compounds like methyl chloroform) especially in desert regions such as Uzbekistan (photoelectrochemistry caused by solar UV irradiation of titanous sand?) but there also is **trifluoroacetic acid** which is unlikely to come from any biosynthesis (though some living organisms, especially fungi, some bacteria and tropical sponges, can create [usually aromatic] C-F bonds they would never procure trifluoromethyl groups nor would be capable of degrading them)[5] and now CF_3COO^- is found at sites where CFCs (the most likely precursor actually in use would be 1,1,1-trifluorodichloroethane) did not get now like in the deep ocean where levels are about half those observed in Mediterranian-region rainwaters.

Agent Orange, the main component of which was the (outdated) synthetic auxin 2,4-dichlorophenoxyacetic acid (2,4-D), is a devastating teratogenic preparation but these effects cannot be attributed to any known component of it, including 2,3,7,8-tetrachlorodibenzo(p)-dioxine (TCDD) which is highly toxic but not teratogenic. Or particle size, being susceptible to erosion or particle growth processes may matter, turning the humble, innocent magnesium silicate into carcinogenic killer asbestos. Eventually, there are synergistic effects and the big challenge of geoengineering, e.g., to counteract global warming which means introducing large amounts of substances into the atmosphere effects of which are not, to say the least, fully understood.

Thus identification of emerging pollutants is less a matter of analytical chemistry—now capable of identifying and quantifying many compounds at ppb[6] to often ppq ($1:10^{15}$) levels—but means thoroughly reconsidering

---

[5] When CF_3 groups are introduced into pharmaceuticals, like in the antidepressive agent fluoxetin (Prozac™) which is a β-phenethyl amine which commonly undergo ready oxidation, its purpose is to **preclude** such degradation. Here the CF_3 group is never attacked as such; moreover it can even be combined with organometal moieties: CF_3Li and trifluoromethyl Grignard reagents are fully stable compounds! It comes as no surprise that fluoxetin, like other, more reactive psychoactive compounds including benzodiazepames or caffeine, now is a regular component of surface waters connected to human settlements. Aliphatic CF bonds also withstand reductive PEC attack in conc. formic acid where CCl- or aromatic CF bonds would be cleaved.

[6] Now, even 'clean' air is known to contain many toxic or corrosive chemicals at ppb- to multi-ppt levels, including HCN, HF, HCHO, higher oxides of nitrogen, O_3, benzene and

and in turn changing our look on existing chemicals as well as those right now making it to the markets. Of course, novel problems might also arise if increased amounts of which are released by agriculture, farming (including aquafarming), medicine (PGM cytostatics), lifestyle and commodity chemicals (e.g., plastic microparticles turning up in freshwater lakes which had been deliberately added to shower gels, besides of airborne deposition of plastics ground down by sand movement and photochemical brittling).

The concept of emerging pollutants means not to introduce novel compounds into the environment (or avoid to do so) but to become aware that a given compound might pose some risk beyond its originally intended use (TBT cations were **meant to kill** larval stages of many invertebrates and algae alike on ship vessel surfaces, and anticonceptive pills, after all, aim at some fully reversible kind of reproduction toxicity in humans by suppressing female ovulation, but a fish or alligator cannot stop this by stopping uptake of water). Now, some 100,000 different commodity chemicals are used in highly developed countries, and detailed ecotoxicological information is available for just a few hundred of them (less than 1%!). Research on the latter data does not at all keep pace with introduction of new agents or formulations for which, as a rule, only improvement of the intended performance is considered, not side-effects of additives present in the formulation.

Thus, there are many different scenarios which might turn some chemical into an emerging pollutant. Some examples were given before, but we can be sure we will not be capable to see all the problems ahead before there is actual damage, let alone **synergistic effects** which until now were not mentioned at all. The latter might turn some combination of compounds which both or all are 'harmless' at ambient levels into something causing ecotoxicological damage—even when these compounds do not react directly with each other!

Global Change Scenario, Sustainability and the Ten Ecological Commandments

The earth system, climate and global change[7]

Over the past few decades, evidence has mounted that planetary-scale changes are occurring rapidly. These are, in turn, changing the patterns

---

Footenote 6 contd. ...

even sulfuric acid but we cannot know whether or how much, e.g., human life expectancy would be increased if levels of all these compounds in the troposphere were exactly zero. Moreover for carcinogenic agents like HCHO or benzene there is no clear-cut no-effect level. We can yet state that another increase of air levels of such compounds would be unpleasant to hazardous now.

[7] Related to IGBP, International Geosphere-Biosphere Programme, in Steffen et al. (2004).

of forcings and feedbacks that characterize the internal dynamics of the Earth System. Key indicators, such as the concentration of CO_2 in the atmosphere, are changing dramatically, and in many cases the linkages of these changes to human activities are strong. It is increasingly clear that the Earth System is being subjected to a wide range of new planetary-scale forces that originate in human activities, ranging from the artificial fixation of nitrogen and the emission of greenhouse gases to the conversion and the loss of biological species. It is these activities and others like them that give rise to the phenomenon of global change.

The term Earth System refers to the suite of interacting physical, chemical and biological processes that transport and transform materials and energy and thus provide the conditions necessary for life on the planet (Steffen et al. 2004). Climate refers to the aggregation of all components of weather—precipitation, temperature, cloudiness, for example—but the climate system includes processes involving ocean, land and sea ice in addition to the atmosphere. The Earth System encompasses the climate system, and many changes in Earth System functioning directly involve changes in climate. However, the Earth System, includes other components, biophysical and human, important for its functioning. Some Earth System changes, natural and human driven, can have significant consequences without involving any changes in climate. With respect to Steffen et al. (2004) global change should thus not be confused with climate change. It is significantly more!

Sustainability

Sustainability is a term or criterion which was originally (18th century) introduced in topics of forestry remove/cut as many trees as can reasonably be expected to re-grew in the same period of time on the site, thus maintaining a kind of equilibrium state (Caradonna 2014; Fränzle et al. 2012; Robertson 2014; Sachs 2015). This criterion or state of affairs obviously cannot be applied one by one to any kind of economical activity which makes use of non-renewable resources, be it e.g., production of energy by fossil fuels or ore processing. Here, the end of a (more) sustainable economy takes either replacement by sustainable sources of matter and energy or a considerable extend of recycling (Fränzle et al. 2012; Haber 2010; Schellnhuber 1998).

Acknowledging that 'ideal' sustainability is not at hand, the alternative then is to compare among different old and new options to do or organize something which are likely to include partial use of regenerative resources and estimates of the irreversible effects one item or one good of service will provide from being produced up to removing the remains left after (cradle to grave analysis) (Fränzle et al. 2012). As a rule, at least if effects far extend the time-scales of political decisions and planning: foreseeing several 10^5 years, even lignite becomes a regenerative energy source, as the lakes left after open pit mining are then going to fill up with sediments and peat later on spontaneously turning into lignite once again. Over which periods of time precautions can or should be calculated concerning expenditures and workforce to safeguard deposits of radioactive waste? Obviously both the estimated 'actual' price of nuclear fission energy/kWh and the aspect of realistic safeguarding critically depend on this (Fränzle et al. 2012).

The ten ecological commandments

Healthy ecosystems and environments are necessary to the survival of humans and other organisms. The Ten Ecological Commandments (Menke-Glückert 1968) are important catalysts for permanent activities to solving global ecological problems.

They were already formulated by Peter Menke-Glückert at OECD in Paris during 1966 to 1970. He firstly presented them at an UNESCO conference 'Man and Biosphere' in Paris on March 9, 1968.

The Ten Ecological Commandments (Table 5.1) support humans to identifying ways of reducing negative human impact by environmental-friendly chemical engineering, environmental resources management and environmental protection. The information is gained from green chemistry, earth science, environmental science and conservation biology. Ecological economics studies the fields of academic research that aim to address human economics and natural ecosystem. Moving towards sustainability is also a social challenge that entails international and national law, urban planning and transport, local and individual lifestyles and ethical consumerism. Especially the international and global, ethically sound, and sustainable information transfer will be catalyzed by permanent sustainable activities to solving global ecological problems.

The Ten Ecological Commandments given in Table 5.1 represent a sustainable decaloque for Earth Citizens bearing responsibility by birth for sustaining planet earth and its exuberant nature creative wisdom.

Explanation in Detail of Commandment 9 'Do Not Pollute Information'

In the following Commandment 9 'Do not pollute information' will be described more in detail—representing an example of the other nine Commandments.

To protect nature, to overcome environmental damage needs courage to tell the truth about effects of certain industrial practices, about consequences of certain political actions and affluent energy-wasting life styles—even

Table 5.1. The Ten Ecological Commandments (Menke-Glückert 1968).

FIRST COMMANDMENT

RESPECT THE LAWS OF NATURE! REALIZE: THERE ARE CLEAR LIMITS FOR THE CITIZENS OF SPACESHIP EARTH SET BY NATURE AND ENFORCED BY NATURE! YOU NEED NATURE AND NATURE NEEDS YOU! LAWS OF NATURE SHOULD BE FIRM UNDISPUTABLE GOALS IN POLICY MANAGEMENT, GIVING LONG TERM ENVIRONMENTAL KEY DATA FOR WORLD POLIITICS!

Have respect and respond in humility to the 4.3 billion years old cycles of the universe which up to now guarantees the survival of the biological species HOMO SAPIENS alone by interdependence highest (bio)diversity of billions of eco and cultural different integrated systems. All these fragile systems are threatened to-day by massive over-exploitation of resources and energy in an over-consumption pattern of carelessness, thoughtlessness, ruthlessness. Therefore, adapt yourself to nature's cycles. Without you nature works with great efficiency day by day for you! Without charging you a bill. But with tremendous economic and survivability surplus.

SECOND COMMANDMENT

LEARN AS RESPONSIBLE EARTH CITIZEN FROM THE WISDOM OF NATURE! REMEMBER: THE WHOLE IS MORE THAN ITS PARTS! SCIENCE MUST AIM AT THE WHOLE—REGARDING DETAILS AS INTEGRAL INGREDIENTS OF SYSTEM EARTH... NATURE WHOLE-NESS IS GLOBAL SYSTEM-RETINITY!

It is your duty to inform yourself about the rules of ecosystems on earth so that you can act with wisdom, foresight and circum-spection. Do not curtail ecosystems in their parts without considering what happens to the whole. The whole must be background of all nature learning activities—establishing day-to-day EARTH-CITIZENSHIP with all its obligations!

THIRD COMMANDMENT

DO NOT REDUCE PLURALITY, RICHNESS, AND ABUNDANCE OF LIVING SPECIES! DO NOT INDUSTRIALIZE NATURE! DO NOT STREAMLINE NATURE! DIVERSITY OF SPECIES IS THE MOST PRECIOUS NATURAL HERITAGE OF MANKIND

Every year ten thousands of species disappear because of industrialization, over fertilization, toxic pollution, careless handling of nature's resources such as the cutting of tropical forests and the sealing-off of larger and larger parts of nature by building roads, industries, cities. Thus the information content of the biosphere, its survivability diminishes. Every blade of grass, every butterfly, every wild creek or hedge is an important part of the global ecosystem and therefore condition for its survival and our survival as human species. The abundance of natural species is a cultural treasure of mankind which by all means must be protected and preserved. International law must be supplemented accordingly. International agreements related to the Third Commandment have been already fixed nowadays, but their implementation or continuing existence will be in parts dramatically blocked by strong forces of the international political scene.

FOURTH COMMANDMENT

DO NOT POLLUTE!

Avoid pollution of water, air, soil by not overusing your car, by not asking for unnecessary packaging material, by not wasting water or energy. Because of the enormous pollution-load there must be personal zero-level-pollution saving resources and energy—wherever zero-alternatives such as walking, bicycling, using renewable resources and renewable energies are possible alternatives. The late Danish ambassador Isaak Seidenfaden to OECD phrased

Table 5.1 contd. ...

...Table 5.1 contd.

already this commandment as Number ONE for all economic activities in 1967. Do not ask for state action—you are responsible for the state of nature as an earth citizen! In OECD language it is coined as **PPP—POLLUTER—PAYS—PRINCIPLE**.

FIFTH COMMANDMENT

FACE EARTH—RESPONSIBILITY EVERY DAY FOR OUR CHILDREN AND OUR CHILDRENS' CHILDREN SO THAT THEY MAY LIVE ON AN UNSPOILT LIVEABLE PEACEFUL EARTH AS EARTH CITIZEN WITH FUNDAMENTAL HUMAN RIGHTS AND ETHICAL EARTH SURVIVABILITY RESPONSIBILITIES!

Try to leave this small planet earth as a more beautiful place, more liveable in, more enjoyable, with enough resources for all mankind. Responsibility for planet earth infers global responsibility taking into consideration causes of nature destruction and economic and social underdevelopment of two thirds of mankind. Find ways and means of achieving global responsibility to fight world poverty by local and United Nations actions. Earth policy/ responsibility for Earth Citizenship and local policy depend upon each other. Think globally as an Earth Citizen and act regionally as a national citizen to follow this Fifth Commandment as proud democratic citizen in your local community and as an Earth Citizen by birth belonging to the huge realm of billions of biological species. Legally everyone is entitled to national citizenship, but by world-ethical standards entitled to Earth Citizenship with rights and obligations according to the Ten Eco-Commandments—being in dual duty as national and as Earth Citizen!

SIXTH COMMANDMENT

FOLLOW THE PRINCIPLE OF NATURE PRECAUTION/SUSTAINABILITY IN ALL ECONOMIC ACTIVITIES!

Do not take more resources out of nature's richness and ecosystems as can be restored and renewed by careful long-term resources planning! Give everything back to nature in substance so that resource losses or damage can be reconstructed to soil, plants, functioning ecosystems. Start with re-thinking all your activities, for instance by composting all green waste in your garden or your neighborhood. Burning of waste means destruction of matter and off-heat-pollution. Do not allow that our earth loses its soil and green renewability! Save natural resources and energy! Take precaution!

SEVENTH COMMANDMENT

ACT AS YOU SPEAK!

EVERY DAY IT IS YOUR VERY OWN PERSONAL RESPONSIBILITY AS EARTH CITIZEN TO EASE ENVIRONMENTAL BURDENS AND NATURE-HARDSHIPS!

Do not lie to yourself by making a distinction between the principal acknowledgement of the environmental cause and your own little sins, which are unimportant in comparison with the big sins of others, such as those in the case of industry. Many, many little steps have tremendous end effects. Start in your own household and community neighborhood!

EIGTH COMMANDMENT

PREFER INTELLIGENT BIO-ECONOMIC SOLUTIONS, SMART GRIDS LEARNING FROM NATURE WISDOM COLLECTED BY BIONIC RESEARCH. PREFER NATURE-INTELLIGENT SOLUTIONS WITH SMART GRIDS SAVING NATURE RESOURCES TO TRADITIONAL THOUGHTLESS WASTE TECHNOLOGIES AND INDUSTRIALIZED

Table 5.1 contd. ...

...Table 5.1 contd.

SOLUTIONS STIMULATING OVERCONSUMPTION. LIFE STYLE PATTERNS NEGLECTING IMPACT ON WORLD CLIMATE. ECONOMIC GROWTH IN THE 7th KONDRATEFFCLE MEANS **QUALITATIVE ECONOMIC GROWTH IN COEVOLUTION WITH NATURE**.

Avoidance of Survivability/Sustainability crises of political systems in the future depends on the respect for other cultural and social traditions and identities than your own. Survivability of future political systems and the success of environmental politics depend on many, many steps, problems solutions which are integrated in existing regional cultures, traditions, life styles. Remember: CLEVER/INTELLIGENT HANDLING IS BEAUTIFUL, comprehensible, enjoyable, motivating. Stay down to earth with all your senses!

NINTH COMMANDMENT

INFORMATION ABOUT ENVIRONMENTAL DAMAGE BELONGS TO MANKIND— NOT TO PRIVILIGED BIG BUSSINESS! NOT TO ANYONE ALONE! AVOID INFORMATION POLLUTION! GENERATE SURVIVABILITY INFORMATION TO LASTING KNOWLEDGE AND DAILY ROUTINE! DO NOT MANIPULATE INFORMATION BECAUSE OF COMMON PRACTICE AND BECAUSE OF POSSIBLE LOSS OF PRESTIGE OR CAREER POSITION OR ECONOMIC LOSSES!

To protect nature, to overcome environmental damage needs courage to tell the truth about effects of certain industrial practices, about consequences of certain political actions and affluent energy-wasting life styles—even if it hurts common beliefs, traditional views and interests. One of the great sins of our Media- and Public-Relations-Culture, is manipulation of information. Therefore resist this distortion of information! Earth Citizens have a fundamental right of access to precise environmental information! Included in the United Nations Charter of Human Rights belongs Freedom of Access to environmental information including environmental assessment of technology innovations and its health results. Industry Advertising in green marketing is not sufficient! The coming knowledge and computer society needs a new Codex of Information-Handling-Ethics, needs a world campaign against 'information pollution', one of the biggest and most dangerous environmental sins, confusing fair, precise, understandable and honest communication.

TENTH COMMANDMENT

LISTEN CAREFULLY WHAT YOUR OWN BODY TELLS YOU ABOUT IMPACT OF YOUR VERY PERSONAL SOCIAL AND NATURAL ENVIRONMENT UPON YOUR WELLBEING! DON'T LOSE DAILY CONTACT WITH YOUR VERY OWN **"EARLY BODY WARNING SYSTEM"**! ALWAYS REMEMBER YOU CAN NEVER STEP OUT OF NATURE'S CYCLES, YOU WILL ALWAYS BE PART OF NATURE WITH YOUR OWN BODY!

Environmental awareness and sensibility start with many signals from your own body which link you with larger ecosystems of our earth, and it is your body early warning system to assess environmental damage first and foremost. Man and nature are of the same origin: We all living on planet earth must obey the same billion-old laws of nature! Body language is much older and more precious than computer language, technical jargon, media slogans!

Nature teaches you lessons every day! Your own body is your well-minded NATURE INTERPRETER and personal health exercise teacher! Body Pain is an important warning signal to change your life style! HOMO SAPIENS BODY AND SOUL contains survivability programs of archaic times thousands of years ago, but still functioning in times of hectic stress, breakdowns, health challenges. Learn by training exercises to follow this survivability body language using properly natural muscle motion and breathing techniques! An example how to do it gives Chinese medicine and body culture.

if it hurts common beliefs, traditional views and interests. The coming knowledge and computer society needs a new Codex of Information-Handling-Ethics, needs a world campaign against 'information pollution', one of the biggest and most dangerous environmental sins, confusing fair, precise, understandable and honest communication (Menke-Glückert 1968).

Processing and transferring information now is crucial for achieving a sustainable, high-level way of living, besides addressing the problems related to use of matter and energy (cp. Lubchenco and Mehra 2004; Huntington 2000; Patrinos and Bamzai 2005; Rosnay 2000; Tiessen et al. 2007; Walker 2006). Information means to convert hitherto unknown into established knowledge (the Latin verb informare means to educate, to give things some shape). Recently these pieces of information use to be 'encoded' by certain symbols or formulas, with humans made familiar with those codes in cultivated societies by explicit and lifelong learning, including professional education and personal engagement.

Figure 5.3 shows how information is gained in a stepwise, open but never really completed multi-stage process (Roots 1992). As a rule, measurements and other kinds of observation produce some set of data, thereafter selecting some of these data to obtain specific pieces of information and finally knowledge and recognition (Nefiodov 1999). If

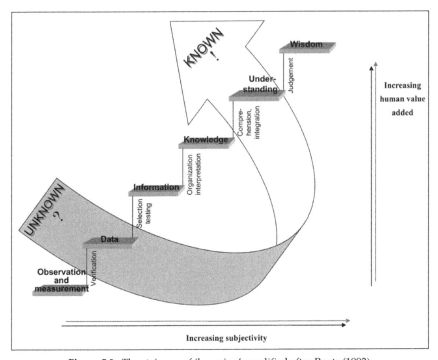

Figure 5.3. The staircase of 'knowing', modified after Roots (1992).

the things thus understood can be judged about also in the later future, both the individual and society end up with a generally valid, secure plus of knowledge.

During the (so far) last steps of evolution, this way and process to gain knowledge at increasing pace was due to the vast (size and complexity) growth of the neocortex in anthropoid primate brains, with culture and learning by trans-generational tradition (by parents, kindergarten, friends, school, etc.) giving mankind an immense gain of knowledge which is going on, now secured (by both acquisition and storage means like writing, electronic storage) and which provides the decisive competitive advantage among societies (Miegel 2005; Biedenkopf 2006; Kirchhoff 2006).

Within about 100,000 years,[8] humans developed patterns of thinking and acting which permitted a most successful spreading across continents almost all over the Earth. Using education in a skillful and intelligent way—that is, selecting and applying solid pieces of knowledge—renders education the principal and crucial factor of future life (Hosang et al. 2005; Markert 2005).

For millenia now there is another pathway of information besides scientific and rational investigation of our environs by (producing) information and knowledge, i.e., religious beliefs (Fig. 5.4). Lieth (2005) coined a term to distinguish between the levels of knowledge based on ratio and faith based on spirituality, namely telling apart real vs. virtual information.

Additional related information on the quality of human values and societies, knowledge, belief, education, analytical chemistry, instrumental measurements, etc. is given in following literature: Costanza et al. 1997; Baydoun 2006; De Bievre 2006; De Bievre et al. 2011; Franziskus 2015; Janajreh et al. 2015; Schwartz 1994; Taylor 1986; Truhaut 1969.

---

[8] The notion that spreading of modern humans all over the Earth (except of Antarctica and southern parts of South America) took just a few thousand years is erroneous. Leaving aside the fact that Homo erectus did the same much earlier being present in Spain, Georgia (Caucasus; Dmanisi excavation site) and Indonesia (Trinil on Java, next to Flores where the last erecti might have survived until a few centuries ago) > 1 mio. years ago, later on also in China ('Beijing man') and Central Europe, the first few (tribes or individuals of) *Homo sapiens* started to migrate off eastern and southern Africa some 200,000 years ago, to arrive in the Middle East more than 100,000 years (Qefna cave, Israel [Mt. Carmel]) ago, making it to Australia some 60,000 y ago and to Northern America 13–15 millenia before now, about the time of onset of agriculture ('Neolithic revolution') in both North Western Asia (Anatolia) and the lower Mekong area (Vietnam, Cambodia). So the average speed of 'migration' of *H. sapiens sapiens* to new territories was about 100 m/y, a group or tribe would thus as a rule not completely relocate 'its' area required for hunting and gathering completely within a generation's time, except when passing open ocean.

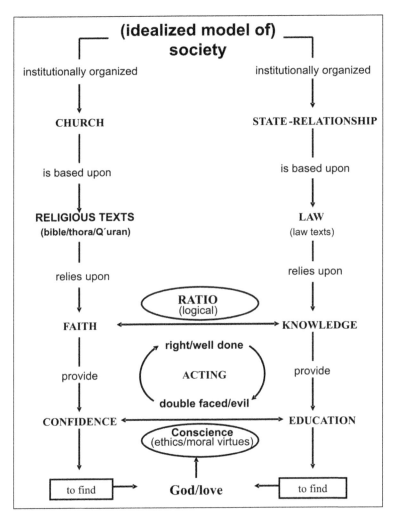

Figure 5.4. An idealizing model description of structures of society: the upper half is constructed by institutions including church(es), religion, state and law while the lower —cultural—part is supported by faith, confidence, knowledge and education (taken from Markert et al. 2005). Convincement is brought about by linking ratio (logical thinking) and conscience (ethics, moral) in a dialogical manner to separate right from wrong and good from evil. Self-confidence and education find their way into people loving each other, considered to be the only acceptable ends of society which may be negotiated. For additional explanation, see text.

It was shown several times, that pieces of information belonging to (either) empirical knowledge or faith must be precisely and meticulously separated (at least here in the secular western world), yet the gain of information and knowledge obtained by either path—empirism and faith

—does not differ too much (Küng 2005; Markert et al. 2006; Ratzinger 2006). Both faith and knowledge are rooted in pieces of observation, experience and hypotheses which are maintained only until they got clearly falsified (Popper 1957). Hence one needs to ask and get criteria whether some piece of information is reliable ('correct') or wrong, prompting us to consider the ways how information is actually processed.

Conclusion

'Information pollution' in respect to emerging pollutants in the global change scenario can be exposed to one of the biggest and most dangerous environmental sins, confusing fair, precise, understandable and honest communication. Concepts of information processing and communication in an (idealized model of) society and the ratio and relationship between subjective and objective factors in processes of recognition are discussed and presented in detail in Markert 2005; Markert et al. 2006; Fränzle et al. 2012. The same literature might be used for being informed of ethical aspects for the society related to a market-based economy and a global working and on a regional level dealing Ethical Consensus (EC). It will be supported by the development and use of a Dialogical Educational Process (DEP).

Acknowledgements

The authors are very thankful to colleagues, friends and others who supported us during discussions according to scientific topics of emerging pollutants in the global change scenario during past decades of years.

References

Adriano, D.C. (ed.). 1992. Biogeochemistry of Trace Metals, Lewis, Boca Raton.
Badran, A. (ed.). 1989. At the Crossroads: Education in the Middle-East, PWPA Book, Paragon House, N.Y.
Baker, A.J.M. 2008. Accumulators and excluders—strategies in the response of plants to heavy metals. Journal of Plant Nutrition 3: 643–654.
Baydoun, E. 2006. Arabic Encyclopaedia on Knowledge for Sustainable Development, Volume 2 (Environmental Dimension). 573 pp. UNESCO publishing.
Biedenkopf, K. 2006. Menschen bei Maischberger. ARD, November 2006.
Caradonna, J.L. 2014. Sustainability: A History, Oxford University Press, USA.
Costanza, R., d'Arge, R., Degroot, R., Farber, S., Grasso, M., Hannon, B., Limburg, K., Naeem, S., Oneill, R.V., Paruelo, J., Raskin, R.G., Sutton, P. and Vandenbelt, M. 1997. The value of the World's ecosystem services and natural capital. Nature 387(6630): 25–260.
De Bievre, P. 2006. On the analogy between the world of measurement and human society. Accred. Quali. Assur. 11: 541–542.
De Bievre, P., Dybkaer, R., Fajgelj, A. and Hibbert, D.B. 2011. Metrological traceability of measurement results in chemistry: Concepts and implementation (IUPA Technical Report). Pure Appl. Chem. 83(10): 1873–1935.

Diatta, J., Chudzinska, E., Drobek, L., Wojcicka-Poltorak, A., Markert, B. and Wünschmann, S. 2015. Multiple-phase evaluation of copper geochemistry. *In*: Sherameti, J. and Varma, A. (eds.). Monitoring and Remediation. Springer, in press.

Fränzle, S., Markert, B. and Wünschmann, S. 2005. Technische Umweltchemie— Innovative Verfahren der Reinigung verschiedener Umweltkompartimente. Ecomed-Biowissenschaften, Verlagsgruppe Hüttig Jehle Rehm GmbH, Landsberg.

Fränzle, S., Markert, B. and Wünschmann, S. 2012. Introduction to Environmental Engineering, Wiley-VCH, Weinheim, New York, Tokyo.

Franziskus, Pope. 2015. Laudato si, die Umwelt-Enzyklika des Papstes. Herder, Freiburg.

Freitas, M.C., Pacheco, A.M.G., Baptista, M.S., Dionísio, I., Vasconcelos, M.T.S.D. and Cabral, J.P. 2007. Response of exposed detached lichens to atmospheric elemental deposition. Ecological Chemistry and Engineering 14(7): 631–644.

Frontasyeva, M., Smirnov, I., Steinnes, E., Lyapunov, S. and Cherchintsev, V. 2004. Heavy metal atmospheric deposition study in the South Ural Mountains. Journal of Radioanalytical and Nuclear Chemistry 259(1): 19–26.

Frontasyeva, M., Pavlov, S., Mosulishvili, L., Kirkesali, E., Ginturi, E. and Kuchava, N. 2007. Accumulation of trace elements by biological matrice of Spirulina platensis. Ecological Chemistry and Engineering 01/2009; 16(S3): 277–285.

Haber, W. 2010. Die unbequemen Wahrheiten der Ökologie. Oekom Verlag, München.

Hooda, P. (ed.). 2010. Trace elements in soil. Wiley and Sons, New York.

Hosang, M., Fränzle, S. and Markert, B. 2005. Die emotionale Matrix—Grundlagen für gesellschaftlichen Wandel und nachhaltige Innovation. Oekom Verlag, München.

Huntington, H.P. 2000. Using traditional ecological knowledge in science: Methods and applications. Ecological Applications 10(5): 1270–1274.

Janajreh, I., Alshrah, M. and Zamzam, S. 2015. Mechanical Recycling of PVC Plastic Waste Streams from Cable Industry: A Case Study. Sustainable Cities and Society 05/2015; 18. doi:10.1016/j.scs.2015.05.003.

Kirchhoff, P. 2006. Das Gesetz der Hydra. Gibt den Bürgern ihren Staat zurück. Droemer Verlag, München.

Knox, A., Seaman, J., Mench, M. and Vangronsveld, J. 2001. Remediation of metal- and radionuclides-contaminated soils by *in situ* stabilization techniques. pp. 21–60. *In*: Iskandar, I. (ed.). Environmental Restoration of Metals-Contaminated Soils. Lewis Publishers.

Lacerda, L.D. and Salomon, W. 2012. Mercury from Gold and Silver Mining: A chemical time bomb? Springer Press, Heidelberg, New York Tokyo.

Lieth, H. 2005. Deutsch-Usbekischer Christlich-Islamischer Dialog: Religion und zivile Gesellschaft. Eine informationstheoretische Betrachtung. Neuthor Verlag, Mittelstadt.

Loppi, S. and Bonini, I. 2000. Lichens and mosses as biomonitors of trace elements in areas with thermal springs and fumarole activity (Mt. Amiata, Central Italy). Chemosphere 41: 1333–1338.

Lubchenco, J. and Mehra, G. 2004. International science meetings. Science 305: 1531.

Marcovecchio, J.E. and Ferrer, L.D. 2005. Distribution and geochemical partioning of heavy metals in sediments of the Bahia Blanca estuary, Argentina. Journal of Coastal Research 21(4): 826–834.

Marcovecchio, J.E., Freije, R.H., De Marco, S.G., Gavio, M.A., Ferrer, L.D., Andrade, S. and Betrame, M.O. 2006. Seasonality of hydrographic variables in a coastal lagoon: Mar Chiquita, Argentina, Aquatic Conversation: Marine and Freshwater Ecosystems 16(4): 335–347.

Maret, W. 2016. The metals in the biological periodic system of the elements: concepts and conjectures. Int. J. Mol. Sci. 17: 66. doi:10.3390/ijms17010066.

Markert, B. 1994. The Biological System of the Elements (BSE) for terrestrial plants (glycophytes). Science of the Total Environment 155: 221–228.

Markert, B., Breure, T. and Zechmeister, H. (eds.). 2003. Bioindicators & Biomonitors. Principles, Concepts and Applications. Elsevier, Amsterdam.

Markert, B. 2005. Bildung befriedet die Welt. Ein MUT Interview geführt von Hermann Bohle. MUT—Forum für Kultur, Politik und Geschichte, Nr. 454: S. 62–69.

Markert, B., Lieth, H., Menke-Glückert, P., Hosang, M. and Fränzle, S. 2006. Zur Existenz eines ganz starken anthropischen Prinzips. Ist Gott ein Perpetuum mobile nullter Art? BOD-Verlag, Norderstedt.

Markert, B., Fränzle, S. and Wünschmann, S. 2015. Chemical Evolution—The Biological System of the Elements. Springer, Heidelberg, New York, Dordrecht, London.

Menke-Glückert, P. 1968. Working paper, Eco-Commandments for World Citizens' presented at UNESCO conference 'Man and Biosphere', March 9 in 1968, Paris.

Miegel, M. 2005. Epochenwende. Gewinnt der Westen die Zukunft? Propyläen Verlag, Berlin.

Nefiodov, L. 1999. Der sechste Kondradieff—Wege zur Produktivität und Vollbeschäftigung im Zeitalter der Information. Rhein Sieg, St. Augustin.

Oehlmann, J. and Schulte-Oehlmann, U. 2003. Molluscs as bioindicators. pp. 577–635. *In*: Markert, B.A., Breure, A.M. and Zechmeister, H.G. (eds.). Bioindicators and Biomonitors, Elsevier Publisher, Amsterdam, New York, Tokyo.

Paoli, L., Fiorini, E., Munzi, S., Sorbo, S., Basile, A. and Loppi, S. 2013. Antimony toxicity in the lichen *Xanthoria parietina* (L.) Th. Fr. Chemosphere. 2013 Nov; 93(10): 2269–75.

Patrinos, A. and Bamzai, A. 2005. Policy needs robust climate science. Nature 438(17): 285.

Popper, K. 1957. Die offene Gesellschaft und ihre Feinde. Franck-Verlag, Bern.

Prasad, M.N.V., Freitas, H., Markert, B., Fränzle, S. and Wünschmann, S. 2010. Knowledge explosion in phytotechnologies for environmental solutions. Environmental Pollution 158: 18–23.

Ratzinger, J.A. 2006. Deus Caritas est, 25.01.2006, 1. Enzyklopädie von Papst Benedikt dem XVI.

Rinklebe, J. and Du Laing, G. 2011. Factors controlling the dynamics of trace metals in frequently flooded soils. pp. 245–270. *In*: Magdi Selim, H. (ed.). Dynamics and Bioavailability of Heavy Metals in the Root Zone CRC Press. Taylor & Francis Group.

Robertson, M. 2014. Sustainability—Principles and Practise, Routledge Publisher, Taylor & Francis, USA.

Roots, E.F. 1992. Environmental information—autobahn or maize. pp. 3–31. *In*: Schroeder, W., Fraenzle, O., Keune, H. and Mandy, P. (eds.). Global Monitoring of Terrestrial Ecosystems. Ernst und Sohn Verlag für Architektur und technische Wissenschaften, Berlin.

Rosnay, D.E. 2000. Homo symbionticus: Einblicke in das 3. Jahrtausend, Gerling, Akademie Verlag, München.

Sachs, J.D. 2015. The Age of Sustainable Development, Columbia University Press, USA.

Schellnhuber, J. and Wenzel, V. (eds.). 1998. Earth System Analysis: Integrating Science for Sustainability. Springer, Heidelberg.

Schüürmann, G. and Markert, B. (eds.). 1998. Ecotoxicology. Ecological Fundamentals, Chemical Exposure, and Biological Effects. Wiley New York, Chichester, Weinheim, Brisbane, Singapore, Toronto and Spektrum Heidelberg, Berlin.

Steffen, W., Sanderson, R.A., Tyson, P.D., Jäger, J., Matson, P.A., Moore, III B., Oldfield, F., Richardson, K., Schellnhuber, H.J., Turner, B.L. and Wasson, R.J. 2004. Global Change and the Earth System. A Planet Under Pressure. Springer, Berlin, Heidelberg, New York.

Steinnes, E., Berg, T. and Uggerud, H.T. 2011. Three decades of atmospheric metal deposition in Norway as evident from analysis of moss samples. Sci. Total Environ. 412-413: 351–8.

Tabors, G. and Lapina, L. 2012. Growth dynamics of the Hylocomium splendens moss. pp. 311–321. *In*: Nriagu, J., Pacyna, J., Szefer, P., Markert, B., Wünschmann, S. and Namiensnik, J. (eds.). Heavy Metals in the Environment. Maralte Publisher, Leiden.

Taylor, P. 1986. Respect for nature. A theory of environmental ethics. Princeton: Princeton University Press.

Tiessen, H., Brklacich, M., Breulmann, G. and Menenzes, R. (eds.). 2007. Communicating global change science to society: an assessment and case studies. SCOPE 68, Island press, London.

Truhaut, R. 1969. Dangers de lére chimique. Pure and Applied Chemistry 18: 11–128.

Van Nevel, L.T., Taylor, Ph., Ornemark, U., Moody, J.R., Heumann, K.G. and De Bievre, P. 1998. The international measurement evaluation programme (IMEP). IMEP 6: Trace elements in water, Accred. Qual. Assur. 3: 56–68.

Walker, P.A. 2006. Political ecology: Where is the policy? Progress in Human Geography 39: 382–395.

Wolterbeek, H.T., Sarmento, S. and Verburg, T. 2010. Is there a future for biomonitoring of elemental air pollution? A review focused on a larger-scaled health related (epidemiological) context. J. Radioanal. Nucl. Chem. 286: 195–210.

Wünschmann, S., Fränzle, S., Markert, B. and Zechmeister, H. 2008. Input and transfer of trace metals from food via mothermilk to the child—an international study in Middle Europe. pp. 555–592. *In*: Prasad, M.N.V. (ed.). Trace Elements: Nutritional Benefits, Environmental Contamination, and Health Implications. Wiley & Sons.

Marine Debris

Problems and Solutions of the Changing Ocean

H.B. Jayasiri

Introduction

Anthropogenic pollution is a major threat to marine life, with the negative effects related not only to chemical contamination from substances such as heavy metals, nutrients and hydrocarbons, but also to marine debris (Santos et al. 2009). Our oceans are increasingly bearing the brunt of direct and indirect impacts from human activities. A wide range of threats such as increasing acidification, coral bleaching, toxins and chemical pollution, nutrient overloading and fisheries depletion including many others are undermining the ocean's ability to sustain ecological functions. Marine debris is a part of this phenomenon. This represents a significant cause for concern, although much of this growing threat to biodiversity and human health is easily preventable with solutions readily available.

This chapter focuses on marine debris and examines its sources, identifies impacts on ecosystems and economies, and possible options and solutions that become marine litter and proposes a framework for responding to marine debris issues in general. The evidence presented on global occurrence in the marine environment, including accumulation, persistence and transboundary sources, movements and impacts on marine

Oceanography Division, National Aquatic Resources Research and Development Agency, Crow Island, Colombo 15, Sri Lanka.
E-mail: hbjayasiri@gmail.com

biodiversity and ecosystems compounded by emerging data on potential impacts and fate makes a strong case for considering marine debris as a global environmental problem.

Defining Marine Debris

Marine debris is any persistent, manufactured or processed solid material discarded, disposed of or abandoned in the marine and coastal environment (UNEP 2009). It consists of items that have been made or used by people, and deliberately discarded or unintentionally lost into the marine environment, including transport of these materials to the ocean by rivers, drainage, sewage systems or by wind (Galgani et al. 2010).

While this definition encompasses a wide range of materials, most items fall into a relatively small number of material types such as glass, metal, paper and plastic. Plastic items are the most abundant type of marine debris on a global scale and plastic is also the most frequently reported material in encounters between debris and marine organisms. Marine debris commonly stems from shoreline and recreational activities, ocean/waterway activities, smoking related activities, and dumping at sea (Ocean Conservancy 2010). Man-made items of debris are now found in marine habitats throughout the world, from the poles to the equator, from shorelines and estuaries to remote areas of the high seas, and from the sea surface to the ocean floor. Some of the most common items are plastic/polystyrene pieces, rope/cord/nets, cotton swabs and food packets (OSPAR 2007). Plastic items consistently rank among the most abundant type of marine debris at a global scale. In fact, recent studies suggest that the 1982 figure of 8 million litter items entering the oceans every day may need to be multiplied several fold (Barnes 2005).

Litter disposal and accumulation in the marine environment is one of the fastest growing threats for the world's oceans health. With an estimated 6.4 million tons of litter entering the oceans each year, the adverse impacts of litter on the marine environment is not negligible. Following global trends, marine debris pollution is global issue and ubiquitous along the coast, and has received some attention in recent years (Hatje et al. 2013). Most of the relevant studies have focused on oceanic beaches (e.g., Ivar do Sul and Costa 2007; Jayasiri 2013a; 2013b), although it is well recognized that estuaries are an important source of oceanic debris; the contribution of which is potentially exacerbated by various environmental events (e.g., high rainfall and associated hydrology; Ivar do Sul and Costa 2007). Few studies have been conducted in the open ocean and sea bottom.

Although the disposal of solid waste at sea was prohibited in 1988 (Annexe V, MARPOL Convention) more and more reports indicate that even the most secluded environments such as polar regions and the deep ocean floor are no longer exempt from contamination with litter (Barnes 2002).

Factors affecting debris accumulation

The geographical distribution of marine debris is strongly influenced by hydrodynamics, geomorphology and human factors. Moreover, there is notable temporal, particularly seasonal, variation with a tendency for accumulation and concentration along coastal and particular geographical areas. Since plastics are floating material, surface water current determines the quantity of plastic particles available on beaches. Over the past 40 years, large items of plastic debris have frequently been recorded in habitats from the poles to the equator (Thompson et al. 2004). Smaller fragments, 'microplastics', a newly recognized and defined threat to environment, have received far less attention.

Strandline surveys are now organized in many countries and provide information on temporal and spatial trends. However, these surveys typically provide data only on coarse trends and larger items. Plastic pellets are one of the major components of plastic debris in the marine environment (Derraik 2002). They are small granules, generally in the shape of a cylinder or disk with a diameter of 1–5 mm. These pellets are widely distributed in the oceans all over the world (Mato et al. 2001) and thus, accumulate on beaches. At high tide, plastic debris moved by waves is commonly stranded along the wrack line of beaches. The location of the wrack line is a function of the swash uprush associated with wave breaking and water level elevation due to wind setup and/or tidal stage. If subsequent waves and tides do not remove the wrack line, trapped debris will not be resuspended (Thornton and Jackson 1998).

Marine debris is composed largely of plastics and may include cigar tips, baby diapers, six-pack rings, beverage bottles and cans, tires, disposable syringes, plastic bags, bottle caps and fishing line and gear. Different types of marine debris are shown in Figs. 6.2 and 6.3. Marine debris can be classified into seven groups.

1) Plastic
2) Metal
3) Leather
4) Glass
5) Foam
6) Clothes
7) Long-life packages

Sources of marine debris

The United Nations Joint Group of Experts on the Scientific Aspects of Marine Pollution (GESAMP 1991) estimated that land-based sources are responsible for up to 80% of marine debris and the remainder was due to sea-based activities (Sheavly 2005; Fig. 6.1). The main land and sea-based sources of marine debris are listed below.

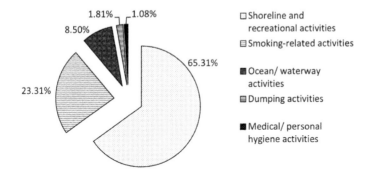

Figure 6.1. Sourcewise distribution of coastal debris collected in 2009 by International Coastal Cleanup Programme (Ocean Conservancy 2010).

Figure 6.2. (a and b) Debris in coastal area of Mahim bay, Mumbai; (c) Plastic debris in sewage outlet entering a creek in Mumbai (Source: Jayasiri et al. 2013).

Figure 6.3. Plastic items found in Mumbai beaches (a) unidentified microplastics (b) Virgin plastic pellets (c) Plastic beads (d) Plastic fragments (Source: Jayasiri et al. 2013).

Land-based sources

Marine debris from land-based sources is blown into the sea, washes into the sea or is discharged into the sea (Sheavly 2005). Land-based sources include the following:

Storm water discharges

Storm drains collect runoff water which is generated during heavy rain events. The drains directly discharge this wastewater into nearby streams, rivers or the ocean. Rubbish from streets can be washed into storm drains and is then discharged straight into the ocean or to streams/rivers which, in turn, may carry the rubbish to the ocean.

Combined sewer overflows

Combined sewers carry sewage as well as storm water. Under normal weather conditions, sewage is carried to a wastewater treatment facility where non-sewage wastes are filtered out. However, during heavy rains the handling capacity of the wastewater treatment system may be exceeded and the sewage plus storm water is then not treated, but is directly discharged into nearby rivers or oceans. This waste can include rubbish such as condoms, tampon applicators, syringes and street litter (US EPA 1992; Sheavly 2005). According to Nollkaemper (1994), waste from combined sewer overflows is one of the major land-based sources of plastic marine debris in the USA.

Littering

Beachgoers may carelessly leave litter at the coast and this will become marine debris. The litter includes items such as food packaging and beverage containers, cigarette butts and plastic beach toys. Fishermen may leave behind fishing gear. Litter from inland areas can become marine debris if it gets into streams or rivers. In this way marine debris may result from rubbish left by workers in forestry, agriculture construction and mining operations (Sheavly 2005).

Solid waste disposal and landfills

Run-off from landfills that are located in coastal areas or near to rivers may find its way into the marine environment. For example, in the USA many estuaries have been contaminated by garbage from nearby solid waste sites (Nollkaemper 1994). In addition to loss from landfills, garbage may be lost to the marine environment during its collection or transportation. Illegal dumping of domestic or industrial wastes into coastal and marine waters is another source of marine debris (USEPA 1992; Sheavly 2005).

Industrial activities

Industrial products may become marine debris if they are improperly disposed of on land or if they are lost during transport or loading/ unloading at port facilities (USEPA 2002). A well known example is small plastic resin pellets, about 2–6 mm in diameter, which are the raw material for the manufacture of plastic products (Derraik 2002). These pellets have been released into the marine environment from accidental spillage during production and processing, transport and handling. Some are buoyant whilst others become suspended or sink. Their presence has been reported in most of the world's oceans (USEPA 1992) and they are found even in more remote, non-industrialized areas in the Southwest Pacific such as Tonga, Rarotonga and Fiji (Derraik 2002). Although plastic pellets are one of the least visible forms of plastic pollution, it is apparent that they have become ubiquitous in ocean waters, sediments and on beaches and are ingested by marine wildlife.

Ocean-based sources

All types of boats and ships and offshore industrial platforms are potential sources of marine debris. The debris may originate from accidental loss, indiscriminate littering or illegal disposal. It may also be the result of waste management disposal practices that were carried out in the past (Sheavly 2005). Ocean-based sources of marine debris include:

Commercial fishing

Commercial fishermen generate marine debris when they fail to retrieve fishing gear or when they discard fishing gear or other rubbish overboard. Debris resulting from commercial fishing includes nets, lines and ropes, strapping bands, bait boxes and bags, gillnet or trawl floats plus galley wastes and household trash (USEPA 1992; Sheavly 2005).

Recreational boaters

Boaters may deposit garbage overboard such as bags, food packaging and fishing gear (Sheavly 2005).

Merchant, military and research vessels

Rubbish from vessels may be accidentally released or blown into the water or may be deliberately thrown overboard. Large vessels with many crew members may carry supplies for several months. They generate solid wastes daily which may end up as marine debris if it is not secured and stored properly (USEPA 1992; Sheavly 2005).

Offshore oil and gas platforms and exploration

Activities on oil and gas platforms may generate items which are deliberately or accidentally released into the marine environment including hard hats, gloves, 55-gallon storage drums, survey materials and personal waste. Undersea exploration and resource extraction also contribute to marine debris (Sheavly 2005). Several studies have demonstrated a positive correlation between marine debris accumulation on beaches and rainfall, particularly when precipitation is strongly seasonal and river flows are greater.

Size classification of debris

The debris can broadly be divided into (Barnes et al. 2009; Ryan et al. 2009; Thompson et al. 2009b) mega-debris (>100 mm), macro-debris (>20 mm), meso-debris (5–20 mm) and micro-debris (<5 mm). Micro-debris, consisting of particles of two main varieties, (1) fragments broken from larger objects, and (2) resin pellets and powders, the basic thermoplastic industry feed stocks, are difficult to trace (Moore 2008). Virgin resin pellets, a common component of debris, enter the oceans routinely via incidental losses during ocean transport or through run-off from processing facilities (Ogata et al. 2009). Although fragments and plastic pellets are usually classified according to size, there is as yet, no consensus nomenclature for describing size. Costa et al. (2010) proposed that the virgin plastic pellets and plastic fragments larger than 1 mm but less than 20 mm be termed small, whereas items that continue to break down into even smaller pieces (Santos et al. 2009) be specifically called microplastics (1 mm or less) based on the existing literature.

Further, the microplastics have been defined differently by different authors (Andrady 2011). Gregory and Andrady (2003) defined micro-litter as the barely visible particles between 0.06 and 0.50 mm in diameter, while particles larger than this are called meso-litter. Others (Betts 2008; Moore 2008) defined the microparticles as being in the size range <5 mm. Ng and Obbard (2006), and Barnes et al. (2009) defined the microparticles as plastics that have dimensions ranging from a few micrometers to 500 µm that are commonly present in sea water. The origins of the microplastics might be attributed to two main sources: (a) direct introduction with runoff and (b) weathering breakdown of meso- and macro-plastic debris (Andrady 2011). The likely mechanism for the generation of a majority of microplastics, however, is the *in situ* weathering of mesoplastics and larger fragments of plastic litter in the beach environment (Gregory and Andrady 2003). Cole et al. (2011) extensively reviewed the microplastics in marine environment with the objectives: (1) to summarize the properties, nomenclature and sources of microplastics; (2) to discuss the routes by which microplastics enter the marine environment; (3) to evaluate the methods by which microplastics are detected in the marine environment; (4) to assess spatial and temporal trends of microplastic abundance; and (5) to discuss the environmental impact of microplastics.

Ivar do Sul et al. (2009) documented the small plastic fragments and pellets on 11 beaches of Fernando de Noronha (Equatorial Western Atlantic) along the strand line and classified as plastic fragments, virgin plastic pellets, nylon monofilaments, polystyrene beads, tar and glass. Marine debris was also classified in eight categories according to size: 2–5, <10, <15 and <20 mm, which were considered small items, and <25, <30, <50 and <100 mm, which were considered medium-size items. Debris of less than 2 mm has not been considered in this classification. In New Zealand beaches, Gregory (1977) found quite considerable quantities, in counts of over 100,000, raw plastic granules per meter of the coast. Martins and Sobral (2011) enumerated 11 size-classes of plastics in sediment from <1 to >10 mm in the Portuguese coastline. The average size of plastic particles in the environment seems to be decreasing, and the abundance and global distribution of micro-plastic fragments have increased over the last few decades. However, the environmental consequences of such microscopic debris are still poorly understood (Barnes et al. 2009). Furthermore, the microscopic fragments of materials used for clothing (polyester, acrylic), packaging (polyethylene, polypropylene) and rope (polyamide) have also been identified from beaches around the United Kingdom (Thompson et al. 2004). For a given size, low-density plastic will float and will be available for uptake by filter-feeders or planktivores, whereas high-density plastics such as polyvinyl chloride (PVC), will tend to sink and accumulate in sediments where they are more likely to be ingested by deposit feeders (Browne et al. 2009).

Plastics

Plastics are synthetic polymers which are typically prepared by the polymerization of monomers derived from oil or gas, and the different types of plastics are usually made from these by the addition of various chemical additives. There are currently some 20 different groups of plastics, each with numerous grades and varieties (APME 2006). The diversity of polymers and the versatility of their properties are used to make a vast array of products that bring medical and technological advances, energy savings and numerous other societal benefits (Andrady and Neal 2009). The common polymers made of plastics are low-density polyethylene (LDPE), high density polyethylene (HDPE), polypropylene (PP) and nylon (Andrady 2011). These are inexpensive, lightweight, strong, durable and corrosion resistant with high thermal and electrical insulation properties (Thompson et al. 2009a). Their low cost, excellent oxygen/moisture barrier properties, bio-inertness and light weight make them excellent packaging materials.

Plastics have transformed everyday life and the usage is increasing. Global production of plastics is estimated at 225 million tonnes per year (APME 2006). Between 1950 and 2009, plastic production has increased from 1.5 million tonnes to 230 million tonnes (Anon 2009). This increase in the usage of plastics has created waste management issues with the end-of-life plastics accumulating in landfills and in natural habitats (Thompson et al. 2009a). The dumping of plastic debris, intentionally or unintentionally, into the ocean is an increasing problem. Since 1990, the dumping of rubbish at sea from ships has been prohibited under the international shipping regulation MARPOL (International Convention for the Prevention of Pollution from Ships) Annex V. Seventy-nine countries have so far ratified the Annex V (CMC 2002) and the signatory countries are required to take steps to fully implement it. A reduction of ship-derived plastic debris should, therefore, be expected, even if the global use of plastics continues to increase (Barnes et al. 2009).

Since plastics are synthetic organic polymers, these are resistant to biodegradation and thus, stay in the environment for a long period of time. Plastics are dumped into the environment unlawfully. Thus, plastic debris has been entering the marine environment in quantities paralleling their level of production over the last half century (Moore 2008). Such debris is composed of fragments of manufactured plastic products (user plastic) and pre-production plastic pellets (industrial pellets, virgin pellets, plastic resin beads or nurdles) that are shipped from manufacturing plants to plastic injection factories to be melted and molded into consumer products (USEPA 1992). Billions of these pellets lost from container ships, down-drains and through other shipping and production mishaps have been reported to be floating in coastal surface waters, in the open ocean and on beaches throughout the world and in sediments (USEPA 1992; Derraik 2002).

Annual global consumption of the major plastic resins is considerable (Andrady and Neal 2009). Plastic films are prone to escape containment as wind-blown debris and are likely to be the major component of terrestrial plastic litter; but plastic litter also includes discarded fishing equipment, food and beverage packaging and many other items that are present in the marine environment (Koutsodendris et al. 2008). Alarming quantities of plastic litter have been found worldwide, both offshore and on beaches. Carpenter et al. (1972) first reported plastic and other man made litter at sea. Plastic debris on beaches is generally a very visible and aesthetically offensive form of pollution (Williams and Tudor 2001). Classifying beach litter into groups based on the material that each piece is made up of has demonstrated that plastic litter predominates (Nakashima et al. 2011).

Characteristics and classification of plastic debris

The size fractions are not normally distinguished in the majority of reports dealing with marine debris, probably due to the sampling difficulties associated with large-scale surveys involving hundreds of square meter transects (Moore 2008). Therefore, their sources, fate and further environmental consequences are poorly understood (Sheavly and Register 2007). Most studies of marine debris have focused on easily visible and identifiable plastic objects (Barnes et al. 2009; 2010; Nakashima et al. 2011). Plastic materials have the ability to proliferate in innumerable sizes, shapes and colors throughout the marine environment worldwide (Moore 2003).

The UK Marine Conservation Society, which organizes annual voluntary beach cleaning on shores all around the UK, reports a 30% increase in the abundance of large fragments (1–50 cm in size) and a 20% increase in the abundance of smaller fragments (1 cm) between 1998 and 2006 (MCS 2007). Thornton and Jackson (1998) classified debris by type, function, degree of fragmentation, length, weight and location on the beach profile at Cliffwood Beach, New Jersey (USA). Turner and Holmes (2011) investigated the abundance and distribution of plastic pellets on the beaches of Malta, and employed various analytical techniques to ascertain the chemical characteristics of the pellets.

Degradation of plastic litter in marine environment

Plastics fragments in the environment are a consequence of photolytic, mechanical and biological degradations. Within the marine environment, plastics also fragment through the combined effects of wave action and abrasion from sediment particles. Plastic debris on beaches degrades more readily than plastic at sea because of higher solar radiation exposure and subsequent increased bulk temperature (Pegram and Andrady 1989). Whereas larger plastic pieces are often removed through active beach cleanups, composition and degradation rates most often dictate retention times of smaller plastic particles on beaches. Degradation occurs faster in virgin pellets that contain no UV stabilizers compared to that in plastic products. The removal of larger pieces of plastic debris from beaches before these are weathered enough to be surface embrittled can have considerable value in reducing the microplastics that end up in the ocean. Beach cleanup, therefore, can have an ecological benefit far beyond the aesthetic improvements of the beaches, and by reducing microplastics, contributes towards the health of the marine food web (Andrady 2011). The physical characteristics of most microplastics show high aging resistance and minimal biodegradation (Moore 2008). The eventual and exact time of biodegradation of plastics in the marine environment still remains unknown (Andrady 2011).

Andrady (2011) documented the process of plastic degradation which is a chemical change that drastically reduces the average molecular weight of the polymer. High-molecular weight plastics used in common applications do not biodegrade at an appreciable rate as microbial species that can metabolize polymers are rare in nature. Over the past decade, there was speculation on the impacts of small plastics, but no specified and uniform results on the subject were obtained (Betts 2008). Only recently, scientists have begun to realize that slowly-decomposing plastic materials may present a long-term threat for marine food chains (Betts 2008).

Debris in Different Marine Settings

Marine debris have been reported in all the marine and coastal ecosystems such as beaches, water column in open and coastal water, oceanic gyres, estuaries, ocean floor.

Beaches

Sandy shores are generally considered important sinks for floating waste, which after stranding generally becomes trapped in/under sand or might be blown farther inland (Williams and Tudor 2001; Kusui and Noda 2003). The factors that determine the amount, type, and distribution of marine debris on beaches are complex and probably interdependent; there are both environmental determinants, such as winds, currents, tides, river flows and beach morphology and socioeconomic variables such as municipal infrastructure (e.g., collection and destination of solid wastes; controlled urban drainage), beach use, social behavior, and level of environmental education among the local and visiting populations (Thiel et al. 2013; Araùjo and Costa 2006).

Debris pollution can also cause aesthetic disturbances, thereby impacting tourism industries that depend on public perceptions of clean beaches (Tudor and Williams 2003). Most solid waste stranded on the shore is allochthonous, brought ashore by waves and currents. Plastic is extremely persistent and dominates the visible litter on sandy beaches worldwide (Derraik 2002). Debris are commonly found at the sea surface or washed up on the shoreline according to wind, tide and current patterns, accumulating temporarily or permanently on beaches and on the seabed at locations known as litter sinks. The visibility of debris as flotsam requires plastics to be positively buoyant in sea water. However, only a few of the plastics typically used in the marine environment has a specific gravity lower than that of sea water.

There are numerous methods used to survey beach litter, depending on the type of litter being sampled; surveying small areas of beach from vegetation to shoreline can bias results, particularly on beaches where litter tends to accumulate in a specific area (Velander and Mocogni 1999). Surveying litter in strandlines gives a high figure for the quantity of fresh and accumulated litter present, making the comparison of studies difficult, which is to be expected since litter tends to accumulate in the strandlines (Velander and Mocogni 1999). The surveys of anthropogenic debris and clean-up operations have generally focused on the larger items along strandlines, and there is a wide geographical variability in the type of data available to examine potential trends. However in the last three decades, it has become apparent that the tiny pellets and micro-plastics have become more numerous and, like larger pieces, these can travel considerable distances (Barnes et al. 2009).

Larger plastic items readily fragment in beach environments, and these fragments have been incorporated in coastal sediments around the world (Barnes et al. 2009). These fragments may remain on beaches longer than larger items because coastal cleanup operations seldom remove them. Andrady (2011) pointed out that there is no acceptable standard procedure presently available for their enumeration in water or sand. However, Frias et al. (2010) and Andrady (2011) suggested similar procedures for the isolation of microplastics in sea water and sand, respectively. It is considered that all of the conventional plastics that have ever been introduced into the environment still remain unmineralized either as whole items or as fragments (Thompson et al. 2005). Barnes et al. (2010) investigated the frequency of plastic pieces on the ocean surface (large) and seabed (large and small fragments) in the most remote areas of the southern Ocean.

Kako et al. (2011) forecasted the quantity of beach litter using a forward in-time Particle Tracking Model (PTM) with the surface currents computed in the ocean circulation model driven by satellite-derived/forecasted wind data and compared the quantity of beach litter determined from the sequential webcam images of the actual beach. Plastic-bottle caps, which showed remarkable seasonal and inter-annual variations, have been chosen as the item for forecasting litter quantities in Goto Islands, Japan. Somerville et al. (2003) assessed the aesthetic quality of 14 beaches in the Firth of Forth, Scotland, using a protocol developed by the UK's National Aquatic Litter Group (NALG).

Nakashima et al. (2011) quantified the total mass of litter on a beach using balloon-assisted aerial photography. The total mass of litter over the beach was derived by multiplying the litter-covered area by the mass of litter per unit area. Light plastics such as polyethelene (PE) made up 55% of all plastic litter on the beach. However, microplastics have not been included in the subsequent analyses. Velander and Mocogni (1999) tested the 10 methods for sampling beach litter on 16 beaches located around the Firth of Forth in order to ascertain the effectiveness of the various methods.

Beaches are the most easily accessible areas for studying plastic debris; yet despite the establishment of many study sites, irregularity of sampling, and differing protocols and observers have led to very few data sets spanning more than a decade (Barnes and Milner 2005). The stranding of larger size fractions of plastics is between one and two orders of magnitude less per length of coastline on remote shores, and at large spatial scales, abundance correlates very strongly with human population (Barnes 2005).

Overall average abundance of 11.6 items m^{-2} and 3.24 g m^{-2} plastics litter has been recorded in Mumbai beaches. More than 80% of plastic particles are within the size range of 20–100 mm in size both by number and weight. Moreover, the colored plastics were predominated with 69% by items and 52% by weight. Probably, the intense use of beaches for recreation, tourism and religious activities has increased the potential for plastic contamination in urban beaches in Mumbai (Jayasiri et al. 2013a).

McDermid and McMullen (2004) quantitatively analyzed the small plastic debris between 1 and 15 mm in size from nine coastal locations on beaches throughout the Hawaiian Islands. The quantification and classification of ship-scrapping waste at Alang-Sosiya, India, has been carried out by Reddy et al. (2003). Monitoring represents an important step towards quantifying spatial and temporal trends in the abundance of all types of debris, including plastics. The temporal trends of macro-plastics on remote islands suggest that the regulations to reduce dumping at sea have been successful to some extent. However, our sustained demand for plastic means that the contamination of the environment by micro-plastic pieces seems set to increase (Barnes et al. 2009).

Cooper and Corcoran (2010) studied the relationship between particle composition and surface textures in natural plastic degradation in five beaches in Kauai, Hawaii. The study has evaluated the effects of mechanical and chemical processes on the degradation of plastic debris, and considered the daily rate of replenishment of plastics on the beaches. Carson et al. (2011) assessed the depth of plastic fragments mixed into beach sediments, determined potential changes to water movement and heat transfer due to these fragments, and speculated on how these changes affect the beach-dwelling organisms. Reddy et al. (2006) documented the accumulation of small plastic debris in the intertidal sediments of the world's largest ship-breaking yard at Alang-Sosiya. Frias et al. (2011) evaluated the types of plastic debris and their size distribution in 10 beaches with two different purposes—Plastic Categorization (CP) and the application of the Clean Coast Index (CCI).

Santos et al. (2008) examined the composition, quantities and distribution of marine debris along ~ 150 km of relatively undeveloped, tropical beaches in Costa do Dendê (Bahia, Brazil), and plastics accounted for 76% of the sampled items, followed by styrofoam. However, the majority of studies on plastic debris were conducted along the temperate coasts in developed countries. The problem seems to be more critical in developing countries where the massive arrival of packaged goods on the market is associated with inefficient waste disposal and also high coastal population density. In spite of these potential problems and having more than 7000 km of coastline, systematic studies concerning marine debris on Indian beaches are still scarce.

The distribution and abundance of marine litter on 26 beaches along the Sea of Japan showed that the most abundant type of stranded litter is plastic, which accounts for 40–80% of the total items in terms of weight and number, and resin pellets have been found on 12 beaches (Kusui and Noda 2003). The most common types of polymers reported on the Ookushi beach were polyethylene (PE), PP, polyethylene terephthalate (PET), PVC, polystyrene (PS), acrylonitrile-butadiene styrene (ABS), acrylonitrile-styrene (AS), polyamide (PA) and polyurethane (PUR) (Nakashima et al. 2011).

Plastics typically constitute approximately 10% of discarded waste, and these comprise 50–80% of the waste items stranded on beaches, floating on the ocean surface and lodged in the seabed (Barnes et al. 2009). The particles consist of both pre-production plastic pellets and post-production plastic fragments of unknown origin.

Plastic litter distribution in beaches

Plastics constitute the majority of marine litter worldwide (Derraik 2002) and there have been many investigations carried out on marine plastics. However, the methods used in the studies are not standardized and hence, cannot be compared. Despite the uncertainty, estimates around the world are reasonably consistent in estimating plastics to comprise approximately 10% of municipal waste mass. In contrast, plastics comprise 50–80% of the waste stranded on beaches, floating on the ocean surface and on the seabed (Gregory and Ryan 1997; Derraik 2002; Barnes 2005). Derraik (2002) carried out an extensive review of the literature on pollution of the marine environment by plastic litter. Horsman (1982) estimated that merchant ships dump 639,000 plastic containers each day around the world; ships are, therefore, a major source of plastic debris (Shaw 1977; Shaw and Mapes 1979). Ribic (1998) carried out beach surveys once a month from 1991 to 1996 at the Island Beach State Park, New Jersey (USA), and demonstrated that the quantity of beach litter had a significant positive trend over that period. Most plastics are less dense than water, but some (e.g., polyamide, polyterephthalate, polyvinyl chloride) are more dense (Ryan et al. 2009).

In general, plastics comprise ~ 20 to 30% of the registered litter on the beaches of Peru and Chile (Thiel et al. 2011). On the beaches in Brazil, plastics account for 76% of the sampled items, followed by styrofoam. Study in the

Tamar Estuary (UK) has shown that the size frequency of plastic debris on the strandline is highly skewed toward smaller debris and that, in terms of abundance, microscopic fragments account for over 80% of the stranded plastic (Browne et al. 2009). Ganesapandian et al. (2011) carried out a survey along a 50-km stretch of shoreline in the northern Gulf of Mannar region of the southeast coast of India to quantify the marine litter and found the predominance of plastic litter.

Ocean floor

Biofouling can play a significant role in controlling plastic debris buoyancy (Ye and Andrady 1991; Moore et al. 2001). Under the weight of fouling by a wide variety of bacteria, algae and animals, accumulated plastics can sink to the seabed (Barnes et al. 2009). Microplastic fragments have already been found in the sediments around the remote Southern Ocean island of South Georgia (Thompson et al. 2009b). On the seafloor, marine litter, particularly plastic, can accumulate in high densities with deleterious consequences for its inhabitants. Yet, because of the high cost involved with sampling the seafloor, no large-scale assessment of distribution patterns is available to date. The denser varieties of plastics such as nylons tend to submerge in the water column and even reach the coastal sediment. Considerable quantities of plastic sink are found on to the continental shelf seabed and even deeper.

Images from the HAUSGARTEN observatory in Arctic deep-sea (79°N) have indicated that litter increased from 3635 to 7710 items km^2 between 2002 and 2011 and reached densities similar to those reported from a canyon near the Portuguese capital Lisboa.

Paranagua Estuarine Complex (PEC) in Brazil located at the northern limit of the Parana coast is among the largest estuaries in Brazil's considerable economic and ecological importance to the entire southern region of the country. Despite being heavily populated, the PEC is considered one of the most preserved Brazilian ecological environments. Among the anthropogenic activities around the bay, port-related industries dominate, followed by tourism, artisanal fisheries (legally restricted to passive gears—i.e., no trawling), agriculture and aquaculture.

The penaeid trawls have been used in terms of their mesh sizes (42- and 26-mm stretched mesh openings in the bodies and codends, respectively), material (0.6 and 1.0 mm diameter polyamide twine, respectively) and design (two seams, with lead-a-head and no sweeps), and only varied slightly in their total opening width (9.44, 9.46 and 9.92 m) to collect the debris in the water column. All trawls were fished in a single-rig configuration (Broadhurst et al. 2013).

The marine debris in seafloor has been examined in Patras Gulf and in Echinadhes Gulf using the trawl nets. The marine debris concentration on the seafloor of Patras and Echinadhes Gulfs is 240 and 89 items/km^2, respectively. The most abundant debris is plastic followed by metal. The high percentage of beverage packaging in Echinadhes Gulf is attributed to shipping traffic, whilst the high percentage of general packaging in Patras Gulf suggests that the source of this material may be on land and it is transported into the gulf by rivers and seasonal streams.

Water column

Visual observations from an aircraft and a variety of airborne sensors (Passive sensors; RGB video, Digital camera, Infrared camera, Hyperspectral cameras and active sensors; Lidar, Radar) have been successfully used for detection of marine debris. Each has specific strengths that could prove the most useful for debris detection in particular applications and areas. None is capable of seeing all debris in any type of sea state, light conditions and turbidity level, so a suite of sensors appropriate to survey conditions will offer the best solution for debris detection with minimal false hits (Veenstra and Churnside 2011). This solution, of course, is more expensive to implement and requires a larger platform than a single small sensor. Lidar is more expensive and precludes operation from small unmanned aircraft, but adds valuable information (Churnside et al. 2010). It is particularly valuable in reducing the number of false detections caused by waves, clouds and sun glint.

The combined survey of both ships has showed that marine debris are more common and abundant in the South Pacific and South Atlantic Oceans than in the Southern Ocean (Barnes and Milner 2005 and Thompson et al. 2009). At the positions, in the Southern Ocean, marine debris has been dominated by plastic. Overall and throughout each sea surveyed, man-made items dominated marine debris and only plastic has been seen south of 63°S. Plastic comprised 43% of the 69 items seen from MV Esperanza and 41% of the 51 items recorded by RRS James Clark Ross. Plastic bags, which have recently been highlighted in a number of countries as a serious issue of environmental concern, only comprised two items in the South Atlantic, both close to the Falkland/Malvinas Islands. Past surveys of marine debris in the South Atlantic Ocean, Scotia Sea and remote Southern Ocean Islands (Barnes and Milner 2005) do not suggest densities of large plastic pieces that are still increasing significantly at highest southern latitudes. Monitoring, accumulation patterns, effects and potential solutions to plastics in the environment are highly complex.

An area of growing concern is the accumulation of buoyant debris such as plastic items and microplastic fragments in the open oceans. Oceanic gyres result from the complex network of currents that circulate water around the oceans, coupled with the effects of wind and the rotation of the globe. They form slowly rotating current systems in which marine debris can accumulate. Due to the durability and persistence of some items of marine debris, once it enters a gyre system it can remain for long periods of time, meaning that the concentration of debris within these systems can be considerably greater than in other areas of the ocean. Moore et al. (2001), for example, found densities of plastic of 334,271 pieces km^{-2}, or 5114 g km^{-2}, within the North Pacific Subtropical Gyre—the largest recording of debris in the Pacific Ocean.

Impacts of marine debris

The impacts of marine debris on marine life are far-reaching, including entanglement in lost or discarded fishing gear and ingestion of plastics and in many cases results in the mortality of the animals. It has been estimated that hundreds of thousands of pounds of lost or discarded fishing gear now foul the marine environment. In addition to being a severe hazard to marine life, human health and navigational safety are also at risk from marine debris. Human impacts range from disease caused by discarded medical waste to navigational hazards for both recreational and commercial boaters.

Effect of marine debris on ecosystems and biodiversity

Marine wildlife is impacted by plastic pollution through entanglement, ingestion, bioaccumulation, and changes to the integrity and functioning of habitats. While macroplastic debris is the main contributor to entanglement, both micro- and macro debris are ingested across a wide range of marine species. The impacts to marine wildlife are now well established for many taxa, including mammals, seabirds, sea turtles, fish and a range of invertebrates. Over 170 marine species have been recorded to ingest human-made polymers that could cause life-threatening complications such as gut impaction and perforation, reduced food intake and transfer of toxic compounds. Although marine debris affects many species (Laist 1997), there is limited data from which to evaluate the collective impact at community and population levels, even for a single species.

Plastic pollution is now recognized worldwide as an important stressor for many species of marine wildlife and their habitats (Moore 2008). The impacts of marine litter on wildlife are recognized around the world and are well documented in literature. Chemical substances such as organic and inorganic pollutants (toxic metals) are contained in plastic products and vary with the types of polymers (Teuten et al. 2009). These chemicals may leach out of the plastics and be taken up into animals via the environment as the plastics are degraded or digested.

The high abundance, lengthy durability and the travel of plastics to even the most remote coasts make them a major potential vector for the dispersal of organisms (Gregory 2009) affecting biodiversity. Robards et al. (1995) have documented the dangers to marine mammals and birds caused by the entanglement in and ingestion of marine litter. Entangled animals and birds tend to exhibit a reduced ability to obtain food, travel and avoid predators, potentially resulting in serious injury or death by starvation, drowning or suffocation (Laist 1997; Anon 2002). Ingested materials (either directly as mistaken prey items or indirectly by bioaccumulation through the food chain) tend to damage and block the digestive tract, and reduce feeding activity due to a false sensation of satiation, potentially resulting in starvation and death (Laist 1987).

PET bottles are a good example of the steps taken to calculate the quantity of metals on beaches. Takahashi et al. (2008) demonstrated that 30.5% of Japanese and 100% of Chinese PET bottles contain more than 10 mg kg^{-1} of antimony, a toxic metal. Likewise, highly toxic organic tin compounds have also been detected in plastic products made from PVC (Takahashi et al. 1999). Toxic lead is also present in plastic bags and occasionally exceeds 10,000 mg kg^{-1} (Sakai et al. 2009). These chemicals are used widely as plasticisers, catalysts, stabilizing additives and pigments. Barnes and Fraser (2003) reported the colonization by a wide range of species, even of higher taxa, on floating plastic in the Southern Ocean.

Plastic litter is a great threat to the marine environment, especially to marine fauna. Plastic debris can attract encrusting organisms as drift plastics (Winston 1982) and easily be ingested or entangled by various marine organisms (Rothstein 1973; Mallory et al. 2006). Documents list at least 267 species worldwide, including 44% of all seabirds, 43% of all marine mammals, 86% of all turtles as well as fish species (Laist 1997). The total number of species is rather difficult to predict or may be underestimated; some affected species may either sink or be eaten by predators and hence, remain undiscovered (Allsopp et al. 2006). The 'three Rs' (reduce, reuse and recycle) have been extensively advocated as solutions to the wasteful nature of our society. These strategies together with the 'fourth and fifth Rs' (energy recovery and molecular redesign) were considered by Hopewell et al. (2009) and Thompson et al. (2009b), who described the current trends and examined the limitations to recycling of plastics.

Entanglement

Marine debris can become entangled around the neck, flippers, tails or flukes of animals and can lead to infection, lower mobility, amputation of limbs and even death. Marine debris is known to have either injured or killed marine mammals, sea turtles and seabirds due to their becoming entangled with it. The most problematic debris are fishing nets and ropes, monofilament lines, six-pack rings and packing strapping bands (Sheavly 2005). Many species are known to have suffered entanglement including 32 species of marine mammals, 51 species of seabirds and six species of sea turtles and 34 species of fish.

Ingestion

Marine debris can also be harmful to marine life if it is ingested. Animals often confuse the debris for food, or accidentally swallow debris when it is in the vicinity of their food source. Debris can cause a physical blockage in the digestive system and/or infection. Through ingestion, they reach the tissues of suspension and deposit feeders (Graham and Thompson 2009) and other biota (Thompson et al. 2004), accumulate through the food web and may enter the human food chain.

Many species of seabirds, marine mammals and sea turtles have been reported to eat marine debris, including plastics. It is thought that this ingestion of marine debris occurs mainly because animals confuse debris for food but may also happen accidentally. Many sorts of plastic items have been ingested by marine organisms including plastic fragments derived from larger plastic items, plastic pellets, which are used as a feedstock material in the plastics industry, plastic bags and fishing line. In some instances the debris may pass through the gut without harming the animal, but in other cases it can become lodged in their throats or digestive tracts. This can lead to starvation or malnutrition if the digestive tract is blocked (USEPA 1992). In addition, debris can accumulate in the gut and give a false sense of fullness, causing the animal to stop eating and slowly starve to death (Sheavly 2005). Ingestion of sharp objects can damage the gut and may result in infection, pain or death.

Spread of alien species by marine debris

The introduction of a non-native species into another habitat is called a biological invasion. The impacts of biological invasions can be devastating for the ecosystem concerned. Natural debris floating in the oceans has always provided 'rafts' which have offered a limited means of travel for certain marine species. Rafts include volcanic pumices, floating marine algae, seagrasses, plant trunks or seeds (Barnes and Milner 2005). However, the introduction of vast quantities of plastic debris into the ocean environment over the past half century has massively increased the amount of raft material and consequently increased the opportunity for the dispersal of marine organisms. This represents an increased potential for alien invasions of new habitats (Barnes 2002; Barnes and Milner 2005).

Plastic debris is long lasting, highly abundant and travels slower than boats, factors which could all favor the survival of rafting organisms (Barnes 2002).

Organisms ranging from algae to iguanas have been observed to raft on rubbish in the marine environment (Barnes and Milner 2005). However, the most commonly found organisms living on plastic waste in the oceans include barnacles, polychaete worms, bryozoans, hydroids and mollusks (Barnes 2002). Plastic encrusted with marine organisms has been found in the Pacific, the Atlantic, the Caribbean and the Mediterranean Sea (Aliani and Molchard 2003).

Plastics are subject to fouling, a process by which algae and other organisms use the plastic as a habitat substrate. Studies have shown that many types of bacterial activity thrive on floating plastics, including harmful algal bloom species. For this reason, plastics serve as vectors for invasive species transport across great distances and to great depths.

Loss of biodiversity and habitat

Marine debris also threatens the biodiversity of the oceans through habitat destruction. Coral reefs are damaged by ghost nets that steamroll through sensitive centers of biodiversity. Habitat destruction also occurs when plastic sheeting covers sea grass beds or other bottom-dwelling species, deadening important feeding and breeding grounds.

Fatality caused by debris has a particularly damaging effect on endangered species. Some species, such as turtles, have long life spans and delayed onset of reproductive maturity, which increases the difficulty of the population's recovery from fatality events. By increasing the fatality rate of already vulnerable populations, marine debris may increase the risk of extinction.

Human impacts of marine debris

Marine debris has the potential to cause many different public health and safety problems. Discarded chemicals, medical waste, radioactive waste, and sewage threaten fishermen, recreational boaters and beachgoers alike. Sewage contains nitrates, phosphates and toxic metal compounds known to cause human health problems. Plastic debris from sewer sludge can include tampon applicators, condoms and disposable diapers, which are a particular problem because they can transmit a variety of gastrointestinal diseases to swimmers. Although sewage-related plastic and other floatable material should be disposed of at treatment sites, these materials enter the ocean when not properly filtered.

Plastics as carriers of toxic chemicals

Plastics can also carry DDE (Dichlorodiphenyldicholorethylene), a breakdown of the pesticide DDT (Dichlorodiphenyltrichloroethane), Nonylphenol, and polychlorinated biphenyls (PCBs), PAH (Polyaromatic Hydrocarbons) which may be transferred to ocean life. Plastics are a known source of polychlorinated biphenyls PCBs in some seabirds and sea turtles. Due to the ubiquitous occurrence on world beaches and their ease of collection and shipment, plastic resin pellets are used by International Pellet Watch (IPW) as passive samplers. IPW is a volunteer-based global monitoring program designed to monitor the pollution status of the oceans and to understand the risks associated with the chemicals in marine plastics. IPW has drawn global pollution maps of POPs and identified hot spots (Ogata et al. 2009; Karapanagioti et al. 2011).

Suggestions and recommendations

Current laws and other government actions do address the issue of marine debris and have for many years, but the inadequacy of these measures is confirmed by the preponderance and persistence of the problem. Insufficient data and research, as well as poor regional and international coordination, make marine debris a difficult problem to address and support. However, by increasing collaboration at all levels of government, improving research and information on the issue, and establishing education and outreach programs to facilitate prevention measures.

The impact of marine debris is far reaching than that it was previously thought. The present widespread use of plastics as an alternative to wood or metal packing allows no room for a complete ban of plastics around the world. Therefore, the option available is to find innovative solution for the collection, segregation, recycling and reuse of plastics to reduce environmental pollution. A number of national and international regulations for preventing pollution in general and plastics in particular exists, such as Annex V of International Convention for the Prevention of Pollution from Ships (MARPOL) prohibits dumping of rubbish at sea from ships, UNEP Global Programme of Action for Protection of the Marine Environment from Land-based activities and the EU Marine Strategy Framework Directive (MSFD). The worrying fact is the extent of implementation of such regulations. Plastics being one of the most persistent pollutants, the strict implementation of the regulations to prevent the plastic pollution and management of plastic debris in a non-detrimental manner is essential at this juncture. Both prevention and cleanup aspects need to be considered in addressing the problem of plastic debris in the coastal and marine environment. As plastic debris of land and sea-based sources is prevalent, the management plans of both marine and terrestrial environment should be integrated.

The plastic industries (manufacturing, packaging and processing) shall accept the responsibility of minimizing the plastic pollution by funding to create awareness among consumers for the reduced use of plastic materials and proper disposal of plastic debris, sorting and collection of plastic debris. The industries should also profusely fund to research programs on developing ecofriendly alternatives to plastics, efficient plastic recycling technologies, and also research on methodologies to monitor and assess the impact of plastics on marine ecosystem and human beings.

Retailers can encourage the consumers to bring their own reusable items, bags, etc. and also have incentive-linked collection point for the discarded packaging material and other plastic debris at each retailer shops to bringing back the discards from the households to the recycling chain. Retailers should be accountable for the quantity of plastic bags and packaging material moved to the consumers and back to the retailers. The retailers who are efficient in bringing back the plastics to recycling should be rewarded.

Littering is a behavioral issue of the consumers. The change in littering behavior can be brought out primarily through education. Educating the consumers on the environmental consequences of litter, would result in generation of less plastic waste as well as proper disposal of the debris. The efforts should be made to bring consistent and sustained behavioral changes among consumers to reduce, reuse, recycle, refuse and rethink to manage the problem of plastic debris.

Researchers should involve themselves in developing eco-friendly alternatives, environment-friendly recycling and energy recovery technologies. In addition, the methodologies for effective monitoring and impact assessment should be researched too. The Governments should effectively implement the regulations and also formulate pro-environment policies with the active participation of the stakeholder to address the menace of plastic debris. The efforts of all the stakeholders (plastic industries, retailers, consumers, researchers and governments) should be harmonized for the effective management of pollution by plastic debris.

These actions are existing measures that are intended to address marine debris in the environment. Many of these actions have been in place for years, while others are relatively new concepts recently put into practice. As a means of organization, actions were classified into four themes:

Marine debris prevention

Preventing the introduction of debris into the marine environment remains the most elusive component of mitigating the impacts of marine debris.

Activities intended to enhance and promote the prevention of marine debris include robust education and outreach campaigns, development and application of appropriate policies, and creation of appropriate incentive programs. Marine waste travels across national borders, thus international cooperation to obtain data and enforce regulatory policies will be necessary to alleviate the effects of marine debris.

Prevention of fishing gear loss is the most fundamental solution to stop ghost fishing. A strategy to prevent loss of fishing gear must include education to increase awareness of the problems of discarded nets together with enforcement of laws that prohibit the dumping of gear at sea. The use of pots/traps with biodegradable parts to permit escape has already been implemented by legislation in some countries but this strategy is needed globally. Finally, retrieval of lost fishing gear can be undertaken to alleviate the problems of ghost fishing.

Response to debris already in the marine environment

Such response activities include nearshore and at-sea cleanups, as well as enforcement of existing environmental laws pertinent to marine debris. Various types of cleanups are necessary in both the coastal regions and the open ocean for complete marine debris removal to be effective in the short and long term. The majority of beach cleanups are facilitated by international and local non-profit organizations. Beach cleanups are generally driven by community involvement and contribute to the education and understanding of marine debris issues.

Research and development of new methods to understand debris impacts and movement

Applying the scientific approach towards an understanding of the causes, effects and cures of marine debris presents a challenging set of variables. The amount and type of debris in the ocean is difficult to measure, and the specific origin of debris and routes by which it travels while at sea are topics that require further research.

In order to maintain a thorough understanding of the marine debris issue, research must also focus on the impact of persistent materials on the marine environment and the development of new technologies for prevention and removal. Although marine debris research to date has not been able to provide a sufficiently comprehensive description of sources, movement or impacts of marine debris, substantial advances have been made in our understanding of marine debris in recent years. Development of new technologies for studying and monitoring marine debris will provide opportunities for improved data collection and better understanding of marine debris and its impacts.

Technologies

Plastic constitutes roughly 80% of the debris that is currently in the oceans. Plastics that are biodegrade, known as bioplastics, are a compelling solution to this problem. Bioplastics are created from natural plant polymers that break down through natural processes such as photodegradation, the interaction with UV radiation, hydrolysis, the interaction with water, and composting, or compositing, interaction with oxygen or oxygenation (Kolybaba et al. 2003).

Marine debris monitoring programs

It is necessary to establish large-scale and long-term monitoring processes across the country and environments (including the sea floor). Since land-based sources provide major inputs for plastic pollution in the marine environment, if we could make the community aware of the environmental consequences of plastic pollution, it would actually make a significant difference.

OSPAR has promoted 'fishing-for-litter' as a practical, simple yet effective means to reduce litter in the marine environment. The project targets the fishing industry by asking fishermen to voluntarily collect marine litter caught up in their nets in large hard-wearing bags provided by the project. The amount and types of litter are recorded onshore before being disposed of in an environmentally friendly way. While most of solid waste management is designed by literates and experts, a major part of the solid waste is managed by illiterates or the least literate or nonexpert. Hence, there will always be shadows between the aspiration and reality—the reason that many solid waste management projects fail. Solid waste management needs more common sense rather than the solution of complicated partial differential equations and financial plans.

We need to address marine debris collectively across national boundaries and with the private sector, which has a critical role to play both in reducing the kinds of wastes that can end up in the world's oceans, and through research into new materials. It is by bringing all these players together that we can truly make a difference.

There are multiple global legal instruments and voluntary agreements aimed at the prevention and management of marine debris. Currently, the most applicable overarching legal framework addressing marine debris is provided by the United Nations Convention on the Law of the Sea (UNCLOS).

Specific agreements regulating different sources of marine debris

The International Convention for the Prevention of Marine Pollution from Ships (MARPOL) and its Annex V prohibiting at-sea pollution by various materials including all plastics and restrictions on at-sea discharge of garbage from ships.

The London Convention for the Prevention of Marine Pollution by Dumping of Wastes and other Matter and its 1996 Protocol.

Conclusion

Marine debris is on the rise, and concerns the politicians, media, scientists, industry and the public. Thus, this chapter discusses the sources and impacts of marine debris and solutions to reduce the impacts in the changing ocean. Marine debris have been reported in all the marine and coastal ecosystems such as beaches, water column in open and coastal water, oceanic gyres, estuaries, ocean floor. Marine debris is composed largely of plastics and may include metal, leather, glass, foam, clothes and long-life packages. Among the debris, plastics are one of the most common and persistent pollutants in the marine environment. There are currently some 20 different groups of plastics, each with numerous grades and varieties. Among them, the common polymers made of plastics are LDPE, HDPE, PP and nylon. The plastic debris reaches coastal and marine environments as their ultimate sink and is accumulating around the world, even in the remote and uninhabited coastal environments. Since plastics are synthetic organic polymers, these are resistant to biodegradation and thus, stay in the environment for a long period of time. Plastics are dumped into the environment unlawfully. Thus, plastic debris has been entering the marine environment in quantities paralleling their level of production over the last half century. The ocean- and land-based sources are discuss disused in detailed. The marine organisms are facing the danger of possible entanglement, ingestion, habitat destruction and bio-invasion. The presence of debris affects the aesthetic and recreational value, causing considerable economic loss. The human health and safety is also threatened by the plastic debris.

Though, a number of national and international regulations for preventing pollution in general and plastics in particular exists, current laws and other government actions are inadequate to manage the issue. It is confirmed by the preponderance and persistence of the problem. By increasing collaboration at all levels of government, improving research and information on the issue, and establishing education and outreach programs to facilitate prevention measures.

The efforts should be made to bring consistent and sustained behavioral changes among consumers to reduce, reuse, recycle, refuse and rethink to manage the problem of marine debris. Researchers should involve themselves in developing eco-friendly alternatives, environment-friendly recycling and energy recovery technologies. In addition, the methodologies for effective monitoring and impact assessment should be formulated. It is necessary to establish large-scale and long-term monitoring processes across the country and environments (including the sea floor). It is by bringing all these players together that we can truly make a difference.

References

Allsopp, M., Walters, A., Santillo, D. and Johnston, P. 2006. Plastic Debris in the World's Oceans. Greenpeace International, Amsterdam, Netherlands, 43 pp.

Andrady, A.L. 2003. Plastics in the environment. *In*: Andrady, A.L. (ed.). Plastics in the Environment. John Wiley & Sons, New Jersey, 762 pp.

Andrady, A.L. and Neal, M.A. 2009. Applications and societal benefits of plastics. Phil. Trans. Roy. Soc. B 364: 1977–1984.

Andrady, A.L. 2011. Microplastics in the marine environment. Mar. Pollut. Bull. 62: 1596–1605.

Anon. 2002. Nationwide Beach Clean and Survey Report. Marine Conservation Society, Herefordshire, 89 pp.

Anon. 2009. The Compelling Facts about Plastics. An Analysis of Plastics Production, Demand and Recovery for 2008. Plastics Europe, Brussels.

APME. 2006. An Analysis of Plastics Production, Demand and Recovery in Europe. Association of Plastics Manufacturers, Brussels.

Barnes, D.K.A. 2002. Man's aid to alien invasions of seas is rubbish. Dir. Sci. 1: 107–112.

Barnes, D.K.A. and Fraser, K.P.P. 2003. Rafting by five phyla on man-made flotsam in the Southern Ocean. Mar. Ecol. Prog. Ser. 262: 289–291.

Barnes, D.K.A. 2005. Remote islands reveal rapid rise of Southern Hemisphere sea debris. Dir. Sci. 5: 915–921.

Barnes, D.K.A. and Milner, P. 2005. Drifting plastic and its consequences for sessile organism dispersal in the Atlantic Ocean. Mar. Biol. 146: 815–825.

Barnes, D.K.A., Galgani, F., Thompson, R.C. and Barlaz, M. 2009. Accumulation and fragmentation of plastic debris in global environments. Philos. Trans. R. Soc. London, Ser. B 364(1526): 1995–1998.

Barnes, D.K.A., Walters, A. and Gonçalves, L. 2010. Macroplastics at sea around Antarctica. Mar. Environ. Res. 70: 250–252.

Browne, M.A., Galloway, T. and Thompson, R. 2009. Microplastic—An emerging contaminant of potential concern? Integrated Environ. Assess. Manag. 3(4): 559–566.

Carpenter, E.J., Anderson, S.J., Harvey, G.R., Miklas, H.P. and Peck, B.B. 1972. Polystyrene particles in coastal waters. Science 178: 749–750.

Carson, H.S., Colbert, S.L., Kaylor, M.J. and McDermid, K.J. 2011. Small plastic debris changes water movement and heat transfer through beach sediments. Mar. Pollut. Bull. 62(8): 1708–1713.

Churnside, J.H., Sharov, A.F. and Richter, R.A. 2010. Aerial surveys of fish in estuaries: a case study in Chesapeake Bay. ICES Journal of Marine Science 67.

Cole, M., Lindeque, P., Halsband, C. and Galloway, T.S. 2011. Microplastics as contaminants in the marine environment: A review. Mar. Pollut. Bull. 62: 2588–2597.

Cooper, D.A. and Corcoran, P.L. 2010. Effects of mechanical and chemical processes on the degradation of plastic beach debris on the island of Kauai, Hawaii. Mar. Pollut. Bull. 60: 650–654.

Costa, M.F., Ivar doSul, J.A., Silva-Cavalcanti, J.S., Araújo, M.C., Spengler, A. and Tourinho, P.S. 2010. On the importance of size of plastic fragments and pellets on the strandline: A snapshot of a Brazilian beach. Environ. Monit. Assess. 168(1-4): 299–304.

Derraik, J.G.B. 2002. The pollution of the marine environment by plastic debris: A review. Mar. Pollut. Bull. 44: 842–852.

Frias, J.P.G.L., Sobral, P. and Ferreira, A.M. 2010. Organic pollutants in microplastics from two beaches of the Portuguese coast. Mar. Pollut. Bull. 60: 1988–1992.

Frias, J.P.G.L., Martins, J. and Sobral, P. 2011. Research in plastic marine debris in mainland Portugal. J. Integrat. Coast. Zone Manage. 11(1): 145–148.

Galgani, F., Fleet, D., Van Franeker, J., Katsanevakis, S., Maes, T., Mouat, J., Oosterbaan, L., Poitou, I., Hanke, G., Thompson, R., Amato, E., Birkun, A. and Janssen, C. 2010. Marine strategy framework directive, task group 10 report: marine litter. *In*: Zampoukas, N. (ed.). JRC Scientific and Technical Reports. Ispra: European Commission Joint Research Centre.

Ganesapandian, S., Manikandan, S. and Kumaraguru, A.K. 2011. Marine litter in the northern part of Gulf of Mannar, southeast coast of India. Res. J. Environ. Sci. 5: 471–478.

GESAMP. 1991. The State of the Marine Environment. Blackwell Scientific Publications, London 146 pp.

Gregory, M.R. 1977. Plastic pellets on New Zealand beaches. Mar. Pollut. Bull. 8: 82–84.

Gregory, M.R. and Ryan, P. 1997. Pelagic plastics and other seaborne persistent synthetic debris: a review of Southern hemisphere perspectives. pp. 49–66. *In*: Coe, J. and Rogers, D. (eds.). Marine Debris: Sources, Impacts and Solutions. Springer Verlag, New York.

Gregory, M.R. and Andrady, A.L. 2003. Plastics in the marine environment. pp. 379–402. *In*: Andrady, A.L. (ed.). Plastics and the Environment. John Wiley and Sons, New York.

Hatje, V., Costa, M.F. and Cunha, L.C. 2013. Oceanografia e química: unindo conhecimentos em prol dos oceanos e da sociedade. Quim. Nova 36: 1497–1508.

Ivar do Sul, J.A. and Costa, M.F. 2007. Marine debris review for Latin America and the Wider Caribbean Region: from the 1970 until now and where do we go from here? Marine Pollution Bulletin 54: 1087–1104.

Ivar do Sul, J.A., Spengler, A. and Costa, M.F. 2009. Here, there and everywhere. Small plastic fragments and pellets on beaches of Fernando de Noronha (Equatorial Western Atlantic). Mar. Pollut. Bull. 58: 1236–1238.

Jayasiri, H.B., Purushothaman, C.S. and Vennila, A. 2013. Plastic litter accumulation on high water strandline of urban beaches in Mumbai, India. Environ. Monit. Assess. 185(9): 7709–7719.

Jayasiri, H.B., Purushothaman, C.S. and Vennila, A. 2013. Quantitative analysis of plastic debris on recreational beaches in Mumbai, India. Mar. Pollut. Bull. 77(1-2): 107–112.

Kako, S., Isobe, A., Magome, S., Hinata, H., Seino, S. and Kojim. A. 2011. Establishment of numerical beach-litter hindcast/forecast models: An application to Goto Islands, Japan. Mar. Pollut. Bull. 62: 293–302.

Kolybaba, M., Tabil, L.G., Panigrahi, S., Crerar, W.J., Powell, T. and Wang, B. 2003. Biodegradable Polymers: Past, Present, and Future. The Society for Engineering in Agricultural, Food, and Biological Systems (ASAE), Paper No. RRV03-0007. St. Joseph, Michigan.

Koutsodendris, A., Papatheodorou, G., Kougiourouki, O. and Georgiadis, M. 2008. Benthic marine litter in four gulfs in Greece, Eastern Mediterranean; abundance, composition and source identification. Estua. Coast. and Shelf Sci. 77: 501–512.

Kusui, T. and Noda, M. 2003. International survey on the distribution of stranded and buried litter on beaches along the Sea of Japan. Mar. Pollut. Bull. 47: 175–179.

Laist, D.W. 1987. Overview of the biological effects of lost and discarded plastic debris in the marine environment. Mar. Pollut. Bull. 18: 319–326.

Laist, D.W. 1997. Impacts of marine debris: Entanglement of marine life in marine debris including a comprehensive list of species with entanglement and ingestion records. pp. 99–119. *In*: Coe, J. and Rogers, D.B. (eds.). Marine Debris: Sources, Impacts and Solutions. Springer Series on Environmental Management, New York.

Martins, J. and Sobral, P. 2011. Plastic marine debris on the Portuguese coastline: A matter of size? Mar. Pollut. Bull. 62: 2649–2653.

Mato, Y., Isobe, T., Takada, H., Kanehiro, H., Ohtake, C. and Kaminuma, T. 2001. Plastic resin pellets as a transport medium for toxic chemicals in the marine environment. Environ. Sci. Technol. 35: 319–324.

McDermid, K.J. and McMullen, T.L. 2004. Quantitative analysis of small-plastic debris on beaches in the Hawaiian archipelago. Mar. Pollut. Bull. 48: 790–794.

MCS. 2007. Beach Watch 2006—The 14th Annual Beach Litter Survey Report. Marine Conservation Society, Ross on Wye.

Moore, C.J., Moore, S.L., Leecaster, M.K. and Weisberg, S.B. 2001. A comparison of plastic and plankton in the north Pacific central gyre. Mar. Pollut. Bull. 42(12): 1297–1300.

Moore, C.J. 2003. Trashed: Across the Pacific Ocean, plastics, plastics everywhere. Nat. Hist. 112(9): 46–51.

Moore, C.J. 2008. Synthetic polymers in the marine environment: A rapidly increasing, long-term threat. Environ. Res. 108: 131–139.

Nakashima, E., Isobe, A., Magome, S., Kako, S. and Noriko, D. 2011. Using aerial photography and *in situ* measurements to estimate the quantity of macro-litter on beaches. Mar. Pollut. Bull. 62: 762–769.

Nollkaemper, A. 1994. Land-based discharges of marine debris: From local to global regulation. Mar. Pollut. Bull. 28: 649–652.

Ocean Conservancy. 2010. Trash Travels. International Coastal Cleanup Report. Washington, DC, 60 pp.

Ocean Conservancy. 2013. 2012 International Coastal Cleanup: Working for Clean Beaches and Clean Water, Washington, DC, 24 pp.

Ogata, Y., Takada, H., Mizukawa, K., Hirai, H., Iwasa, S., Endo, S., Mato, Y., Saha, M., Okuda, K., Nakashima, A., Murakami, M., Zurcher, N., Booyatumanondo, R., Zakaria, M.P., Dung, L.Q., Gordon, M., Miguez, C., Suzuki, S., Moore, C., Karapanagioti, H., Weerts, S., McClurg, T., Burres, E., Smith, W., Velkenburg, M.V., Lang, J.S., Lang, R.C., Laursen, D., Danner, B., Stewardson, N. and Thompson, R.C. 2009. International Pellet Watch: Global monitoring of persistent organic pollutants (POPs) in coastal waters. 1. Initial phase data on PCBs, DDTs and HCHs. Mar. Pollut. Bull. 58: 1437–1446.

OSPAR. 2007. OSPAR Pilot Project on Monitoring Marine Beach Litter: Monitoring of marine litter on beaches in the OSPAR region. London: OSPAR Commission.

Pegram, J.E. and Andrady, A.L. 1989. Outdoor weathering of selected polymeric materials under marine exposure conditions. Polym. Degrad. Stabil. 26: 333–345.

Reddy, M.S., Basha, S., Kumar, V.G.S., Joshi, H.V. and Ghosh, P.K. 2003. Quantification and classification of ship scraping waste at Alang-Sosiya, India. Mar. Pollut. Bull. 46: 1609–1614.

Reddy, M.S., Basha, S., Adimurthy, S. and Ramachandraiah, G. 2006. Description of the small plastics fragments in marine sediments along the Alang-Sosiya ship-breaking yard, India. Estuar. Coast. Shelf Sci. 68: 656–660.

Ribic, A. 1998. Use of indicator items to monitor marine debris on a New Jersey beach from 1991 to 1996. Mar. Pollut. Bull. 36: 887–891.

Robards, M.D., Piatt, J.F. and Wohl, K.D. 1995. Increasing frequency of plastic particles ingested by seabirds in the subarctic North Pacific. Mar. Pollut. Bull. 30: 151–157.

Ryan, P.G., Moore, C.J., Franeker, J.A. and Moloney, C.L. 2009. Monitoring the abundance of plastic debris in the marine environment. Phil. Trans. Roy. Soc. B 364: 1999–2012.

Sakai, S., Aasari, M., Sato, N. and Miyajima, A. 2009. Lead contained in plastic shopping bags and its substance flow. J. Environ. Chem. 19(4): 497–507.

Santos, I.R., Friedrich, A.C. and Ivar do Sul, J.A. 2008. Marine debris contamination along undeveloped tropical beaches from northeast Brazil. III Congresso Brasileiro de Oceanografia—CBO'2008 I Congresso Ibero-Americano de Oceanografia—I CIAO Fortaleza (CE), 20–24 May 2008.

Santos, I.R., Friedrich, A.C. and Ivar do Sul, J.A. 2009. Marine debris contamination along undeveloped tropical beaches from northeast Brazil. Environ. Monit. Assess. 148: 455–462.

Shaw, D.G. 1977. Pelagic tar and plastic in the Gulf of Alaska and Bering Sea: 1975. Sci. Total Environ. 8: 13–20.

Shaw, D.G. and Mapes, G.A. 1979. Surface circulation and the distribution of pelagic tar and plastic. Mar. Pollut. Bull. 10: 160–162.

Sheavly, S.B. 2005. Sixth Meeting of the UN Open-ended Informal Consultative Processes on Oceans & the Law of the Sea. Marine debris—an overview of a critical issue for our oceans. June 6–10, 2005. http://www.un.org/Depts/los/consultative_process/consultative_process.htm.

Sheavly, S.B. and Register, K.M. 2007. Marine debris & plastics: Environmental concerns, sources, impacts and solutions. J. Polym. Environ. 15: 301–305.

Somerville, S.E., Miller, K.L. and Mair, J.M. 2003. Assessment of the aesthetic quality of a selection of beaches in the Firth of Forth, Scotland. Mar. Pollut. Bull. 46: 1184–1190.

Teuten, E.L., Saquing, J.M., Knappe, D.R.U., Barlaz, M.A., Jonsson, S., Björn, A., Rowland, S.J., Thompson, R.C., Galloway, T.S., Yamashita, R., Ochi, D., Watanuki, Y., Moore, C., Viet, P.H., Tana, T.S., Predente, M., Boonyatumanond, R., Zakaria, M.P., Akkhavong, K., Ogata, Y., Hisashi, H., Iwasa, S., Mizukawa, K., Hagino, Y., Imamura, A., Saha, M. and Takada, H. 2009. Transport and release of chemicals from plastics to the environment and to wildlife. Phil. Trans. Roy. Soc. B 364: 2027–2045.

Thiel, M., Bravo, M., Hinojosa, I.A., Luna, G., Miranda, L., Núñez, P., Pacheco, A.S. and Vásquez, N. 2011. Anthropogenic litter in the SE Pacific: An overview of the problem and possible solutions. J. Integr. Coast. Zone Manage. 11: 115–134.

Thompson, R.C., Olsen, Y., Mitchell, R.P., Davis, A., Rowland, S.J., John, A.W.G., McGonigle, D. and Russell, A.E. 2004. Lost at sea: Where is all the plastic? Science 304: 838.

Thompson, R.C., Moore, C.J., Saal, F.S. and Swan, S.H. 2009a. Plastics, the environment and human health: Current consensus and future trends. Phil. Trans. Roy. Soc. B 364: 2153–2166.

Thompson, R.C., Swan, S.H., Moore, C.J. and vom Saal, F.S. 2009b. Our plastic age. Phil. Trans. Roy. Soc. B 364: 1973–1976.

Thompson, R., Moore, C., Andrady, A., Gregory, M., Takada, H. and Weisberg, S. 2005. New directions in plastic debris. Science 310: 1117.

Thornton, L. and Jackson, N.L. 1998. Spatial and temporal variations in debris accumulation and composition on an estuarine shoreline, Cliffwood Beach, New Jersey, USA. Mar. Pollut. Bull. 36(9): 705–711.

Tudor, D.T. and Williams, A.T. 2003. Public perception and opinion of visible beach aesthetic pollution: The utilization of photography. J. Coast. Res. 19: 1104–1115.

Turner, A. and Holmes, L. 2011. Occurrence, distribution and characteristics of beached plastic production pellets on the island of Malta (Central Mediterranean). Mar. Pollut. Bull. 62: 377–381.

UNEP. 2009. Marine Litter: A Global Challenge, pp. 232. Nairobi: UNEP.

United States Environmental Protection Agency (US EPA). 1992. Plastic pellets in the aquatic environment: Sources and recommendations. Environmental Protection Agency: Oceans and Coastal Protection Division Final Report 842-B-92-010. Washington, DC.

Veenstra, T.S. and Churnside, J.H. 2011. Airborne sensors for detecting large marine debris at sea. Marine Pollution Bulletin.

Velander, K. and Mocogni, M. 1999. Beach litter sampling strategies: Is there a 'Best' Method? Mar. Pollut. Bull. 38(12): 1134–1140.

Williams, A.T. and Tudor, D.T. 2001. Litter burial and exhumation: Spatial and temporal distribution on a Cobble Pocket beach. Mar. Pollut. Bull. 42(11): 1031–1039.

Winston, J.E. 1982. Drift plastic-An expanding niche for a marine invertebrate? Mar. Pollut. Bull. 13(10): 348–351.

7

Global Shipping, Ballast Water and Invasive Species

Sami Souissi,[1,*] *Olivier Glippa*[2] *and Hans-Uwe Dahms*[3]

Introduction

Aquatic systems such as the marine environment and inland waters are faced with a number of threats including water pollution, over-fishing, degradation of habitats and climate change. Aquatic systems are also affected by the introduction of alien species, which are called invasive species once they make a severe impact on the ecosystems they invade.

Aquatic invasions are closely linked to the globalization of ship transport. Generally, globalization refers to the international integration of various interchanged items (O'Rourke et al. 2000). Advances in transportation (from steam locomotive to container ship) and in telecommunication infrastructure (from telegraphy to the internet) enhanced the global exchange of economic and cultural activities. Environmental threats like air, soil, water pollution, global change, and overharvesting are linked to globalization (Bridges 2002). Globalization affects and has effects on natural environments, economies and socio-cultural resources (Ritzer 2008).

Among the different means of transport, ocean shipping turns out to be the most energy efficient mode of long-distance transport for large quantities of goods (Rodrigue et al. 2006). As much as 90% of the world

---

[1] Univ. Lille, CNRS, Univ. Littoral Cote d'Opale, UMR 8187 LOG, Laboratoire d'Océanologie et de Géosciences, F-62930 Wimereux, France.
[2] Novia University of Applied Sciences, Raseborgsvägen 9, Ekenäs, Finland.
[3] Department of Biomedical Science and Environmental Biology, Kaohsiung Medical University, Kaohsiung, Taiwan, R.O.C. & Department of Marine Biotechnology and Resources, National Sun Yat-sen University, Kaohsiung, Taiwan, R.O.C.
* Corresponding author: sami.souissi@univ-lille1.fr

trade is hauled by ships (International Maritime Organization 2016). Global shipping networks have a critical economical role since maritime traffic is one of the most important drivers of global trade. The trade volume exceeded 30 trillion ton-miles and is growing faster than the global economy before 2007 (UNCTAD 2015). There are spatiotemporal differences in maritime trafficking. A most frequented sea route is between the US and China with unbalanced orientation since goods from China to the US are four times higher than in the opposite direction. New shipping routes are expected to open. An example is the Northeast Passage north of the Russian territory which is expected to provide a faster passage between Europe and China (Kaluza et al. 2010). The shipping network shares differences in the movement patterns of different ship types. Oil tankers and bulk carriers generally move less regularly between ports than container ships. This is an important result regarding the spread of invasive species because these ships often go empty and exchange larger quantities of ballast water than other ship types.

The worldwide maritime network plays a critical role in today's spread of invasive species. Invasive species are recognized as one of the greatest threats to the world's oceans since they provide a major menace to global biodiversity (EC 2008). In the course of globalization, the probability of successful invasion becomes greater and accompanying threats are getting even larger. Two major pathways for marine bioinvasions are discharged water from ballast water tanks and hull fouling. According to UNCTAD (2015) ten billion tons of ballast water are transferred globally. Even terrestrial species such as insects are sometimes transported as larvae in shipping containers or as freshwater larvae in ballast water if this is retrieved from riverine harbors (Lounibos 2002). In several parts of the world, invasive species have caused dramatic levels of landscape change and the alteration of species. Damaged ecosystems and their functioning and hazards for human livelihood, health and local economies are some of the consequences (Mack et al. 2000). The financial loss caused by invasive species is estimated in a not so recent account to be US$ 120 billion per year in the US alone (Pimentel et al. 2005). Despite this it seems that the shipping industry is far less perceived as a cause of environmental concern than other industrial sectors by the public.

The Global Cargo Ship Network

Ship journeys provide us with a view of their actual network. The combined cargo capacity of ships calling at a given port (measured in gross tonnage) follows a heavy-tailed distribution (Fig. 7.1).

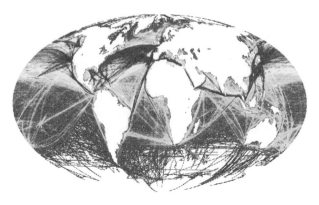

Figure 7.1. Map of world commercial shipping activity (from NCEAS 2008 [https://www.nceas.ucsb.edu/globalmarine2008/impacts]).

How Can a Species be Transported

Several pathways of species introduction were suggested: ballast water, sediment, fouling, aquaculture, floating objects, canals or unintentional introductions (aquarium, aquaculture) (Panov and Caceres 2007; Alekseev et al. 2010; Gollasch 2011).

Cruise ships, large tankers, and bulk cargo carriers use a huge amount of ballast water, which is often taken on in coastal waters in one region after ships discharge wastewater or unload cargo. The ballast water is then discharged at the next targeted port (Fig. 7.2). Ballast is essential to ensure that ships operate safely and effectively, while providing balance and stability when the ship lacks cargo. At the same time can ballast water with its accompanying living organisms pose serious ecological and economic changes? Thousands of species, small enough to pass through the water pumps that pump ballast water from outside into the tanks, get transported between two harbors. It is estimated that about 10,000 marine species are transported around the world in ballast water every day (NOAA 2007).

Ballast water discharge typically contains a variety of organisms, including plants, animals, viruses, bacteria and their resting stages (Panov and Caceres 2007; Gollasch 2011). Resting eggs and cysts of algae and aquatic invertebrates have been frequently reported in ballast tank sediments. Ballast water tanks contain settled accumulations that vary from a few cm to more than 30 cm layers of sediment (Hamer 2002), resembling the upper layers of lake, estuarine or sea bottom sediments (Fig. 7.2).

Since diapausing organisms (Figs. 7.3, 7.4) were selected to overcome adverse periods, they have the advantage of retaining this characteristic in the man-made ballast water environment as well. Diapausing organisms

Figure 7.2. Mud extracted from a ship ballast water tank (photo Olivier Glippa; GIP Seine-Aval BIODISEINE project).

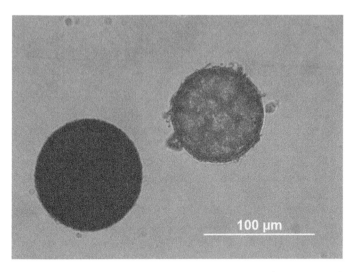

Figure 7.3. Copepod resting eggs from a ship ballast water tank (photo Olivier Glippa; GIP Seine-Aval BIODISEINE project).

are more protected from invasive antagonistic species (competitors and predators) that have no resting stages (Alekseev et al. 2001; Panov et al. 2004; Alekseev et al. 2007).

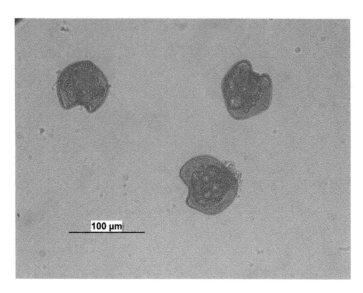

Figure 7.4. Dinoflagellate cysts (photo Olivier Glippa; GIP Seine-Aval BIODISEINE project).

How Does an Introduced Species Become Invasive?

A species is introduced when it is relocated to outside the limit of its original area of distribution by human activity, either accidentally or deliberately. An invasive species is a plant, fungus or animal species that is not native to a specific location (an introduced species), and which has a tendency to spread. It may disturb its new environment to a degree believed to cause damage to ecosystems, human economy or even human health. A species becomes invasive if it proliferates (Fig. 7.5), and disrupts the functioning of ecosystems, leads to the disappearance of other, indigenous or endemic species and may cause serious economic impacts (Mack et al. 2000; Pimentel et al. 2005; Strayer et al. 2006).

Species- and ecosystem factors should be considered among the mechanisms that when combined, establish invasiveness of a newly introduced species. Invasive species appear to have specific traits or specific combinations of traits that allow them to outcompete native species, e.g., fast growth, rapid reproduction, high dispersal ability, phenotypic plasticity (the ability to alter growth characteristics to become better adapted to the local conditions), tolerating a wide range of environmental conditions (ecological competence), using a wide range of food types (generalism), and associations with humans.

A species that gets introduced to a new location will have to maintain itself as a small size founder population before it can successfully establish and becomes invasive in the new environment (Tilman 2004). It may

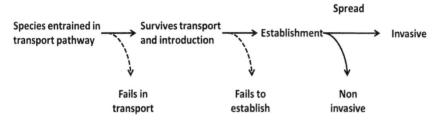

Figure 7.5. From the introduction to the invasive species (modified from Kolar and Lodge (2001)).

be difficult for an introduced species to survive and reproduce at low population density, so a species might need to reach a location for several times before it can establish and becomes invasive (Grosholz 2005). Repeated introductions, such as through routine ship movements between ports with the same founding species provide repeated possibilities of immigration for a given species (Verling et al. 2005).

Examples of Marine Invasions

The introduction of non-native species is not a new phenomenon. For example, the American Soft-shell clam *Mya arenaria* is the oldest species being introduced to European waters. Even the Vikings might have been in the position to transfer this species from North America to Europe in 1245–1295 (Petersen et al. 1992). In the Black Sea, its introduction (probably with ballast water) is more recent and dates back to 1966 (Gomoiu et al. 2002).

As mentioned will ballast water taken up at port or sea and released in a port of even another continent by transoceanic vessels be the largest source of non-native aquatic species invasions (Xu et al. 2014). Around the world, more than 3,000 different species of aquatic organisms may be transported daily by such vessels (Moretzsohn et al. 2013). Those passive migrants establish at other places as alien or non-indigenous species. Ruiz et al. (2015) documented for marine and estuarine waters of North America a total of 450 Non Indigenous Species (NIS) of invertebrates and algae that have been established there.

For example, freshwater Zebra mussel *Dreissena polymorpha*, native to the Black, Caspian and Azov Seas, probably reached the Great Lakes as larvae via ballast water from a transoceanic vessel (Molnar et al. 2008). *D. polymorpha* has infested over 40% of inland waterways and more than US$5 billion were necessary for its removal and treatment, since 1989, added to the expenses for the control of its expansion.

Some of the transported species have natural resting stages (in dormancy or diapause stage of metabolic arrest). Resting stages are characterized by a range of adaptations such as reduced metabolic activity, accumulation of protective materials (stress enzymes, antifreezing compounds), energy reserves. This makes them strongly resistant to adverse factors. They are pre-adapted to be transported with ship ballast water and sediments without losing substantial viability for long periods of time, from months to even decades (Alekseev et al. 2010). It has been demonstrated that diapause allows a species to colonize new environments and facilitate passive transport of the resting forms, e.g., *Cladoceran* ephippia (Alekseev and Starobogatov 1996). For example, Bailey et al. (2003) found invertebrate diapausing eggs in residual sediments from transoceanic vessels. The authors studied the viability of eggs collected from ballast tanks of vessels operating on the North American Great Lakes and found egg viability ranging between 0 and 92% for individual species.

Panov and coworkers (2004) showed that the cladocerans *Cercopagis pengoi* and *Bythotrephes longimanus* have an extended diapause that allows them to disperse and invade successfully. Both species originated from the eastern Baltic Sea from where they initially invaded the Great Lakes in North America which was shown by genetic markers (Berg et al. 2002). This likely happened through transport of diapausing resting stages in ballast water tanks which is generally assumed to be the primary pathway of intercontinental dispersal of aquatic organisms (Hebert and Cristecu 2002).

What Effects May Invasive Species Have?

Species that become invasive may alter their ambient environment and the functioning of the ecosystems they are invading by modifying abiotic or biotic factors (Mack et al. 2000; Molnar et al. 2008).

Biotic invasions are considered as one of the top five drivers for global biodiversity loss and its frequency is increasing as a result of tourism and globalization (Millennium Ecosystem Assessment 2005).

Invasive species may drive local native species to extinction via competitive exclusion, niche displacement, or hybridization with related native species. This way, introduced species may result in extensive changes in the community structure, dispersal, and worldwide distribution of species which ultimately causes a loss of biodiversity (Odendaal et al. 2008).

Since it is difficult to correlate extinctions with species invasions in every case, we need to evaluate possible ecological risks with possible economic benefits of species introductions (Mooney and Cleland 2001). The majority of reports, however, indicate negative effects of invasive species.

An example for a major negative economic impact is provided by an efficient feeder on zooplankton and fish larvae alike, the comb jelly

Mnemiopsis leidyi which disturbs whole food webs wherever it invades (Javidpour et al. 2006). This comb jelly originates from coastal waters of the western Atlantic. *M. leidyi* was found in the Black Sea in the 1980s, and in the Caspian Sea in 1999. Thereafter, it was found throughout the Mediterranean and in the northwestern Atlantic. In the North Sea it was recorded first in 2006 and since then it has successfully colonized the Baltic Sea (Javidpour et al. 2006). It is unknown how this comb jelly invaded the North and Baltic Seas. The immigration might have happened by drifting individuals, or with ballast water. This may have taken place either from its natural habitat along the western Atlantic coastline or from the Black Sea. From the Black Sea there are two possible pathways, via the Mediterranean and eastern Atlantic or from the Black Sea to the Rhine Estuary through the Rhine-Main-Danube Canal in ballast water tanks (Oliveira 2007). The latter is reasonable since *M. leidyi* has a wide salinity tolerance. The riverine pathway from the Ponto-Caspian area into continental Europe is also known for other freshwater invaders like the zebra mussel *Dreissena polymorpha* (Oliveira 2007). This example also shows that dispersal pathways can be diverse or unknown in several cases. This makes a control of invaders more difficult.

In several countries, consumption of some suspension feeding shellfish, such as oysters can cause paralysis and even death in humans. Such effects can be due to introduced microscopic algae, which are responsible for Harmful Algal Blooms (HABs), such as red tide causing protists (Holtcamp 2012).

Native species can be threatened with extinction through the process of genetic pollution. Genetic pollution through unintended hybridization leads to the homogenization or replacement of local genotypes as a result of competitive advantage of the introduced species (Aubry et al. 2005). Genetic pollution depends on the introduction or habitat modification that allows previously isolated species to contact. Hybrids resulting from rare species that interbreed with abundant species can then dominate the gene pool of the rarer species (Rhymer and Simberloff 1996).

There are not only disturbing effects of invasive species. Facilitation may occur where one species physically modifies a habitat in ways that are advantageous to other species. For example, zebra mussels which are a menace throughout North America and the Great Lakes, increase habitat complexity on lake floors, and provide crevices in which invertebrates live. This increases structural complexity and adds nutrition provided by the waste products of the suspension feeding mussel, which in turn increases the abundances and diversity of benthic invertebrate communities (Silver-Botts et al. 1996). Economics also plays a major role in exotic species introduction. High demand for the valuable Chinese mitten crab *Eriocheir sinensis* is one explanation for the possible intentional release of the species in foreign waters of North America as well as Europe.

The Pacific oyster *Crassostrea gigas* provides another example of an economically valuable marine invertebrate that has invaded much of the Atlantic and apparently also due to climate change and the warming of ocean water appears to invade wide areas of Scandinavian coastal waters (Nehring 2011; Laugen et al. 2015). In addition of being a resource for commercial fishing, oyster reefs also provide diverse ecosystem services such as, among others, water quality improvement, and seashore stabilization (reviewed by Grabowski et al. 2012). These authors estimated an economic value that oyster reef services provide (not even including the harvest and sale of oysters) ranging between US$55,000 and US$99,000 per hectare per year.

In middle Europe oysters may have become 'naturalized' meanwhile. At long term, there is a high risk that *C. gigas* develops a bio-invasion here. Site-specific strategies for conservation makes different approaches necessary. One strategy would be to accept the northward spread of the species in the ecosystem and its impact on coastal habitats. A different conservation approach is to control the density control of *C. gigas* in areas with the highest risk for dispersal and bio-invasion. Cooperation with local groups of volunteers that handpick the species may allow the control of a smaller area whereas cooperation with commercial fisheries may allow for controlling larger areas.

Economic Impacts of Marine Invasive Species

To the ecological cost, an economic cost is added that is difficult to assess though since indirect effects such as human and agricultural product health effects are difficult to estimate. In Europe, the economic costs of invasive species are estimated to be more than Euros 12 billion per year (EC 2008). In the Great Lakes region, the cost of controlling zebra mussels at its 381 water treatment facilities exceeds US$100 million annually (Roasen et al. 2012).

Treatment Methods/Ballast Water Management/Ship Hull Treatment

Once a species became invasive to an ecosystem, attempts to eradicate them usually fail (Elton 1958). Ecosystem changes that follow an introduction rather quickly then may become irreversible. Therefore, the only way to prevent invasions is to control the transport of potential invaders.

Canada was the first country to recommend that all ships should discharge their ballast in the deep ocean (>2000 m) in 1989 to reduce the threat of ballast water introductions (Endresen et al. 2004). Even if most of the coastal organisms present in ballast water should not survive in this oceanic conditions, Minton et al. (2005) showed that some organisms can persist. Meanwhile, there were several summits on Ballast Water (see for a recent event: [http://www.wplgroup.com/aci/event/15th-ballast-water-management-summit/]) and several International Agreements were made that are currently under validation. Recently, Ghabooli et al. (2016) studied the population attenuation in zooplankton communities during trans-oceanic transfer in ballast water. They used molecular tools to assess the biodiversity and confirmed the persistence of some groups during the transfer.

Conclusions

Ever increasing volumes of ballast water are transported from port to port between countries and continents with the present increase of shipping intensity. The resulting discharge of ballast water often contains a multitude of alien organisms (viruses, microorganisms, fungi, plants and animals). The non-native species can cause major environmental and ecological changes—which are mainly disturbing—in aquatic systems. If pathogens or parasites are among the invading species, they can directly lead to health issues among live stock, wild life and humans. Several organisms are transported as early planktonic stages of their life cycle. At this stage, most marine planktonic organisms are very sensitive to oxygen conditions and pollution of ship ballast water. Natural resting stages show a strong resistance to adverse factors and may be transported with ship ballast water and sediments for long periods of time without losing their viability. These resting stages are particularly difficult to eradicate from ballast water tanks and from ship hulls although there are smart technologies emerging for this purpose. Generally, however, resting stages are difficult to remove/or to monitor in the different compartments of ballast water tanks.

The development of larger and more powerful ships, combined with a rapid increase in world trade and global ship networking are reducing natural barriers for the dispersal of oceanic species. Therefore, the species composition of whole ecosystems gets changed, from primary producers to commercial species, and the economic impact can be huge due to a change of biodiversity patterns, the threatening of native species, and a change or even disturbance of ecosystem functioning. We have to study these processes in more detail and not only focus on stable habitat theory, like the ones we are trying to develop in global warming or restoration ecology. A few models include these aspects in the dynamics of biodiversity as we have little knowledge about these processes.

But we can conclude that species introductions become accelerated with global change, particularly climate change. Recent studies were focusing on the risk of invasions in the Arctic as the intensity of shipping traffic will increase in the future and this mainly due to global warming. But respective studies are only at their early stage and the projection of future changes in biodiversity at global scale should necessarily combine both effects of climate change as well as the intensification of shipping routes in the future. It is probable that the future environmental conditions as projected by the IPCC (IPCC 2007; 2013) will be favorable for the geographical extension of many species. Moreover, the rate of alien species naturalization may increase in the context of global change. Consequently, we may expect changes in marine biodiversity, mainly of tiny organisms to be not only impacted by climate change but also by species introductions. The prediction of marine biodiversity changes under global warming scenarios may not be sufficient to take into account the unpredictable impact of species introductions. In fact, most studies on the impact of climate change on marine biodiversity neglected the processes based on species introduction. The analysis of species invasions in aquatic ecosystems as discussed in this chapter confirms the key importance of this phenomenon. We therefore recommend accurate monitoring and understanding all mechanisms of species introductions via ballast water as a key process in biodiversity dynamics.

Finally, regarding the major risk engendered by ballast transportation a global solution is needed. In this context we have to mention the global initiatives (see several summits in Singapore) from a legislation and management (human) point of view. In order to minimize the environmental risk of transfer via ballast water discharges, international and national rules have been developed (see: https://www.bimco.org/About/Viewpoint/02_Ballast_Water_Management.aspx; http://www.worldshipping.org/industry-issues/environment/vessel-discharges/ballast-water).

However, the success of such initiatives requires more scientific research to understand the best technologies to reduce the risk of alien species introduction.

Acknowledgements

We thank the GIP Seine-Aval who provided financial support on the project BIODISEINE to study the role of resting eggs in the biodiversity patterns in the Seine estuary, France (http://www.seine-aval.fr/projet/biodiseine/). Special thanks to the authorities of Le Havre Harbor for their precious help during the sampling of ship ballasts. We are grateful to Victor Alekseev for all fruitful discussions and collaboration in the field of the role of resting eggs in plankton dynamics and invasions. H.-U. Dahms acknowledges the support of a grant from the Research Center for Environmental Medicine, Kaohsiung Medical University (KMU), the Asia-Pacific Ocean Research Center of the Department of Oceanography (No. 76211194) in the frame of the KMU/NSYSU cooperation, and MOST104-2621-M-037-001 to T.H. SHIH.

References

Alekseev, V.R. and Starobogatov, Y.I. 1996. Types of diapause in the Crustacea: definitions, distribution, evolution. Hydrobiol. 320: 15–26.

Alekseev, V.R., Makrushin, A. and Hwang, J.S. 2010. Does the survivorship of activated resting stages in toxic environments provide cues for ballast water treatment? Mar. Poll. Bull. 61: 254–258.

Aubry, C., Shoal, R. and Erickson, V. 2005. Grass cultivars: their origins, development, and use on national forests and grasslands in the Pacific Northwest. USDA Forest Service, 44 pages, plus appendices; Native Seed Network (NSN), Institute for Applied Ecology, 563 SW Jefferson Ave, Corvallis, OR 97333, USA. Nativeseednetwork. org. Retrieved 2011-05-17.

Bailey, S.A., Duggan, I.C., Van Overdijk, C.D., Jenkins, P.T. and MacIsaac, H.J. 2003. Viability of invertebrate diapausing eggs collected from residual ballast sediment. Limnol. Oceanogr. 48: 1701–1710.

Berg, D.J., Garton, D.W., MacIsaac, H.J., Panov, V.E. and Telesh, I.V. 2002. Changes in genetic structure of North American *Bythotrephes* populations following invasion from Lake Ladoga, Russia. Freshwater Biol. 47: 275–282.

Botts, P.S., Patterson, B.A. and Schlosser, D.W. 1996. Zebra mussel effects on benthic invertebrates: Physical or biotic? J. North Am. Benthological Soc. 15: 179–184.

Bridges, G. 2002. Grounding globalization: the prospects and perils of linking economic processes of globalization to environmental outcomes. Econ. Geogr. 78: 361–386.

Drake, J.M. and Lodge, D.M. 2004. Global hot spots of biological invasions: evaluating options for ballast-water management. Proc. Roy. Soc. Lond. B. 271: 575–580.

Drake, J.M. and Lodge, D.M. 2007. Hull fouling is a risk factor for intercontinental species exchange. Aquat. Invasions 2: 121–131.

EC. 2008. Towards an EU Strategy on Invasive Species. COM/2008/789, European Commission, Brussels.

Elton, C.S. 1958. The ecology of invasions by animals and plants. London: Methuen 18.

Endresen, O., Behrens, H.L., Brynestad, S., Andersen, A.B. and Skjong, R. 2004. Challenges in global ballast water management. Mar. Poll. Bull. 48: 615–623.

Ghabooli, S., Zhan, A., Paolucci, E., Hernandez, M.R., Briski, E., Cristescu, M.E., MacIsaac, H.J. 2016. Population attenuation in zooplankton communities during transoceanic transfer in ballast water. Ecology and Evolution, 1–8 p, doi:10.1002/ece3.2349.

Gollasch, S. 2011. Chapter 17. Global shipping and the introduction of alien invasive species. pp. 293–314. *In*: The World Ocean in Globalisation: Climate Change, Sustainable Fisheries, Biodiversity, Shipping, Regional Issues.

Gomoiu, M.T., Alexandrov, B., Shadrin, N. and Zaitsev, Y. 2002. The Black Sea—a recipient, donor and transit area for alien species. pp. 341–350. *In*: Leppäkoski, E., Gollasch, S. and Olenin, S. (eds.). Invasive Aquatic Species of Europe. Distribution, Impacts and Management, Kluwer Academic Publishers, Dordrecht, The Netherlands.

Grabowski, J.H., Brumbaugh, R.D., Conrad, R.F., Keeler, A.G., Opaluch, J.J., Peterson, C.H., Piehler, M.F., Powers, S.P. and Smyth, A.R. 2012. Economic Valuation of Ecosystem Services Provided by Oyster Reefs. BioSci. 62: 900–909.

Grosholz, E.D. 2005. Recent biological invasion may hasten invasional meltdown by accelerating historical introductions. Proc. Natl. Acad. Sci. USA 102(4): 1088–1091.

Hamer, J.P. 2002. Ballast tank sediments. pp. 232–234. *In*: Leppäkoski, E. and Olenin, S. (eds.). Invasive Aquatic Species of Europe.

Hebert, P.D.N. and Cristescu, M.E.A. 2002. Genetic perspectives on invasions: the case of the Cladocera. Can. J. Fish. Aquat. Sci. 59: 1229–1234.

Holtcamp, W. 2012. The emerging science of BMAA: do cyanobacteria contribute to neurodegenerative disease? Environ. Health Perspect 120: 110–116.

IPCC. 2007. Climate Change 2007: the Physical Science basis. pp. 1–996. *In*: Solomon, S., Qin, D., Manning, M., Chen, Z., Marquis, M., Averyt, K.B., Tignor, M. and Miller,

H.L. (eds.). Contribution of Working Group I to the Fourth Assessment Report of the Intergovernmental Panel on Climate Change. Cambridge Univ. Press, London.

IPCC. 2013. Climate Change 2013: The Physical Science Basis. Contribution of Working. Group I to the Fifth Assessment Report of the Intergovernmental Panel on Climate Change. *In*: Stocker, T.F., Qin, D., Plattner, G.K., Tignor, M., Allen, S.K., Boschung, J., Nauels, A., Xia, Y., Bex, V. and Midgley, P.M. (eds.). Near-term Climate Change: Projections and Predictability. Cambridge University Press, Cambridge, New York, NY, UK, USA. 1535 pp.

James, P. and Steger, M.B. 2014. A Genealogy of 'globalization': The career of a concept. Global. 11: 417–434.

Javidpour, J., Sommer, U. and Shiganova, T.A. 2006. First record of *Mnemiopsis leidyi* A. Agassiz 1865 in the Baltic Sea. Aquat. Invasions 1: 299–302.

Kaluza, P., Kölzsch, A., Gastner, T. and Blasius, B. 2010. The complex network of global ship movements. J. R. Soc. Interface 7: 1093–1103.

Kolar, C.S. and Lodge, D.M. 2001. Progress in invasion biology: predicting invaders. Trends Ecol. Evol. 16: 199–204.

Laugen, A.T., Hollander, J., Obst, M. and Strand, A. 2015. The pacific Oyster (*Crassostrea gigas*) invasion in Scandinavian coastal waters: impact on local ecosystem services. *In*: Canning-Clode, J. (ed.). Biological Invasions in Changing Ecosystems. Vectors, Ecological Impacts, Management and Predictions.

Lounibos, P. 2002. Invasions by insect vectors of human disease. Annual Review of Entomology 47(1): 233–266.

Mack, R.N., Simberloff, D., Lonsdale, W.M., Evans, H., Clout, M. and Bazzaz, F.A. 2000. Biotic invasions: causes, epidemiology, global consequences and control. Ecol. Appl. 10: 689–710.

Millennium Ecosystem Assessment. 2005. Ecosystems and Human Well-being: Biodiversity Synthesis (PDF). World Resources Institute.

Minton, M.S., Verling, E., Miller, A.W. and Ruiz, G.M. 2005. Reducing propagule supply and coastal invasions via ships: effects of emerging strategies. Frontiers in Ecology and the Environment 3(6): 304–308.

Molnar, J.L., Gamboa, R.L., Revenga, C. and Spalding, M.D. 2008. Assessing the global threat of invasive species to marine biodiversity. Front. Ecol. Environ. 6: 485–492.

Mooney, H.A. and Cleland, E.E. 2001. The evolutionary impact of invasive species. Proc. Natl. Acad. Sci. USA 98: 5446–5451.

Moretzsohn, F., Sánchez Chávez, J.A. and Tunnell, J.W., Jr. (eds.). 2013. Invasive Species, GulfBase: Resource Database for Gulf of Mexico Research. Harte Research Institute for Gulf of Mexico Studies at Texas A&M University-Corpus Christi.

Nehring, S. 2011. NOBANIS—Invasive Alien Species Fact Sheet—*Crassostrea gigas*. From: Online Database of the European Network on Invasive Alien Species—NOBANIS www. nobanis.org, Date of access 31/01/2013.

NOAA. 2007. Coastal Services Center, National Oceanic and Atmospheric Administration. Ballast Water: Michigan takes on the Law. Coastal Services http://www.csc.noaa.gov/magazine/2007/04/article2.html.

Odendaal, L.J., Haupt, T.M. and Griffiths, C.L. 2008. The alien invasive land snail *Theba pisana* in the West Coast National Park: Is there cause for concern? Koedoe. 50: 93–98.

Oliveira, O.M.P. 2007. The presence of the ctenophore *Mnemiopsis leidyi* in the Oslofjorden and considerations on the initial invasion pathways to the North and Baltic Seas. Aquat. Invasions 2: 185–189.

O'Rourke, K.H. and Williamson, J.G. 2000. When Did Globalization Begin? NBER Working 7632.

Panov, V.E., Krylov, P.I. and Riccardi, N. 2004. Role of diapause in dispersal and invasion success by aquatic invertebrates. J. Limnol. 63: 56–69.

Panov, V.E. and Caceres, C. 2007. Chapter 12. Role of diapause in dispersal of aquatic invertebrates. pp. 187–195. *In*: Alekseev, V., de Stasio, B. and Gilbert, J.J. (eds.). Diapause in Aquatic Invertebrates.

Petersen, K.S., Rasmussen, K.L., Heinemeier, J. and Rud, N. 1992. Clams before Columbus? Nature 359: 679.

Pimentel, D., Zuniga, R. and Morrison, D. 2005. Update on the environmental and economic costs associated with alien-invasive species in the United States. Ecol. Econ. 52: 273–288.

Roasen, A.L. 2012. Anderson economic group 2012: The costs of Aquatic Invasive Species to Great Lakes States 51.

Rhymer, J.M. and Simberloff, D. 1996. Extinction by hybridization and introgression. Annu. Rev. Ecol. Syst. 27: 83–109.

Ritzer, G. (ed.). 2008. Studying globalization: Methodological issues. *In*: George Ritzer. The Blackwell Companion to Globalization. John Wiley & Sons 146.

Rodrigue, J.P., Comtois, C. and Slack, B. 2006. The geography of transport systems. London, UK: Routledge.

Ruiz, G.M., Rawlings, T.K., Dobbs, F.C., Drake, L.A., Mullady, T., Huq, A. and Colwell, R.R. 2000. Global spread of microorganisms by ships. Nature 408: 49–50.

Ruiz, G.M., Fofonoff, P.W., Steves, B.P. and Carlton, J.T. 2015. Invasion history and vector dynamics in coastal marine ecosystems: a North American perspective. Aquat. Ecosyst. Health. Manag. 18: 299–311.

Strayer, D.L., Eviner, V.T., Jeschke, J.M. and Pace, M.L. 2006. Understanding the long-term effects of species invasions. Trends Ecol. Evol. 21: 645–651.

Tilman, D. 2004. Niche tradeoffs, neutrality, and community structure: A stochastic theory of resource competition, invasion, and community assembly. Proc. Natl. Acad. Sci. USA 101: 10854–10861.

UNCTAD. 2015. Review of maritime transport 2015. United nations conference on trade and development, 108. http://unctad.org/en/PublicationsLibrary/rmt2015.

Verling, E., Ruiz, G.M., Smith, L.D., Galil, B., Miller, A.W. and Murphy, K.R. 2005. Supply-side invasion ecology: characterizing propagule pressure in coastal ecosystems. Proc. Roy. Soc. Lond. B 272: 1249–1256.

Xu, J., Wickramarathne, T.L., Chawla, N.V., Grey, E.K., Steinhaeuser, K., Keller, R.P., Drake, J.M. and Lodge, D.M. 2014. Improving management of aquatic invasions by integrating shipping network, ecological, and environmental data. Proceedings of the 20th ACM SIGKDD international conference on Knowledge discovery and data mining—KDD 14: 1699–1708.

Xu: NCEAS 2008 [https://www.nceas.ucsb.edu/globalmarine2008/impacts]; https://www.bimco.org/About/Viewpoint/02_Ballast_Water_Management.aspx; http://www.worldshipping.org/industry-issues/environment/vessel-discharges/ballast-water.

8

High Seas Deep-Sea Fisheries under the Global Changing Trends

Gui Manuel Machado Menezes[1] and *Eva Giacomello*[2,*]

Introduction

Oceans covers about 361 million km² of the earth's surface of which about 219 km² million (~ 60%) are in Areas Beyond National Jurisdiction (ABNJ) or 'High Seas' areas (Fig. 8.1). Under the United Nations Convention on the Law of the Sea (UNCLOS), adopted in 1982 and implemented in 1994, several maritime zones were defined in which coastal states were able to exercise varying degrees of sovereignty and jurisdiction and certain rights and responsibilities. These zones include the Internal Waters, Territorial Sea, Contiguous Zone, the Exclusive Economic Zone (EEZs), the Continental Shelf, the Area and the High Seas (Fig. 8.1).

The high seas, is essentially an area of 'deep sea' since more than 90% of this area is deeper than 200 m. This means that in what concerns the exploitation of living resources in the high seas it is not possible to dissociate it from the problems of exploitation of deep-sea resources.

The deep ocean is a vast, complex and diverse space that still remains poorly known. The average depth of the global oceans is ca. 3,850 m,

---

[1] Universidade dos Açores (UAç) – Departamento de Oceanografia e Pescas (DOP) - MARE – Marine and Environmental Sciences Centre, Rua Prof. Dr. Frederico Machado, 4; 9901-862 Horta, Azores, Portugal.
E-mail: gui.mm.menezes@uac.pt

[2] MARE – Marine and Environmental Sciences Centre, IMAR - Instituto do Mar, Rua Prof. Dr. Frederico Machado, 4; 9901-862, Horta, Azores, Portugal.

* Corresponding author: evagiacomello@gmail.com

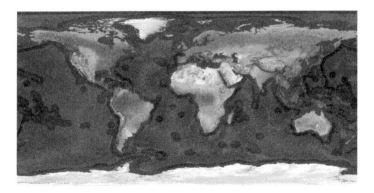

Figure 8.1. World EEZs (dark grey) and high seas areas. R. Medeiros©ImagDOP.

about 88% of the oceans beyond the continental shelves are deeper than
1,000 m and 76% have depths of 3,000–6,000 m. The deep ocean is therefore
the world's largest connected biome but only a fraction of it has been
investigated as only about 0.0001% of the deep sea has been sampled
biologically (Rogers 2015).

The high seas hold among others several types of ecosystems like
fracture zones or oceanic ridges, abyssal plans, hydrothermal vents or
seamounts. Seamounts, which support a large number of fish species, and
high biodiversity (Rogers 1994) are ubiquitous topographic features of the
world's oceans, being estimated that more than 100 thousand (of seamounts
higher than 1,000 m) may exist on the seas and oceans of the planet (Wessel
et al. 2010). However, if we consider small seamounts (> 100 m) the number
may exceed 1 million (Wessel 2007).

In this chapter we will treat only high seas fisheries of demersal/
deep-sea fishes (for definition of terms used in this chapter please see
Table 8.A1). Deep-sea fishing exploration in the high seas acts mostly up
to 1,500–1,800 m depth, although in some cases fishing can occur down
to 2,000 m (Morato et al. 2006; Clark et al. 2007; Rogers and Gianni 2010).
Hence we may consider that all areas in the high seas within those depth
limits are potentially exploitable. These presumed fishable areas (between
200 to 2,000 m) correspond however to only about 9% of all oceans seafloor
(Fig. 8.2). Additionally, considering that the bottom trawl make up about
80% of high seas bottom fishing and taking into account that the habitats or
topographic features on which it occurs (e.g., rocky substrates of mid-ocean
ridges, seamounts and submarine canyons, Morgan et al. 2005; Norse et al.
2012) are rare and occupy less than 5% of the seafloor (based on Costello et

Table 8.A1. Definition of terms used in the chapter.

Term	Definition	Source
Deep-sea fisheries	Fisheries in which the total catch includes species that can only sustain low exploitation rates, where the fishing gear is likely to contact the seafloor during the normal course of fishing. They target deepwater benthic, demersal and benthopelagic resources at depths greater than 200 metres exploited primarily on the high seas.	FAO 2008, 2009
Vulnerable marine ecosystem (VME)	A marine ecosystem is described as vulnerable based on the characteristics that it possesses, including; uniqueness or rarity; functional significance of the habitat; high susceptibility to degradation by anthropogenic activities; life-history traits of component species that make recovery difficult; structural complexity.	FAO 2009 (paragraph 42)
Significant Adverse Impacts (SAIs)	Impacts that compromise ecosystem integrity (i.e. ecosystem structure or function). The scale and significance of an impact are determined based on factors related with the type of impact (its intensity, spatial extent, duration and frequency), and the characteristics of the ecosystem affected (its sensitivity/vulnerability, ability and rate of recovery, extent of alteration in ecosystem function).	FAO 2009 (paragraph 17-20)
Deep sea	Ocean lying below 200 m depth, generally beyond the depth of continental shelves and with insufficient light for net primary production by photosynthesis.	Rogers 2015
Mesopelagic zone	Area between 200-1,000 m depth in the pelagic realm, where sunlight is detectable but is insufficient for photosynthesis.	Rogers 2015

Term	Definition	Reference
Bathyal zone	Area between 200-2,000 m depth in the benthic ecosystem.	Rogers 2015
Oxygen Minimum Zones (OMZs)	Areas where level of oxygen is very low (<20–45 μmol kg^{-1}), due to elevated surface Production,and weak deep circulation of oxygen-rich water.	Gilly et al. 2013
Demersal fish	Fish that live and feed on or near the seabed.	Rogers 2015
Deep-sea fish	Definitions of deep-sea fishes can be based on a common upper depth of occurrence (e.g. 200 or 400 m) or on shared life history traits affecting species vulnerability to exploitation.	Bergstad 2013
Seamount(s)	A discrete (or group of) large isolated elevation(s), greater than 1000 m in relief above the sea floor, characteristically of conical form (length/width ratio <2).	Harris et al. 2014
Guyot(s)	A seamount having a flat top >10 km^2 in areal extent and with a gradient of $<2°$	Harris et al. 2014
Ridge(s)	An isolated (or group of) elongated (length/width ratio >2), narrow elevation(s) of varying complexity having steep sides, $>1,000$ m in vertical relief.	Harris et al. 2014
Canyon(s)	Steep-walled, sinuous valleys with V-shaped cross sections, axes sloping outward as continuously as river-cut land canyons and relief comparable to even the largest of land canyons.	Shepard 1963 (in Harris et al. 2014)

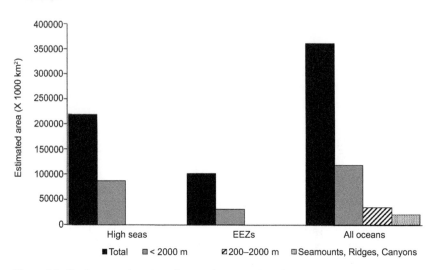

Figure 8.2. Total area and total seafloor surface areas less than 2,000 m depth estimated for the high seas and the EEZs, and the estimated surface areas between 200–2,000 m depth and of seamounts, ridges and canyons in all oceans (numbers originally from Costello et al. 2010; 2015; Harris et al. 2014).

al. 2010; 2015; Harris et al. 2014, see Fig. 8.2), the suitable exploitable areas represent even a smaller fraction of the vast ocean space.

Taking into account the surface areas of those topographic features where deep-sea fishing potentially occurs in each of the main oceans regions, the Pacific Ocean comprises about 61% of the available area of seamounts, ridges and canyons. If we only consider seamounts and guyots, where the majority of the high seas deep-sea fisheries occurs (Clark et al. 2007; Pitcher et al. 2010), again the Pacific Ocean represents almost 70% of the total area (8.7 million km²) containing those features in all regions of the oceans (Fig. 8.3). In terms of management and protection of seamount ecosystems it is also important to note that approximately 50% of potential seamounts worldwide occur within EEZs (Kitchingman et al. 2007). In the Pacific, where 63% of the large seamounts occur (8955 out of 14287) about 60% of the large seamounts lie within EEZs (Kitchingman et al. 2007). The Atlantic and Indian Oceans comprise about 18% (2704) and 11% (1658) of the world seamounts respectively, with large seamounts occurring within EEZs in the 28% of cases in the Atlantic and 35% in the Indian Ocean (Kitchingman et al. 2007). Mid-ocean ridges, which share most of the exploitable species with those of seamounts (Bergstad et al. 2008; Vecchione et al. 2010), occupy relatively important and similar areas in the Atlantic, Pacific and Indian Oceans, totaling about 6.7 million km² (Fig. 8.3).

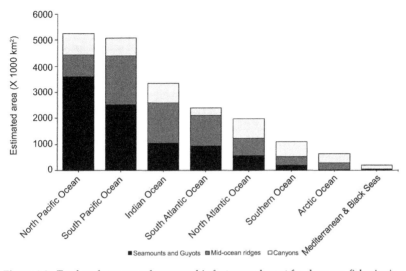

Figure 8.3. Total surface areas of topographic features relevant for deep-sea fisheries in each ocean region (numbers originally from Harris et al. 2014).

Deep-sea fisheries in the high seas

Deep-sea fisheries expanded mainly in the late 1960s and during the 1970s with the former USSR and Japanese distant-water fleets, and from the 80s other nations involved in these fisheries in several regions of the planet (Troyanovsky and Lisovsky 1995; Merret and Haedrich 1997; Gordon et al. 2003; Clark et al. 2007; Bensch et al. 2008; Rogers and Gianni 2010). The first large-scale seamount fisheries started in the Pacific Ocean previous to the development of this type of fisheries in the Atlantic and Indian Oceans (Clark et al. 2007).

With the decrease of the fishing resources on the traditional shallower fishing grounds of the continental platforms the deep-sea fishing expanded and dispersed to vast areas of the global oceans with an increasing fishing efficiency (Devine et al. 2006; Morato et al. 2006; Rogers and Gianni 2010; Watling et al. 2011; Weaver et al. 2011). The increasing fishing restrictions within national waters (Gordon et al. 2003; Morgan et al. 2005), and the simultaneous increase in the demand of fish by the developed countries (Sumaila et al. 2015) also promoted the expansion of fisheries in the high seas. Stock declines and the loss of fishing opportunities in traditional shallower shelf areas boosted the development of new and more robust fishing gears (Roberts 2002; Gordon et al. 2003). Larger vessels with improved equipment and efficiency (e.g., more powerful winches, stronger cables and rockhopper trawls) made trawling on rough bottom seamounts more feasible (Clark et al. 2007), greatly expanding the range of the deep-sea fishing (Roberts 2002). Besides the capability to stay at sea for longer periods reaching distant and remote fishing grounds, and the capacity to process large amounts of fish on the onboard fish-processing plant (from about four to six tows each day), these vessels are well-equipped with electronics for navigating, determining the topography and finding fish (Merret and Haedrich 1997). By the end of the 1990s, McAllister et al. (1999) estimated that 40% of the world's trawling grounds were operating in waters deeper than the continental shelves.

In the Northeast Atlantic, trawl fisheries for deep-water fishes developed along the western Atlantic continental margins and the Reykjanes Ridge and subsequently the continental margins of Europe (Gordon et al. 2003). In the mid-1980s French trawlers targeting deep-water species initiated major commercial operations greatly boosting these fisheries (Gordon et al. 2003), and Icelandic and Spanish trawlers were also regularly involved in deep water fishing activities (Merret and Haedrich 1997). Areas of the northern Mid-Atlantic Ridge (MAR) have been, and still are, exploited for deep-sea fish species such as the redfish (*Sebastes* spp.) (Clark et al. 2007) as well as several other ridge features in the Indian and Pacific Oceans like the Southwest Indian Ridge or the Emperor Ridge (see Clark et al. 2007 for details). By the mid-1990s most of the fisheries in which the large distant fleet trawlers participated had collapsed (Merret and Haedrich 1997).

Bensch et al. (2008) provide a comprehensive review of bottom fisheries in the high seas, characterizing, for different ocean regions, catch and effort and the main species, fleets and gears involved in these fisheries. In this chapter we use some of the data referred in Bensch et al. (2008), to identify main species and gears used in the high seas fisheries of different ocean regions (Table 8.1). This information not only was used to generally describe these fisheries (see below), but also to identify target species for further analysis. Based on the target species listed by Bensch et al. (2008), we obtained time series catch data from Sea Around Us (SAU) project database (Pauly and Zeller 2015). Figure 8.4 provides reconstructed catch data, in the EEZs and high seas areas, clearly showing the initial impulse of the deep-sea fisheries and in the high seas in the 1970s.

Since the end of the 1980s it is also evident that there was a substantial increase of both the deep-sea fishing and the high seas deep-water fishing (Fig. 8.4), although from the mid 1990s the world catches of deep-sea fishes decreased within the EEZs. On the contrary, between 1990–2010 the estimated catches from the high seas stayed more or less constant, with annual average volumes of about 26,000 t, probably supported by the continuous search and exploitation of new areas, new species, and the involvement of new nations in these fisheries. The move to deep waters

Table 8.1. Description of deep-water high seas fisheries. Based on Bensch et al. 2008.

Scientific name	Common name	Depth range (m)	NEA	NWA	SEA	SWA	Med	NP	SP	IO	SO	IVI
Class Osteichthyes – bony fishes												
Alepocephalus bairdii	Baird's slickhead (Baird's smoothhead)	365-1700	T									70
Allocyttus niger	Black oreo	560-1300							T	T/L		69
Allocyttus verrucosus	Warty oreo	0-1800						G				69
Anarhichas spp.	Wolffishes	1-1700		T								78
Anoplopoma fimbria	Sablefish	175-2740						T/G/P/L				61
Antimora rostrata	Blue antimora (morid)	350-3000									L	66
Aphanopus carbo	Black scabbardfish	200-1700	T					T/G/P/L				73
Benthodesmus tenuis	Slender frostfish (cutlassfish)	200-850										66
Beryx splendens	Alfonsino	25-1300	T					T/G/P/L	T			57
Beryx splendens and B. decadactylus (Beryx spp.)	Alfonsino	25-1300			T			T/G	T			72
Brosme brosme	Tusk	18-1000	L									65
Caproidae	Boarfishes nei	na								T/L		16
Centrolophus niger	Rudderfish	40-1050								T/L		85
Conger conger	European conger	0-1171					T/G/L					86
Coryphaenoides acrolepis	Pacific granadier	300-3700						T/G/P/L				79
Coryphaenoides rupestris	Roundnose grenadier	180-2600	T	T								67
Dissostichus eleginoides	Patagonian toothfish	50-3850			L	L						71
Dissostichus spp. (D.eleginoides and D.mawsoni)	Patagonian and Antarctic toothfish	0-2200				L					L	81
Emmelichthys nitidus	Cape bonnethmouth	86-500										56
Epigonus denticulatus	Pencil cardinal (cardinal fish)	130-830						T/G/P/L		T/L		32
Epigonus spp.	Cardinal fishes	5-3000										70
Epigonus telescopus	Black cardinal fish	75-1200		T					T	T/L		70
Erilepis zonifer	Skilfish	0-680						L				85
Etelis coruscans	Deepwater longtail red snapper (ruby snapper)	90-400								L		45
Gadus morhua	Atlantic cod	0-600	T/L/G									67
Gempterus blacodes	Pink cusk eel	22-1000				T						79
Glyptocephalus cynoglossus	Witch flounder	18-1570		T								68
Helicolenus avius	Rockfishes nei	450-600						L				63
Helicolenus dactylopterus	Blackbelly rosefish (blue-mouth redfish)	50-1100	T/L/G				T/G/L					67
Hippoglossoides platessoides	American plaice	10-3000		T								66

Table 8.1 contd. ...

...Table 8.1 contd.

Scientific name	Common name	Depth range (m)	NEA	NWA	SEA	SWA	Med	NP	SP	IO	SO	IVI
Hippoglossus hippoglossus	Atlantic halibut	50-2000	T									88
Hoplostethus atlanticus	Orange roughy	180-1809	T	T	T				T	T		73
Hozukius guyotensis	Rockfishes nei	500-1000						L	nT	T/L		69
Hyperoglyphe antarctica	Bluenose warehou (blue-eye trevalla)	40-1500			T							51
Hyperoglyphe japonica	Pacific barrelfish (butterfish/medusafish)	150-1537						T/G/P/L				58
Lepidorhombus boscii	Four-spot megrim	7-800					T/G/L					52
Lethrinus mahsena	Sky emperor (dame berri)	2-100								T/L		65
Limanda ferruginea	Yellowtail flounder	27-364		T								37
Lophiodes miacanthus	Goosefish	110-960						T/G/P/L				25
Lophius americanus	American angler (anglerfish)	0-668		T								77
Lophius piscatorius and L. budegassa	Monkfish	20-1013	G						T			71
Macrouridae	Grenadiers, rattails nei	0-0										75
Macrourus berglax	Roughhead granadier	100-1000	T/L/G	T								75
Macrourus carinatus	Macrourids or grenadiers	200-1200									L	45
Macrourus holotrachys	Macrourids or grenadiers	300-1400									L	63
Macrourus whitsoni	Macrourids or grenadiers	400-3185									L	64
Macruronus magellanicus	Patagonian grenadier	30-500				T						71
Melanogrammus aeglefinus	Haddock	10-450	T			T						47
Merluccius australis (Merluccius polylepis)	Southern hake	28-1000				T						52
Merluccius hubbsi	Argentine hake	50-800	G			T						58
Merluccius merluccius	European hake	30-1075					T/G/L					65
Micromesistius australis	Southern blue whiting	50-900				T						52
Micromesistius poutassou	Blue whiting	150-3000					T/G/L					33
Molva dypterygia	Blue ling	150-1000	T									75
Molva molva	Ling	100-1000	L									77
Mora moro	Common mora (ribaldo)	450-2500							T			61
Nemadactylus spp. (*N.macropterus*)	Morwongs (king tarakihi)	0-220							nT			50
Neocyttus rhomboidalis	Spiky oreo	200-1240							T	T/L		70
Nototheniidae	Antarctic rockcods noties nei	na										65
Oreosomatidae	Oreo dories nei	na										49
Pagellus bogaraveo	Blackspot seabream	na-700	T/L/G									57
Paristiopterus labiosus	Giant boarfish (sowfish)	20-170							T			55
Patagonotothen ramsayi	Longtail Southern cod	50-500										36

Scientific name	Common name	Depth (m)	Codes	%
Phycis blennoides	Greater forkbeard	10-1047	T/L/G	39
Phycis phycis	Forkbeard	13-614	T/G/L	45
Phycis spp.	Forkbeards nei	na	T	49
Physiculus spp.	Morid cods	na		49
Plagiogeneion rubiginosum	Rubyfish	50-600	T/G/P/L	41
Polyprion americanus	Wreckfish	40-600	nT, T/L	76
Polyprion spp. (oxyginius)	Hapuka (wreckfish)	na	T, T/L	58
Pseudocyttus maculatus	Smooth oreo dory	400-1500	T, T/L	72
Pseudopentaceros richardsoni	Pelagic armourhead (boarfish)	0-1000	T/L	44
Pseudopentaceros wheeleri	Slender armourhead (pelagic armourhead)	146-800	T/G	65
Reinhardtius hippoglossoides	Greenland halibut	1-2000	L	73
Schedophilus velaini (Schedophilus labyrinthica)	Violet warehou (black butterfish)	40-500	nT, T/L	61
Sebastes fasciatus	Acadian redfish	128-366	L, T	44
Sebastes marinus	Golden redfish (Giant redfish)	100-1000	L, T	71
Sebastes mentella	Beaked redfish	300-1441	T	56
Seriola lalandi	Yellowtail amberjack (yellowtail kingfish)	3-825	nT	69
Setarches guentheri	Channeled rockfish (scorpionfish)	150-780		13
Trachyrincus scabrus	Roughsnout granadier	395-1700	L	34
Urophycis tenuis	White hake	180-1000	T/G/P/L	72
Zenopsis nebulosus	Mirror dory (mirror perch)	30-800	T	64
Class Chondrichthyes - cartilaginous fishes				
Bathyraja eatonii	Rajids or skates	na	L	62
Bathyraja irrasa	Rajids or skates	565-1218	L	70
Bathyraja maccaini	Rajids or skates	167-500	L	70
Bathyraja meridionalis	Rajids or skates	na -800	L	70
Bathyraja spinicauda	Spinetail ray	140-1463	T	85
Centrophorus squamosus	Leafscale gulper shark	145-2400	T	86
Centroscymnus coelolepis	Portuguese dogfish	150-3700	T	63
Dalatias licha	Kitefin shark (seal sharks)	37-1800	L, T, T/L	81
Deep-sea sharks	Deep-sea sharks	na	L, T/L	80
Raja georgiana	Rajids or skates	20-350	L	60
Raja hyperborea	Arctic skate	140-2500	T	61
Raja radiata (Amblyraja radiata)	Starry ray (thorny skate)	20-1000	T	59
Raja spp.	Skates	na	T	72

Table 8.1 contd. ...

Table 8.1 contd. ...

Scientific name	Common name	Depth range (m)	NEA	NWA	SEA	SWA	Med	NP	SP	IO	SO	IVI
Rajidae	Rays and skates nei	na			T							72
Selachimorpha	Sharks (deep-sea) nei	na			T							80
Squalus mitsukurii	Shortspine spurdog	29–600						T/G/P/L				61
Class Holocephali - chimaeras												
Chimaeridae (mainly *Hydrolagus mirabilis* and *Chimaera monstruosa*)	Rabbit fish	40–1200	T/L/G									74
Class Malacostraca, order Decapoda (crabs, shrimps, lobsters)												
Aristaeomorpha foliacea	Giant red shrimp	na					T					na
Aristeus antennatus	Blue and red shrimp	na					T					na
Chionoecetes tanneri	Deep-sea crabs	na						P				na
Geryon spp.	Deep-sea (red or king) crabs (*Geryon* nei)	na	G		P		T/G/L	P				na
Nephrops norvegicus	Norway lobster	na										na
Palinurus barbarae	Deep-sea lobster	na								T/L		na
Pandalus borealis	Northern shrimp	na		T								na
Paralomis spp.	Deep-sea crabs	na						P				na
Parapenaeus longirostris	Deep-water rose shrimp	na					T/G/L					na
Plesionika martia	Golden shrimp	na					T/G/L					na
Class Cephalopoda (squid and octopuses)												
Eledone cirrosa	Horned octopus (curled octopus)	na					T/G/L					na
Illex argentinus	Argentine shortfin squid	na				T						na
Illex illecebrosus	Northern shortfin squid	na		T								na
Loliginidae	Squid	na			T							na
Loligo gahi	Patagonian squid (common squid)	na				T						na
Octopodidae	Octopus	na			T							na

Main species (target species in dark grey and non-target species in light grey) and gears used in the high seas fisheries are listed per ocean region, following Bensch et al. (2008). Species/taxa in bold are selected species for which reconstructed catch data were obtained from Pauly and Zeller (2015); among the target fishes listed in the Table, for which catch data were available, we selected those that, according to Froese and Pauly (2015) and to our knowledge, exhibit an usual depth range deeper than 200 m. The category "non-target species" includes both types "other" and "associated species" (species retained if they are of commercial value, otherwise discarded), as defined by Bensch et al. (2008). Depth range of fishes was obtained by Froese and Pauly (2015). The index of intrinsic vulnerability of fish species/taxa was obtained by Froese and Pauly (2015) and Cheung et al. (2007). Abbreviations: NEA—North-East Atlantic; NWA—North-West Atlantic; SEA—South-East Atlantic; SWA—South-West Atlantic; Med—Mediterranean Sea; NP—North Pacific; SP—South Pacific; IO—Indian Ocean; SO—Souther Ocean; IVI Index of intrinsic vulnerability. Fishing gears abbreviated as: T—trawl; G—gillnet; P—pot/traps; L—longline; nT—non trawl (includes longline and Dahn lines, trot lines and other line gears, pots, traps and Danish seines). na–not available.

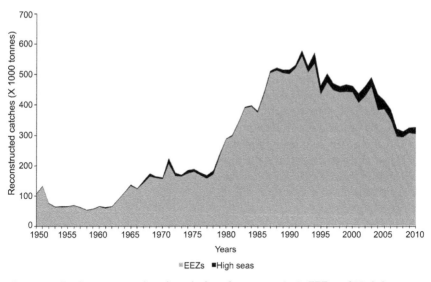

Figure 8.4. Total reconstructed catches of selected target species in EEZs and High Seas areas in all ocean regions (numbers originally from Pauly and Zeller 2015).

was encouraged by government subsidies as a way to alleviate the collapse due to overfishing of the shallower-water fish stocks (Koslow et al. 2000). In certain areas the observed catch trends were also influenced by the EEZs declarations in late 1970s and by the consequent more restrictive management measures imposed within EEZs (Clark et al. 2007).

Estimated landings and discards from bottom trawling comprise more than 80% of landings from high seas bottom fishing. When compared to the reconstructed catches from deep sea fishing inside the EEZs, from 1950 to 2010, the high seas deep-sea fishing only accounts annually for about 4% of the EEZs catch, corresponding to an annual average of about 12,000 t (Fig. 8.4). In the same period those catches corresponded to an annual average of only about 5% of the value of the EEZs catches, delivering around US$ 28 million.

Based on the best available information, Gianni (2004) estimated that the majority of the high seas bottom trawl fishing occurs in the North Atlantic, Southern Indian Ocean and Southwest Pacific Ocean, and that 60% of the world's high seas bottom trawl catch comes from the Northwest Atlantic.

Bensch et al. (2008) list 83 species of teleosts, 17 elasmobranchs, 10 crustaceans and 6 cephalopods that are caught by the high seas fisheries in different ocean regions (see also Table 8.1). Target species vary according to the different oceans and, from the reconstructed catches, for some species volumes caught are relatively small. The ocean region with the highest number of species involved is the North-East Atlantic (26 species), and that with the lowest number of species is the Southern Ocean (10 species). The majority of species caught are fish species (86% of total no. of species, of which >80% are bony fishes), and similarly most of target species are fishes. Most of the species (78%) are target or by-catch species of fisheries in only one ocean, and only one species (the orange roughy *Hoplostethus atlanticus*) is the target species in four out of nine ocean regions (North-East and South-East Atlantic, South Pacific, Indian Oceans).

The by-catch issue in these fisheries is visible: five out of seven ocean regions have 70% or more of non-target species involved in their high seas fisheries. The extreme case is represented by the Southern Ocean, in which 90% of the fish species are non-target species of the longline fishery targeting the toothfish *Dissostichus* spp. (Fig. 8.5).

The fishing gears used in the high seas bottom fishing include trawl, gillnet, longline and pot/traps. The same species can be exposed to different gears, even in the same ocean region. The gear to which most of the species are exposed is trawl (90 out of 116 species, 78%). Only 22% of the species are not exposed to trawl gear, of which 38% are exclusively listed in the Southern Ocean. Considering each ocean region, trawl gears are those used to fish more than 40% of species in five out of seven ocean regions; and only in one region, Southern Ocean, a different gear—longline—is used to catch 100% of species. The regions Mediterranean Sea and North Pacific show a more varied pattern, with three or four gears almost equally distributed between the listed species (Fig. 8.6).

Based on available catch data series obtained by SAU database the following selected target species account for the majority of catches in the period 1950–2010: *Reinhardtius hippoglossoides, Coryphaenoides rupestris, Sebastes mentella, Dissostichus eleginoides, Sebastes norvegicus, Dissostichus*

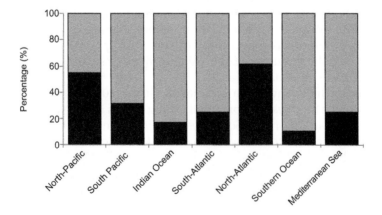

Figure 8.5. Percentage of target (black) and non-target (gray) species involved in the high-seas fisheries by ocean region (based on information referred in Table 8.1 and Bensch et al. 2008).

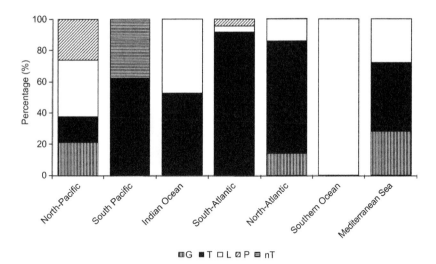

Figure 8.6. Percentage of type of gears used in the high-seas fisheries, by ocean region (based on information referred in Table 8.1 and Bensch et al. 2008). The same species can be fished by different gears. Gillnets (G); Trawl nets (T); Longlines (L); Pots and Traps (P); Non-trawl gears (nT).

mawsoni, Hoplostethus atlanticus, Centroscymnus coelolepis, Molva dypterygia, Alepocephalus bairdii, Beryx splendens, Aphanopus carbo, Polyprion americanus, Centrophorus squamosus, Epigonus telescopus. The accumulated catches of these species on the period corresponded to about 752,000 t which contrasts with the 16 million tonnes caught of the same species in the EEZs (Fig. 8.7). It is also noteworthy that about 71% of the deep sea high seas catches were conducted in the period 1990–2010 (Fig. 8.8). Examples of the most emblematic fish species caught in the high seas in the Atlantic Ocean include *Coryphaenoides rupestris, Molva dypterygia, Aphanopus carbo,* the alfonsinos (*Beryx splendens*), the orange roughy (*Hoplostethus atlanticus*) and several deep-water shark species. Range of depth inhabited by the species is reported in Table 8.1.

The depth inhabited by the species involved in the high seas fisheries span a range of almost 4000 m, from 0 to 3850 m; the largest range-3800 m—is attributed to *Dissostichus eleginoides* and the smallest—of around 100 m—to *Lethrinus mahsena*. The average range is 1194 m, the narrowest average range (851 m) is attributed to the species fished in the Indian Ocean, and the widest to those in the Southeast Atlantic (1817 m).

Bensch et al. (2008) estimated that in 2006 about 285 fishing vessels were actively involved in the deep-sea fisheries in the high seas worldwide. They estimated that the total catch (considering all species) of these fleets amounts to about 250,000 t, representing 0.3% of the world fisheries and about €450 million in value. Around 80% the fishing vessels engaged on these fisheries

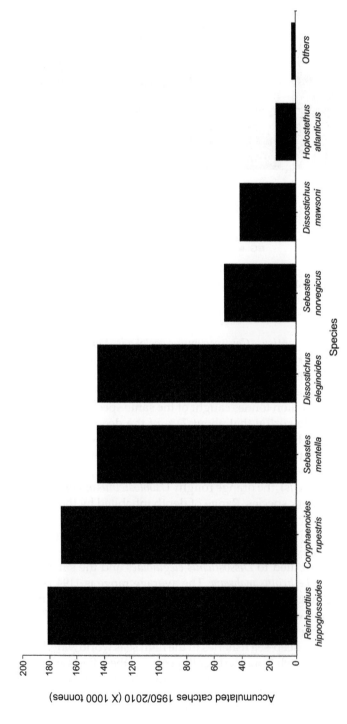

Figure 8.7. Total accumulated reconstructed catches for the period 1950–2010 of the selected group of target species (based on Table 8.1 and Bensch et al. 2008) in all high seas areas (numbers originally from Pauly and Zeller 2015).

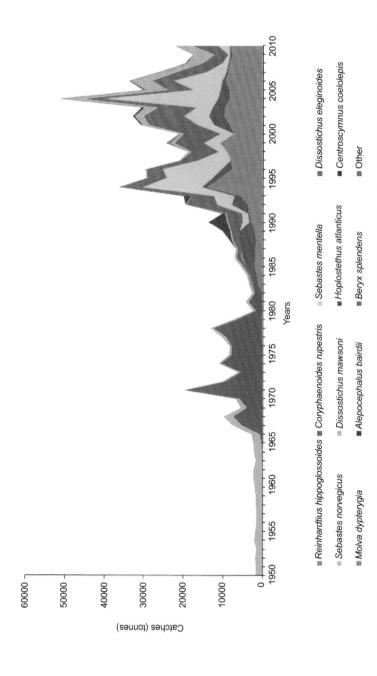

Figure 8.8. Reconstructed catch data of the selected group of target species (based on Table 8.1 and Bensch et al. 2008) in all high seas areas (numbers originally from Pauly and Zeller 2015).

had flags from Spain, Korea, New Zealand, Russian Federation, Australia, France, Portugal, Belize and Estonia. It is worth noting that about one third were estimated to have European flags and more than half of the catches were caught by European fleets (Bensch et al. 2008). Based on reconstructed catches (Pauly and Zeller 2015) we see that the top 10 countries involved on high seas fisheries changed their importance along time (Fig. 8.9) with a larger number of nations being involved on the high seas fisheries since the 1990s. In the period 1950–1990 about four to 10 countries were involved on these fisheries, and from 1990 to 2010 about 13 countries in average were involved on the high seas deep sea fishing annually.

Catch data of the above selected target deep sea species between 1950–2010 indicate that Spain, Russian Federation and Uruguay rank first with about 64% of the accumulated catches. Significant catches were also done by vessels with flags from the United Kingdom, Norway, France, Ukraine, USA and Sierra Leone (Fig. 8.10).

It is also interesting to note an increase in catch ratios between the High Seas/EEZs fisheries, especially after the 1960s and the 1990s, as well as in the number of target species, particularly since the 1990s (Fig. 8.11).

This may result from the increasing technological efficiency (i.e. after the 1960), and from the shift of exploitation from coastal to deep-sea resources of the distant and remote ecosystems of the high seas, mainly over seamounts. This led to successive depletion of species and the continuous search for new species and fishing grounds, to keep high catch levels and yields.

The failures of an effective regulation on the high seas fisheries led also to an increase of exploratory fishing in all oceans being likely that commercial fishing has had an impact on almost all of the known seamounts with summits shallower than 1,000 meters (Stone et al. 2004). Seamounts were probably the most affected because they provide lodging for fish aggregations of several species easily detectable and exploitable (Koslow et al. 2000; Clark et al. 2007; Pitcher et al. 2010) in the vast and impoverished deep ocean. Pitcher et al. (2010) estimate that about 3 million tonnes of the annual catches come from seamounts, and many of them are located in high seas areas. These ecosystems provide habitat for a large number of organisms, some of them distinctive and frequent inhabitants, such as cold-water corals (CWC) and sponges, which by their nature are known to be extremely vulnerable to fishing impacts, mainly those caused by the trawl gears (Koslow et al. 2001; Roberts 2002; Althaus et al. 2009; Williams et al. 2010). These animals and the benthic communities they belong to are recognized as Vulnerable Marine Ecosystems (VMEs) (FAO 2009). Once the fishing gear is dragged over the seafloor, they may suffer major damages, and taking into account their biological characteristics such as slow growth and high longevity (such as millennial life span of some CWC, Roark et al. 2006) they are unlikely to be able to recover from this kind of

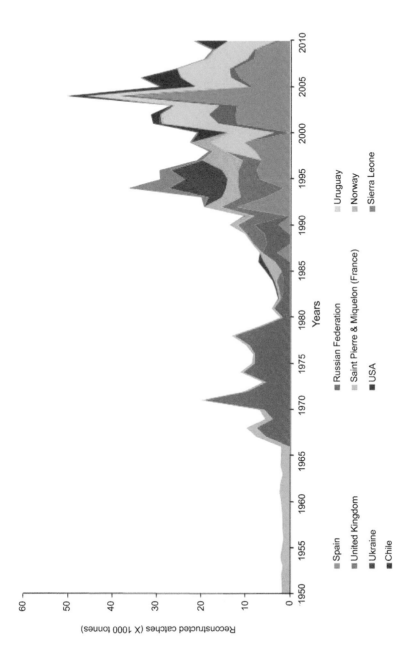

Figure 8.9. Reconstructed catch data the selected group of target species (based on Table 8.1 and Bensch et al. 2008) by the top 10 countries involved in high seas fisheries (numbers originally from Pauly and Zeller 2015).

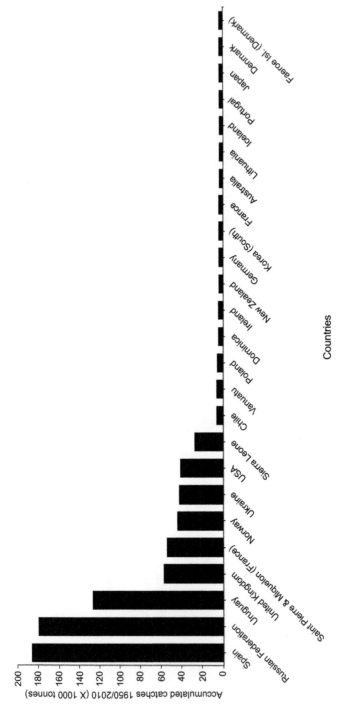

Figure 8.10. High seas accumulated catches by countries in the period 1950–2010 of the selected group of target species (based on Table 8.1 and Bensch et al. 2008) by the top 10 countries involved in high seas fisheries (numbers originally from Pauly and Zeller 2015).

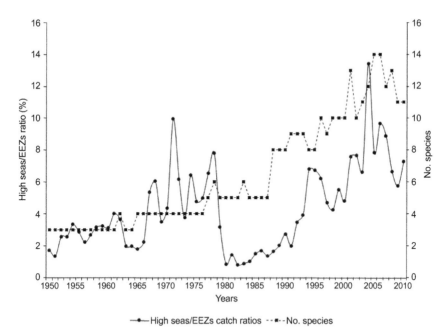

Figure 8.11. High seas vs. EEZs annual catch ratios, and the number of species caught per year of the selected group of target species (based on Table 8.1 and Bensch et al. 2008) by the top 10 countries involved in high seas fisheries (numbers originally from Pauly and Zeller 2015).

impact (Roberts 2002; Clark et al. 2006; Clark and Koslow 2007; Waller et al. 2007; Althaus et al. 2009; Clark and Rowden 2009; Williams et al. 2010; Watling et al. 2011). A recent study on trawl fishing impacts on seamounts off Australia and New Zealand considers that the damaged ecosystems will hardly recover and if they do it will be in a time frame corresponding to several human generations (Clark and Rowden 2009). Following these concerns several initiatives have been proposed to protect VMEs, at the regional and international level.

Most of the deep sea catches are made over spawning aggregations, as it is the case of orange roughy (*Hoplostethus atlanticus*), the alfonsinos (*Beryx splendens*) or the blue ling (*Molva dipterygia*) (Rogers 1994; Koslow 1997; Roberts 2002; Clark et al. 2007). Catches from those aggregations can be extraordinary, 60 tonnes from a 20-minute trawl (Roberts 2002). Catch levels of this scale on relatively small and isolated or discrete topographic features, and over aggregations of fish species already biological vulnerable quickly led to sharp local depletions likely increasing fragmentation of populations and thus decreasing their ability to recover.

In general, deep-sea fisheries are considered not sustainable and even at low exploitation levels not economically viable (Morato et al. 2006; Clark et al. 2007). An example of extremely vulnerable species is the orange roughy (*Hoplostethus atlanticus*), due to its biological features such as low productivity, high longevity (it may attain about 150 years, Fenton et al. 1991) and late maturation (c. 30 years, Francis and Horn 1997; Horn et al. 1998). In New Zealand and south of Australia the fishery for orange roughy started in the 1980s with the discovery of spawning grounds. Catches were initially very high and this was the largest and most persistent orange roughy fisheries in the world. By the 1990s the landings sharply declined and the populations collapsed to less than 20% of pre-exploitation abundance along with a general decline of other incidental species in trawl surveys (Merrett and Haedrich 1997; Clark 1999; Koslow et al. 2000). In 1994 four aggregations of this species were discovered in Namibia and only five vessels fished them, but it took only six years to overexploit these populations to about 10% of their original biomass (Branch 2001). In the Northeast Atlantic the fishery for orange roughy also started in the 1980s but even with relatively modest catches, by the 1990s the catches showed signs of depletion and a series of management measures were taken following the successive advices of the International Council for the Exploration of the Sea (ICES). At present, orange roughy catches are not allowed in the Northeast Atlantic in all ICES areas (catch quotas are zero).

Deep water elasmobranchs are another example where deep-water fisheries can have great impacts on fish populations. Many deep-water elasmobranchs have very wide, often global distributions (Moura et al. 2014) with minimal population structure (Cunha et al. 2012 but see Catarino et al. 2015) being difficult to assess their population status with certainty (Neat et al. 2015). The high vulnerability of this species to exploitation arises from their life-history traits (Dulvy and Forrest 2010) because many of these are long-lived, slowly reproducing species that cannot be expected to recover for many years following overfishing (Simpfendorfer and Kyne 2009). In the North-East Atlantic Ocean industrialized deep-water fisheries established in the mid-1970s, but deep-water sharks—often landed together and known as siki shark—began to be landed in quantity in the 1990s as the fishery spread (Gordon 2001). The main commercial species were *Centrophorus squamosus* and the Portuguese dogfish *Centroscymnus coelolepis*, but a number of other species made up landings in smaller quantities (Neat et al. 2015). Elasmobranchs are among the most misreported and under-reported of deep-water fishes (Hoffmann et al. 2010); however, it is clear that they have been affected in recent decades with the expansion of deep-water fisheries (Neat et al. 2015). Graham et al. (2001) observed that the mean catch rate of elasmobranchs in the 1990s was around 20% of the catch rate in the 1970s in continental slope of south-east Australia, and Neat et al. (2015) found severe declines of about 10% of earlier catch rates for species of the genus Centrophorus on the Rockall Trough (North-East Atlantic Ocean). Although the assessment of the status of the stocks in the North-East Atlantic Ocean is uncertain due to the lack of robust catch and landings data, Basson et al. (2001) found that stocks of the main commercially exploited species had declined significantly between the end of 1980s and the end of 1990s. Since 2010 no catches of deep-water shark are allowed in the North-East Atlantic in all ICES areas (catch quotas are zero). As a result of fishing regulations put in act by the European Union (EU) in the last decade the overall scale of deep-water fisheries reduced considerably in recent years, but still many species of sharks continue to be taken as by-catch and discarded in unknown quantities (Neat et al. 2015).

Figure 8.12 shows the vulnerability of fish species per ocean region, obtained averaging the index of intrinsic vulnerability elaborated by Cheung et al. (2005). The index (from 1 to 100, higher values represent greater vulnerability) is based on life history traits such as longevity, fecundity, growth, natural mortality, geographic range and varies between habitats (Cheung et al. 2005). Deep water demersal and benthopelagic fish assemblages, especially those around seamounts, are more intrinsically vulnerable to fishing and the highest average value is represented by the group of seamount-aggregating fish (Cheung et al. 2007). Since 1950 the vulnerability of global fish catches at seamounts has declined, as more vulnerable species became exploited and serially depleted (Cheung et al. 2007). In line with what was found by Cheung et al. (2007), the index of the species included in the deep-water high seas fisheries in our chapter ranges from high to very high vulnerability for all ocean regions (Fig. 8.12). Considering both target and non-target species, the ocean region with the highest vulnerability is the North Atlantic, followed by the Southern Ocean,

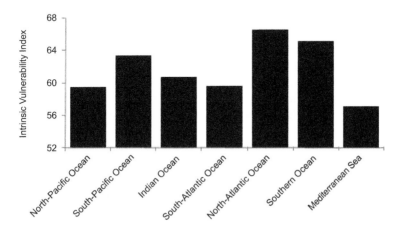

Figure 8.12. Average index of intrinsic vulnerability of target and non-target species of fishes, per ocean region (based on Table 8.1 and Bensch et al. 2008). The same species can be fished in different ocean regions.

and that with the lowest is the Mediterranean Sea, although this figure may be influenced by the fact that this is the region with the highest portion of invertebrates—50%—for which no vulnerability index is available.

Besides the known biological vulnerability of most of the deep sea fish species the failure to account for spatial patterns or fragmentation of fish and fisheries contributes to the growing incidence of fishery collapses around the world (Norse et al. 2012). This is particularly important for the high seas fisheries which mainly operate in fragmented or discrete habitats and patchy populations (as is the case of the seamounts fisheries), following a serial depletions exploration pattern as described by Berkes et al. (2006). Serial depletions result from repeating a sequence of fishing on profitable resource patches until each patch becomes unprofitable and successively moving to new patches until there are no more profitable patches to exploit. As noted by Norse et al. (2012) this model is particularly relevant for the deep-sea fisheries and in the high seas because vessels are generally larger, and catch more quantities of each patch, also adding to the fact that target species are slower to recover from increasing mortality. On the other hand, serial depletions appear almost inevitable because it is economically rational to reduce each stock to unprofitability until no more can be taken and then reinvest the capital to obtain higher return investments (Norse et al. 2012).

The lack of sufficient data and information to technically assess and manage high seas fish populations (Gianni 2004), is also a serious threat to the sustainable management of these fisheries around the world. Unfortunately, too often critical information is not gathered until the target fisheries have peaked and collapsed (Haedrich et al. 2001), and most deep-sea fisheries started before there was adequate scientific knowledge of the fish stocks for management purposes and without adequate consideration of their impacts on the ecosystem (Rogers 2015). This should be particularly true in the high seas were the lack of robust catch data is still a reality, figures from different sources are frequently not coincident, and the status of exploited populations is still unknown.

Catches can be underestimated due to the Illegal, Unreported and Unregulated (IUU) fishing, which is known to be particularly important in high seas fisheries. IUU fishing can have serious economic, social and environmental impacts (MRAG 2005). According to Agnew et al. (2009) annual IUU fishing is estimated to amount between 11–26 million tonnes worth between US$ 10 and US$ 23 billion (EEZs and the high seas). In 2006 the Ministerially-led Task Force on IUU Fishing on the High Seas estimated the likely level of annual value from high seas IUU catches of groundfish to be around US$ 318 million (High Seas Task Force 2006). Agnew et al. (2009) found a positive correlation between regional estimates of IUU fishing and the regional number of depleted stocks, indicating how the IUU fishing contributes to overexploitation of fish stocks. Lacking controls on unreported catches can also prevent recovery of exploited fish populations (Agnew et al. 2009). It is also known that IUU fishing can cause serious damages to marine environment, since IUU vessels may use destructive fishing practices or ignore management actions, having far-reaching impacts on sensitive habitats like seamounts (MRAG 2005).

Climate Change, Deep-Sea Ecosystems and Fisheries

Bottom fisheries targeting deep-sea low productivity species and destroying the sea bottom and fragile habitats such as cold-water corals (Rogers 2015; Ramirez-Llodra et al. 2011) are considered at present the most serious human-related activities impacting the deep sea. Among the other impacts from human activities (as dumping, pollution, mechanical and acoustic disturbance, bioprospecting, oil and gas industry or the deep seabed mining, scientific research), climate change has been predicted as the most impacting on the deep-sea in the near future (Ramirez-Llodra et al. 2011).

Climate change impacts natural and human systems, through changes in the atmospheric and oceanographic conditions that affect organismal physiology, populations and communities, food webs and ecosystems (i.e., IPCC 2014). Global changes in the oceans include warming temperatures, increased stratification and more acidic waters, decrease of sea ice, increase of sea level, deoxygenation and decreased productivity (IPCC 2014). The vast literature on climate change includes evidence of effects already acting on organisms, ecosystems and ecosystem services (Perry et al. 2005; Tresher et al. 2007; Dulvy et al. 2008; Daufresne et al. 2009; Poloczanska et al. 2013), as well as projected scenarios of the future effects of changes in climate on marine systems (Cheung et al. 2010; 2012; Guinotte et al. 2006; Tittensor et al. 2010; Jones et al. 2014). Deep-sea ecosystems make no exception, and the functions and services they provide, such as deep-sea fisheries (Thurber et al. 2014) respond to variations in climate (i.e., Lehodey and Grandperrin 1996; Thresher et al. 2007) and are expected to be affected by future changes in ocean biogeochemistry driven by climate change (Mora et al. 2014). The effects of climate change are expected to interact with the impacts from other stressors, as exploitation and disposal, in a synergistic way, ending up in a magnified effect on the deep-sea ecosystems (Ramirez-Llodra et al. 2011).

The capacity of predicting the potential effects of future environmental changes on ecosystems is particularly limited for those systems for which knowledge on the basic functioning and structure is scarce, lack of basic biological data hampers a credible assessment (Christiansen et al. 2014), and the nature of linkages between physical and biological process is uncertain (Croxall 1992). The technological challenges and elevated costs needed to study in detail and on the long-term the deep-sea, to have a good spatial and temporal coverage, are limiting our understanding of deep ecosystems structure and functioning (Rowden et al. 2010; Hughes and Narayanaswamy 2013), consequently our scientific knowledge of environmental baselines of deep-sea ecosystems is still scanty (Rogers 2015). Projecting impacts of climate change on the deep sea, with only limited information on the natural variability of deep-sea communities, is therefore challenging (Hughes and Narayanaswamy 2013) although several studies explored the effects of climate change on deep ecosystems.

Although the biogeochemical shifts in the deep sea, such as changes in temperature and pH, are expected to be at smaller magnitude than shallower waters (Mora et al. 2014), responses of deep-sea fauna may be stronger. In general, deep-sea communities are adapted to stable conditions and to narrow ranges of environmental variation, i.e., in temperature (Danovaro et al. 2004; Hughes and Narayanaswamy 2013; Mora et al. 2014), therefore they show low tolerability to even small changes in environmental parameters as water temperature, salinity and currents and may not be able to respond well to general simultaneous exposure to changes in all parameters, requiring multiple physiological adjustments. For example, changes in temperature as reduced as 1/10 of degree Celsius (0.05–0.1°C) can affect diversity, structural and functional aspects of community of deep-sea nematodes (Danovaro et al. 2004). The responses of deep-sea ecosystems to changes in climate vary on a wide range of temporal scales, occurring quickly, as in the case of deep-sea bacteria and benthic fauna in the eastern Mediterranean (Danovaro et al. 2001), or on the long-term, as feedbacks from the deep sea may be on a millennial scale (Ruhl et al. 2008).

The deep-sea oceans are especially energy-deprived systems, due to low temperatures and virtually no *in situ* productivity (McClain et al. 2012), making the deep-sea communities generally dependent upon processes taking place in the upper ocean that affect supplies to the seafloor. Under a range of climate change scenarios, open-ocean net primary production is projected to redistribute and to decrease globally by 2100 (from the stringent mitigation scenario to the very high greenhouse gas emission scenario, IPCC 2014), causing a reduced export of organic matter to the deep sea, altering quantity and quality of food for deep-sea organisms (Ruhl et al. 2008; Hughes and Narayanaswamy 2013) on which they rely (Gage and Tyler 1991; Ramirez-Llodra et al. 2010). Changes in the flux of Particulate Organic Carbon (POC) from the surface to the seafloor are expected to represent the primary impact on benthic ecosystems (Jones et al. 2014). The flux of POC is estimated to reduce up to 13% of current values (Mora et al. 2014), with decreases at low to mid latitudes, and possible increase at high latitudes depending on alteration in seasonality (Jones et al. 2014; Rogers 2015).

Long-term *in situ* experimental studies demonstrate a correlation between climate change indicators, food supply and community structure and abundance of abyssal benthic communities (Ruhl and Smith 2004; Smith et al. 2008). Since deep-sea organisms are highly vulnerable to variations in space and time availability of food (Smith et al. 2008), such changes will impact deep and abyssal ecosystem structure and function, altering the provision of ecosystem services (Smith et al. 2008; McClain et al. 2012; Hughes and Narayanaswamy 2013; Thurber 2014; Jones et al. 2014; Rogers 2015) and patterns of biodiversity (Danovaro et al. 2004). The flux of particulate organic matter to the deep-sea can be quite variable, and the way that the quantity and quality of food reaching the seafloor influences deep-sea benthos dynamics may vary (Wigham et al. 2003; Rex et al. 2006). In a global-scale synthesis across deep-sea benthic communities, Rex and colleagues (2006) showed that the significant decline with depth of abundances and biomasses of meiofauna, macrofauna and megafauna is related with the exponential decrease in the rate of nutrient input from sinking phyto-detritus with increasing depth and distance from productive coastal waters. Due to food limitations, some species may not be capable of maintaining their distribution. Large macrofauna and megafauna organisms are more affected by the reduction in food input with depth, compared to smaller size groups, probably because although they have a more efficient metabolism, they need more energy than small organisms to perform their activities and at great depth they may experience a decrease in density to jeopardize their reproductive activities (Rex et al. 2006).

The impact of changes to deep-water habitats due to the effects of climate change on ocean primary production and the export of POC to the seafloor are projected to be significant. By 2100, under a severe climate change scenario (RCP8.5), the total global seafloor biomass is projected to decrease by 5.2%, with a general shift to smaller sizes; decline of megafauna biomass, including large epifauna and demersal fishes, ranges from 2% (under a moderate scenario) to 5% (under severe scenarios) (Jones et al. 2014). Although the greatest changes are projected at depths >3,000 m, negative changes in biomass are expected in deep-water fishing grounds, and in general in habitats of vital importance for deep-sea fisheries (such as in >80% of canyons, seamounts and cold-water coral reefs) (Jones et al. 2014). Despite projected declines of biomass of potential exploitable resources (e.g., epifauna and demersal fishes) are smaller than those of meiofauna and macrofauna, we still consider them significant for future deep sea fisheries. For example, in areas suitable for fishing (e.g., seamounts, canyons) or important for fishery resources (such as cold-water corals), the percentage of biomass change are projected to vary between –3.62 and –8.09% respectively for the moderate and very high greenhouse gas emission scenario (Table 8.2).

Changes in climate include ocean warming, one of the main driving forces for species composition in marine ecosystems, since large-scale

Table 8.2. Percentage projected changes in POC flux and megafauna biomass including epibenthic invertebrates and demersal fish) between 2006–2015 and 2091–2100 under moderate (RCP4.5)/very high (RCP8.5) greenhouse gas emission scenarios (adapted from Jones et al. 2014).

Area	POC flux to seafloor		Megafauna biomass	
	RCP4.5	**RCP8.5**	**RCP4.5**	**RCP8.5**
Globe	−4.44	−11.40	−2.05	−5.15
Atlantic	−6.86	−15.40	−2.92	−6.86
Pacific	−1.84	−5.51	−1.06	−2.64
Indian	−1.41	−3.98	−0.71	−1.94
Arctic	−4.95	−12.60	0.58	0.94
Southern	3.76	9.64	1.86	3.84
Bathyal	−4.42	−10.90	−1.88	−4.87
Area with fishing	−6.11	−14.50	−2.56	−5.90
Area with seamounts	−5.44	−13.80	−2.45	−6.15
Area with canyons	−4.85	−11.20	−2.23	−5.36
Area with cold- water corals	−10.70	−20.90	−3.62	−8.09

geographical distribution of marine animals is defined by temperature (Pörtner and Peck 2010). Marine fauna has specialized climate-related temperature windows that change with latitude within and among species, and/or with ontogeny, with particularly sensitive phases being represented by early life and spawning stages (Pörtner and Peck 2010). Under global warming, marine ectotherms will suffer a rise in metabolic rate, and less oxygen will make it difficult to meet increased metabolic needs, especially in large individuals (Pörtner and Peck 2010). The ecological responses to warming include the shift of species toward higher latitudes, changes in community composition and abundance of species (Perry et al. 2005; Dulvy et al. 2008; Poloczanska et al. 2013), changes in phenology (Walther et al. 2002) and reduced body size, as observed across ectothermic aquatic taxa from the community to the individual level (Daufresne et al. 2009; Sheridan and Bickford 2011).

The effects of increasing temperatures on demersal fishes and fisheries in shallow waters have been studied. In the North Sea, demersal fishes responded to increased temperatures across a 25-year time frame, with many species shifting their distribution northward and/or in deeper waters (Perry et al. 2005; Dulvy et al. 2008). Shifting species were smaller, matured faster and at smaller size at maturity than those not shifting, suggesting that species with slower life histories may not be able to compensate rapidly for warming, adding to their high vulnerability (Perry et al. 2005). A meta-analysis on the effects of warming on fish body size in the North and Baltic Seas indicate that shrinking body size resulted in the expansion of geographical range of small species and reduction of that of large species, in the decrease of size-at-age in commercial species, potentially adding to the effect of fisheries (Daufresne et al. 2009). When disentangling the effects of changes in climate and fishing pressure on the North Sea demersal fish community, the effects of warmer waters—higher species richness of warm-favoring species and a decreased body size—can also act independently from fishing pressure, although fish community respond to both fishing and climate pressure (ter Hofstede and Rijnsdorp 2011). Organisms with broad ecological niche (diverse prey, wide thermal tolerance) should be able to maintain their sizes or get larger, whereas organisms that are not capable to adapt to quick changes in climate could go extinct (Sheridan and Bickford 2011). The effects of changing temperature can differently affect individual species performance and their productivity, as climate sensitivity differs among species (Pörtner and Peck 2010). The variability in response rate among organisms across the trophic web will likely alter ecological interactions leading to effects at higher level in the ecosystem and disrupting provision of ecosystem services (Sheridan and Bickford 2011). As an example, fish stock dynamics of commercial species feeding on zooplankton may be altered as a consequence of regime shifts among copepods due to warmer temperatures (Beaugrand et al. 2002; Möllmann et al. 2009).

Increasing temperatures are expected to have an impact on deep-sea species, both through direct and indirect effects. Although it is expected that increased temperatures will have an effect on the distribution and diversity of deep-sea species, the information on the effects of temperature on benthic and pelagic deep-sea communities is still limited (Rogers 2015). McClain et al. (2012) investigated the roles that carbon flux and temperature play in influencing several aspects of deep-sea benthic organisms, across multiple scales of organization, of a range of deep-sea taxa. They concluded that the effects of thermal (temperature) and chemical (POC) energy vary across scales of biological organization, the former prevailing at lower levels of organization (like organisms, limited in their individual metabolic rates by temperature and body size) the latter gaining importance at higher levels (driving patterns in biomass, abundance and biodiversity). The impacts of shifts of ranges in organisms that are linked through trophic interactions are difficult to predict and how they may act on the deep sea through altered flux of food, both as POC and actively transported through migrating animals, is not well understood (Rogers 2015).

The impacts of changing climate on demersal and deep-water fishes and fisheries have been documented for some species. A study on the effects of long-term changes in climate on long-lived fish species in the Southwest Pacific showed that fish growth rates changed across time reflecting long-term changes in temperature and differed—in the scale and direction of changes—between species, depending on the depth ranges inhabited. They increased in shallow water species (inhabiting waters

< 250 m), decreased in the deep species (> 1000 m), and slightly decreased in intermediate depth species (500–1000 m) (Thresher et al. 2007). At least for the species studied in this work, on a century time frame, these long-lived species showed no compensation (ecological or physiological) to changing temperatures, confirming a pattern of slow ecological responses to changes in climatic conditions. Another, less recent, study on the effects of changes in temperature driven by El Niño Southern Oscillation on the demersal commercial species *Beryx splendens*, goes along with this trend, indicating a strong correlation between fish growth and temperature of intermediate water layers (100–500 m) and faster fish growth as a consequence of El Niño events (Lehodey and Grandperrin 1996).

Temperature interacts with other large-scale stressors, such as deoxygenation and acidification of the oceans. Changes in climate are also expected to affect ocean circulation, with projected weakening of the Atlantic Meridional Overturning Circulation over the 21st century and a general decline in deep water formation (Gregory et al. 2005; Canals et al. 2006; IPCC 2014). Major changes in circulation can affect deep-ocean ecosystems. For example, under climate change scenarios dense shelf water cascading events are expected to reduce both in intensity and frequency. These are periodical events occurring on high and low latitudinal continental margins, in which cooler and heavier seawater sink into the deep sea, funnelling sediment and organic material through canyons, from shallow to deep-ocean (Canals et al. 2006; Company et al. 2008). The decline in deep water formation is expected to impact on transport of nutrients and larvae, and to alter off-shelf carbon fluxes, and the functioning of deep-sea ecosystems (Canals et al. 2006; Nellemann et al. 2008). In the Mediterranean Sea, these cascading events are suggested to represent regenerative mechanism through which contrasting the effects of an intense and prolonged trawl fishing pressure on the deep-sea shrimp, *Aristeus antennatus*. Through large lateral inputs of particulate organic matter to deep regions, such climate-driven phenomenon enhances the recruitment of the deep-sea shrimp, apparently mitigating fishery overexploitation (Company et al. 2008).

Increased temperatures, stratification of the oceans and changes to global circulation patterns, as effects of global climate change, result in anoxic conditions. Under climate change scenarios, dissolved oxygen is projected to decrease globally of 1 to 7% by the year 2100 (reviewed in Keeling et al. 2010) and under scenarios of centuries of protracted global warming the deep ocean oxygen concentrations generally decline by between 20 and 40% (Matear and Hirst 2003). The decrease of oxygen content in the oceans and in the tropical Oxygen Minimum Zones (OMZs), and the expansion and shoaling of OMZs, particularly between 200–700 m depth, have been already documented (i.e., Keeling et al. 2010; Stramma et al. 2010; Koslow et al. 2011). Declines in oxygen, leading to expansion of OMZs, are expected to have widespread consequences, ranging from biogeochemical cycling to effects on organisms (Keeling et al. 2010; Pörtner and Peck 2010).

Upwelling-driven hypoxia events are known to have deleterious effects on organisms in coastal systems (Chan et al. 2008; Stauffer et al. 2012), as well as the expansion of OMZs is already having an impact on pelagic fishes (Stramma et al. 2011). In general, mobile animals are more susceptible to variations in oxygen than pH (Melzner 2009; Kroeker et al. 2013) and crustaceans and fish are considered to be the most sensitive organisms to low levels of oxygen (Keeling et al. 2010; Pörtner and Peck 2010). In a study on deep-sea fauna of submarine canyons on oceanic islands—areas of enhanced fish abundance and species richness—De Leo et al. (2012) show that low dissolved oxygen (in OMZ) can be the factor limiting diversity and abundance of deep-sea fishes and that demersal fishes have a lower tolerance than invertebrate megafauna for low oxygen concentrations. The effects of decreased oxygen content may be especially negative for those organisms that inhabit the mesopelagic zones, and paleogeographic evidence indicate that the expansion of OMZ in the past caused rearrangements in biodiversity patterns and biogeography of midwater fishes (White 1987). Koslow and colleagues (2011) found a strong relationship between the decline of mesopelagic fish abundance and decreased oxygen in the OMZ southern Californian current—of around 20% since the 1980s—suggesting it may be due to the shoaling of the hypoxic boundary layer, leading the mesopelagic organisms to be more exposed to light and visual predators. The mesopelagic fish represent a vital link between the plankton and higher predators, between the pelagic and the benthic environment, playing a major role in the ocean food webs (Koslow et al. 2011; Colaço et al. 2013). Under climate change scenarios, changes in oxygen content at intermediate waters are expected to affect the distribution of mesopelagic organisms, probably impacting the carbon transport to the deep-sea and the pelagic food webs (Netburn and Koslow 2015).

Increasing level of atmospheric CO_2 results in increasing CO_2 in the water, and acidification of the oceans (i.e., Pörtner and Peck 2010). Among the negative effects on biological responses of marine taxa, the largest are on the survival and growth of calcifying organisms, especially for those organisms, such as corals and calcifying phytoplankton taxa, playing an important function as habitat builders and foundations of food webs (Kroeker et al. 2010). The effects of CO_2 on marine organisms depend on concentration and time scale (Pörtner 2008) and variations in the life-history may cause some species to be more resilient than others (Kroeker et al. 2010). Reduced pH can disrupt the acid-base status of extracellular body fluid and since regulation of acid-base is related to energy turnover and dependent on phylogeny and life style, organismal responses may be less or more able to compensate the acid-base disturbance (Pörtner 2008). Those organisms with lower capacities to regulate acid-base mechanism (such as invertebrates), are more sensitive to CO_2, whereas mobile organisms with developed regulatory mechanisms (such as fish) may be more resilient to acidification. On the other hand, acidification, warming and increased frequency of hypoxia co-occur in the oceans and sensitivity of organisms to temperature, CO_2 and O_2 integrate, reducing the scope for performance at ecosystem level, with implications on the ranges of the species and interaction between species (Pörtner 2008).

Changes in sea water chemistry, in addition to rising sea temperatures and variation in salinity and circulations patterns, are expected to also affect deep-sea calcifying organisms such as Cold-Water Corals (CWC), many of which provide habitat and refuge for many associated species (Roberts et al. 2009). Under acidification scenarios, deep-sea stony corals are projected to redistribute, as well as the organisms that depend on them, impacting on their physiology, calcification rates and source of food (Guinotte et al. 2006). The suitable habitat for cold-water stony corals is projected to be strongly reduced, fragmenting populations, decreasing resilience and having an impact on larval recruitment (Tittensor et al. 2010). While still limited, there is some evidence of the functional relation between CWC and deep-water fish species (Baillon et al. 2012; Henry et al. 2013; Pham et al. 2015), being likely that future projected changes on CWC distribution have an effect on the distribution of some commercial deep-sea species.

Climate change can have an effect on fisheries, adding to other non-climatic threats such as overfishing and pollution, affecting the stocks and their resilience (i.e., IPCC 2014; Mora et al. 2014). The projected global changes in climate by the mid-21st century are expected to cause a redistribution of marine species at the global level, with a reduction of biodiversity in sensitive regions, challenging the provision of fisheries and other ecosystem services (IPCC 2014). Under a 2°C global warming relative to pre-industrial temperatures scenario, species richness and fisheries catch potential are expected to increase at mid and high latitudes and to decrease at tropical latitudes and semi-enclosed areas (Cheung et al. 2010; IPCC 2014). Projected impacts of climate change on global fisheries productivity by 2055 vary among ocean depth zones, with the average catch potential increasing (+ 10–20%) in offshore regions (> 200 m depth), and decreasing (−4–5%) at the continental shelf (< 200 m depth), probably as a consequence of inshore-to-offshore shift of species as water temperature increase (Cheung et al. 2010). Cheung et al. (2010) also observed that some high-latitudes regions in the northern hemisphere are expected to gain, whereas others, located in the tropics/subtropics, are expected to lose in catch potential, with probable large implications for food security, especially in those countries particularly vulnerable to global changes in climate. Moreover, modeling the integrated changes in ecophysiology and distribution of > 600 species of exploited demersal fishes under scenarios of temperature and oxygen levels predicted for the future indicates a reduction in body size in both the tropics and temperate regions, of up to 25% from 2001 to 2050. The projected reduction in body size is due equally to changes in individual weight and changes in species composition (the poleward shift of fish populations) (Cheung et al. 2012). These projections include many of the fish species commercially valuable for the high-seas fishery included in this chapter. The modeling studies inevitably bear some uncertainties, linked to the necessarily adopted assumptions and simplifications (Cheung et al. 2012), and the effects of changes such as distributional shifts acting on the strength of ecological interactions inside the whole community, altering trophic dynamics, and the links between the surface, water column and deep seafloor remain difficult to predict.

Management and Conservation Legal Framework

By definition the High Seas are "all parts of the sea not included in the exclusive economic zones, in the territorial sea or in the internal waters or in the archipelago waters of an archipelago state" (UNCLOS art. 86). In accordance with the UNCLOS all States have right to exercise the freedom of the high seas under conditions laid down in the Convention. Subsequent legal developments have however reduced these freedoms under the domain of the fisheries in particular those over the straddling fish stocks and the highly migratory fish stocks, setting out principles for their conservation and management and establishing that such management must be based on the precautionary approach and the best available scientific information (UNFSA—UN Fish Stocks Agreement—Conservation and Management of Straddling Fish Stocks and Highly Migratory Fish Stocks in force December 2001). The Agreement elaborates on the fundamental principle, established in the Convention, that States should cooperate to ensure conservation and promote the objective of the optimum utilization of fisheries resources both within and beyond the EEZs. Another important resolution in the context of the high seas fisheries is the UN Food and Agriculture Organization's Code of Conduct for Responsible Fisheries (CCRF, FAO 1995), which adopts the precautionary principle as a basis for fisheries management, along with other provisions of mutual control by fishing nations. The CCRF sets out "principles and international standards of behaviour for responsible practices with a view to ensuring the effective conservation, management and development of living aquatic resources, with due to respect the ecosystem and biodiversity" (FAO 1995). These provisions incentivise the cooperation among States to avoid conflicts and to ensure the conservation and the appropriate management of fishing resources. In spite of these efforts it is known that most of these provisions were not implemented or ratified by many States imposing limitations to their effectiveness. In practical terms the high seas is still loosely regulated and the little commitment of countries in managing it led to a continuous overexploitation of living resources in EEZs and in the high seas and resulted in negative environmental impacts related to habitat destruction and by-catch of non-target species negative impacts (Mortensen et al. 2005; Clark and Koslow 2007; Weaver et al. 2011; White and Costello 2014; Rogers 2015). This situation led in recent years to several posterior agreements, voluntary guidelines and codes of practice to be implemented by States and the establishment of several Regional Fisheries Management Organizations (RFMOs) (see selected key events timeline on Fig. 8.13) with mandatory areas that include important zones of the high seas.

The relevant institutions for the regulation of the high seas activities are the UNGA (United Nations General Assembly) and their advice institutions, in particular the International Seabed Authority (ISA) which regulates mineral resources extraction and the International Maritime Organization (IMO) regulating shipping and waste disposal. Other relevant international entities include the Intergovernmental Oceanographic Commission (IOC-UNESCO) (United Nations Educational, Scientific and Cultural Organization) which among others promotes international cooperation and coordinates ocean research programs; the International Union for Conservation of Nature (IUCN) which among others deal with Marine Protect Areas (MPAs) implementation including those in the high seas through the World Commission on Protected Areas (WCPA) High Seas MPA Task Force. Issues dealing with fisheries advice and management, living and non-living resources, conservation and monitoring of the oceans fall within the scope of FAO (Food and Agriculture Organization), of the Convention for the Biological Diversity (CBD) and the Regional Fishery Bodies (RFBs) and RFMOs.

The regionalization of international environmental law and policy has emerged as one of the most important legal trends in recent years following the recognition that not every international environmental problem needs to be dealt with on a global level (Rochette et al. 2015a). Regional mechanisms that deal with international oceans governance include the Regional Seas programs (RS), the Large Marine Ecosystems (LME) (see Rochette et al. 2015a for details), and the Regional Fisheries Bodies (RFBs and RFMOs), being the RFBs and RFMOs the most relevant in the context of the high seas fisheries including the deep-sea fisheries. The different typology of RFBs is due to their diverging mandates. The main distinction is between advisory RFBs—with management advice only—and RFMOs—with competence to establish legally binding measures for conservation and management (Rochette et al. 2015a). At present there are 41 marine RFBs worldwide, 21 of which are RFMOs and 20 are advisory

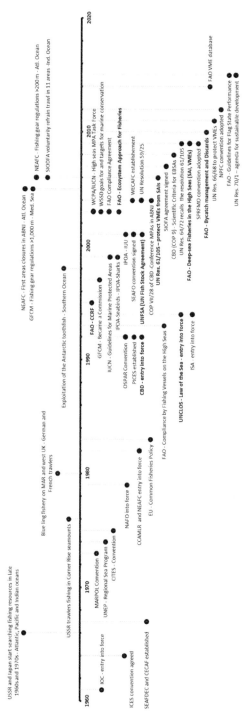

Figure 8.13. Selected key events, initiatives and organizations with some direct and indirect relevance to the high seas living resources exploitation and management. Abbreviations : UN Res.—United Nations Resolution; Atl. Ocean—Atlantic Ocean; Med. Sea—Mediterranean Sea; Ind. Ocean—Indian Ocean; WSSD—World Summit on Sustainable Development. The other acronyms are defined in the text. (last updated in January 2016).

RFBs (three scientific and 17 management) (http://www.fao.org/fishery/rfb/search/en, Rochette et al. 2015a). From those that deal with high seas deep-sea fisheries issues, eight are RFMOs (GFCM—General Fisheries Commission for the Mediterranean, NAFO—Northwest Atlantic Fisheries Organization, SEAFO—South East Atlantic Fisheries Organization, SIOFA—Southern Indian Ocean Fisheries Agreement, SPRFMO—South Pacific Regional Fisheries Management Organization, CCAMLR—Commission for the Conservation of Antarctic Marine Living Resources, NEAFC—North-East Atlantic Fisheries Commission, NPFC—North Pacific Fisheries Commission), two are scientific advisory RFBs (ICES—International Council for the Exploration of the Sea and PICES—North Pacific Marine Science Organization) and three are management advisory RFBs (WECAFC—Western Central Atlantic Fisheries Commission, SEAFDEC—Southeast Asian Fisheries Development Center, CECAF—Fishery Committee for the Eastern Central Atlantic).

Despite the RFMOs being the preferred entities for regulating fisheries at a regional level, the process of covering all high seas areas by an RFMO is still taking place, and is far from being efficient in implementing the UNGA resolutions on deep ocean biodiversity protection (Wright et al. 2015). Geographic gaps exist between RFMOs in parts of the Atlantic, Indian and Pacific Oceans, particularly for those in charge of managing bottom fisheries and species other than tuna (Ban et al. 2014). Notwithstanding the gaps, RFMOs regulatory areas cover large areas of international waters, and therefore have to play an essential role in the preservation of resources and ecosystems in the high seas (Rochette et al. 2015a).

In contrast with the seabed mining in international waters, which are overseen by ISA as previewed in the UNCLOS, the impacts of the deep-sea fishing on the international seabed were not initially anticipated and were therefore less legally constrained (Probert et al. 2007). The deep sea is however of interest for an increasing number of nations, targeting the high seas resources at a global level and the concerns regarding the conservation and management of deep-sea ecosystems have been growing (Berkes et al. 2006; Probert et al. 2007).

International concerns about the negative impacts of the exploitation of living resources in the high seas, particularly in the last decade, stimulated an intense debate about the effects of deep-sea fisheries on deep-sea vulnerable ecosystems, the management of deep-sea fisheries and conservation of deep-sea ecosystems in the high seas. Despite the issue of protecting biodiversity in the deep sea in ABNJ has been gaining momentum in the political and scientific fora, and has been debated extensively by the international community in recent years, these areas are still poorly managed/protected. A study reviewing MPAs worldwide well illustrates the lack of protection of these ecosystems: out of

5880 MPAs, covering about 4.2 million km^2 (e.g., only about 1.17% of the ocean area), most of them are located within the 12-mile coastal area and none on the high seas (Toropova et al. 2010). Although in recent years some high seas MPAs have been implemented (i.e., OSPAR high seas MPAs, see Table 8.3), they can still be considered as isolated cases in a general situation of fragmented or no management. Despite the importance of existent regional mechanisms, concern has been raised about their effective coordination and efficiency and in some circumstances the overlap in their aims (Rochette et al. 2015a), and some authors have called for multilevel governance institutions operating at diverse levels to implement measures and take action (Berkes et al. 2006). A recent review on bottom fisheries closures in ABNJ implemented by RFMOs and arrangements to protect VMEs addresses the shortcomings of such organizations in conservation of marine biodiversity, calling for a broader framework for integrated conservation (Wright et al. 2015). Adopting an integrated approach beyond national jurisdictions is made even more challenging by the fragmented management of the high seas, dictated by UNCLOS, that separates governance of the water column from that of the seabed (Ban et al. 2014), and by the UNCLOS environmental provisions not taking into account global threats such as climate change or vulnerability of deep-sea ecosystems (Gjerde 2012). The result of the regional and sector-by sector approach, the overlapping in the same areas of different management bodies, and the lack of management bodies for some other areas is a fragmented management of the high seas, as well as poorly ecosystem-based and targeted to biodiversity conservation, lacking to address cumulative impacts (Ban et al. 2014).

As a result of the debates in recent years, in addition to the UNCLOS, CCFR, and UNFSA provisions, various regulatory initiatives at the international level have emerged. These discussions have taken place within the framework of various international organizations such as the UNGA, the EU, the ICES, the PICES, the Oslo-Paris Convention for the protection of the Northeast Atlantic Marine Environment (OSPAR) and within many RFMOs such as the NEAFC, the NAFO, the SEAFO. Conservation and environmental organizations, scientists and several States called for the urgent action of UNGA to protect deep-sea biodiversity on the high seas, including measures such as a moratorium on high seas bottom trawling (Probert et al. 2007).

After a long debate the UNGA recognized that the provisions of the UNFSA should also apply to discrete fish stocks including the deep-sea fisheries in high seas (Rogers 2015), and adopted in 2004 and 2006 several resolutions that addressed those concerns calling for a review of the state of conservation of the high seas, and regional action for protecting VMEs, although not including a global moratorium (UNGA Resolutions 59/25 in 2004 and 61/105 in 2006).

Table 8.3. Conservation spatial measures implemented to date in the high seas. (last updated in January 2016).

Region/Management Body/Area	Period in force	Conserv. Measure	Main Aim	Notes/No. of measure
Southern Ocean - CCAMLR				
South Orkney Islands Southern Shelf MPA	2009-	MPA	BD	CCAMLR/CM-91-03 (2009)
Four areas in the Convention Area	2012-	AC	VME	CCAMLR/CM -22-09 (2012)
Convention Area–toothfish fisheries	2009-	FGR	CH	Fishing prohibited in depths < 550 m - CCAMLR/CM-22-08 (2009)
Convention Area	2010-	FGR	PA	Deep-sea gillnetting prohibited - CCAMLR/CM -22-04 (2010)
Mediterranean Sea -GFCM				
Pelagos Sanctuary	2002-	MPA	BD	Sanctuary agreement
Capo Santa Maria di Leuca; Eratosthenes Seamount; Nile Delta Area Cold Hydrocarbon Seeps	2006-	AC	CH	GFCM/REC.CM-30/2006/3
Areas with depth >1,000 m	2005-	FGR	CH/PA	Towed dredges and trawlnets prohibited - GFCM/REC.CM-29/2005/1
North-West Atlantic Ocean -NAFO				
Corner Seamount; New England Seamount; Newfoundland Seamount;Orphan Knoll	2007-2020	AC	PA-EAF	NAFO/FC Doc. 15/01
30 Coral Protection Zone;	2008-2020	AC	VME	NAFO/FC Doc. 15/01
Fogo Seamount 1; Fogo Seamount 2	2008-2020	AC	VME	NAFO/FC Doc. 15/01
High Sponge and Coral Concentration Areas [a]	2010-2020	AC	VME	NAFO/FC Doc. 15/01
All seamount areas in the Regulatory Area	2016-	AC	CH/EAF	37th Annual Meeting 21-25 September 2015
North-East Atlantic Ocean - NEAFC				
Altair Seamount; Antialtair Seamount	2005-2017	AC	VME	NEAFC/ Rec.19/2014 as amended by Rec.09/2015
Faraday seamount; Hekate seamount; Reykjanes Ridge	2005-2009	AC	CH	NEAFC/ Rec.04/2005
Hatton Bank 1; Logachev Mounds; West Rockall Mounds	2007-2017	AC	VME	NEAFC/ Rec.19/2014 as amended by Rec.09/2015
Middle MAR Area (Charlie-Gibbs Fracture Zone and sub-Polar Frontal Region); Northern MAR Area; Southern MAR Area	2009-2017	AC	VME	NEAFC/ Rec.19/2014 as amended by Rec.09/2015
Edora's bank	2013-	AC	VME	NEAFC/ Rec.19/2014 as amended by Rec.09/2015
Hatton-Rockall Basin; Southwest Rockall Bank; Hatton Bank 2	2014-2017	AC	VME	NEAFC/ Rec.19/2014 as amended by Rec.09/2015
Rockall Bank	2015-2017	AC	VME	NEAFC/ Rec.19/2014 as amended by Rec.09/2015

Rockall Haddock Box	2013-	AC	FM	Protect juvenile haddock - NEAFC/ Rec. 03-2013;
Reykjanes Ridge- blue ling Closure	2013-2016	sAC	FM	Protect blue ling spawning ground - NEAFC/ Rec. 05/2013
Areas with depth > 200 m	2006-	FGR	FM	Gilnets, entangling and trammel nets prohibited - NEAFC/Rec. 3/2006
North-East Atlantic Ocean - OSPAR				
Charlie-Gibbs North	2012-	hsMPA	-	OSPAR Decision 2012/1
Altair Seamount; Antialtair Seamount; Charlie-Gibbs South; Josephine Seamount; Mid-Atlantic Ridge north of the Azores; Milne Seamount complex	2010-	hsMPA	BD	OSPAR Decision 2010/1 to 6
South-East Atlantic Ocean - SEAFO				
Africana Seamount; Malakhit Guyot; Schabenland and Herdman Seamounts; Schmidt-Ott and Erica Seamounts; Vema Seamount; Wüst Seamount	2007-	AC	CH	SEAFO/CM18-10
Dampier Seamount; Discovery, Junoy and Shannon Seamounts; Molloy Seamount; Panzarini Seamount	2007-2010	AC	CH	SEAFO CM06-06
Kreps Seamount; Unnamed Seamount 14; Unnamed Seamount 15; Unnamed Seamount 17; Unnamed Seamount 18;	2011-	AC	CH	SEAFO/CM18-10
South Indian Ocean - SIODFA				
Agulhas Plateau; Atlantis Bank; Bridle Seamount; Coral Seamount; East broken Ridge; Fools Flat; Gulden Draak Seamount; Mid-Indian Ridge; Rusky Seamount; South Indian Ridge North/South; Walters Shoal	2006-	vAC	FM	Voluntary closures upon agreement of SIODFA members; the management body for this area (RFMO SIOFA) came into force in 2012 and did not implement spatial measures to date.
Middle of What Seamount; Banana Seafloor Feature	2013-			
North Pacific Ocean - NPFC				
no formal closures declared yet				

Table 8.3 contd. ....

…Table 8.3 contd.

Region/Management Body/Area	Period in force	Conserv. Measure	Main Aim	Notes/No. of measure
South Pacific Ocean - SPRFMO[b]				
Convention Area	2013-	FGR	CH/FM	Large-scale pelagic driftnets and deepwater gillnets prohibited - SPRFMO CMM 1.02 Annex M
Closures implemented by New Zealand	2007-	AC	VME/PA	Implementation of SPRFMO interim measure (2007)[d]
South –West Atlantic Ocean[c] 9 areas containing VMEs	2011	VAC	VME	6-months closure implemented by Spain following a EU regulation[e]

Conservation Measures include: MPA—Marine protected area; hsMPA—high seas MPA; AC—Area closure (bottom fishing, deep-sea fishing or fishing activities prohibited); sAC—Seasonal area closure; vAC—Voluntary area closure; FGR—Fishing gear regulation. Main aim for conservation measures include: BD—preserve or restore marine biodiversity; CH—protect deep-sea communities, habitats and ecosystems; EAF—apply and/or maintain the Ecosystem Approach to Fisheries management; FM—fisheries management measure, to protect deep-water stocks, and maintain sustainable fisheries; PA—adopt a precautionary approach (to avoid detrimental effects on marine environment and species); VME—protect VMEs and prevent SAIs on VMEs. No. of measure refers only to the last document referring the area. [a]including the following areas 1. Tail of the Bank; 2. Flemish Pass/eastern canyon; 3. Beothuk Knoll; 4. Flemish Cap; 5. Northeast Flemish Cap; 6. Sackville Spur; 7. Northern Flemish Cap; 8. Northern Flemish Cap; 9. Northern Flemish Cap; 10. Northwest Flemish Cap; 11. Northwest Flemish Cap; 12. Northwest Flemish Cap; 13. Beothuk Knoll; [b]the management body for this area (RFMO NPFC) is not yet functioning, but negotiations on the proposed closure of some areas are ongoing (Wright et al. 2015); [c]there is no RFMO managing this area. [d](SPRFMO 2007). [e](Council Regulation (EC) No. 734/2008). VME—Vulnerable Marine Ecosystem; SAI—Significant Adverse Impact

The UNGA resolution 59/25 in 2004 invited nations with deep-sea fisheries and RFMOs to take urgent action to protect VMEs from destructive fishing practices such as bottom trawling. Subsequent discussions in 2006 related to biodiversity, sustainability and equity, generated several proposals that included moratorium on bottom trawling on the high seas, the prohibition of bottom fishing in areas not covered by RFMOs, or the reverse of the 'burden of proof' for example in the allocation of fishing licenses where fishing companies had to demonstrate that their activity did not affect the resources or habitats (Rogers and Gianni 2010). It was in the same year—2006—that the UNGA created an "Ad Hoc Open-ended Informal Working Group to study issues relating to the conservation and sustainable use of marine biological diversity beyond areas of national jurisdiction" (BBJN WG). Especially relevant was the UNGA Resolution 61/105 (UNGA 2006), which committed the states with fisheries on the high seas to take a series of more stringent measures, such as the need to undertake assessments prior to assess the potential Significant Adverse Impacts (SAIs) of fishing activities; and then to ensure that if there are SAIs they will be managed in order to be avoided, including the possibility of closing some areas to bottom fishing where VMEs are known or suspected to occur (Rogers and Gianni 2010). The resolution also determined the implementation and use of protocols to cease fishing when VMEs are encountered during fishing activities (rules of 'displacement' or 'move-on-rules'). In this sequence, a series of technical guidelines were negotiated and published (such as the International Guidelines for the Management of Deep-Sea Fisheries in the High Seas—FAO 2009), including the criteria for conducting impact studies, for identifying VMEs and for benchmarking the SAIs that deep-sea fisheries would have on the VMEs.

The imposition of international rules by UN led some fishing companies to take the initiative and organize themselves, to follow the International Guidelines for the Management of Deep-Sea Fisheries in the High Seas. One example is the Southern Indian Ocean Deepsea Fishers Association (SIODFA), which implemented, on a voluntary basis, advanced monitoring systems for the alfonsino and orange roughy fisheries taking place on some seamounts in the Indian Ocean. In addition to imposing on their members the fulfillment of specific rules in order to maintain long-term sustainable levels of exploitation, they also agreed on the closure of some seamounts in the areas where they operate (Warner et al. 2012).

In 2009, to overcome the difficulties in implementing resolution 61/105, the UNGA adopted additional measures (Resolution 64/72, 2009) which reaffirm the need for flag States and RFMOs to implement resolution 61/105 in accordance with FAO guidelines defined in 2009, before allowing or permitting any bottom fishing in international waters. UNGA Resolution 64/72 puts particular emphasis on conducting impact assessments on the high seas and warns that any boat or country can start fishing on the high seas without the impact studies being carried out. The resolution also points out that the evaluations of deep-sea stocks should be carried out to ensure its sustainable management.

After a decade, despite several conservation and management initiatives, the UNGA resolutions for protecting VMEs from high seas bottom trawl are still far from being really effective (Wright et al. 2015)—following the usual pattern out of sight out of mind—and high seas nations have been called for stronger protect deep-sea ecosystem and species.

In addition to UNGA resolutions, since 2004 many initiatives took place concerning the conservation and management of deep-sea ecosystems in the high seas (see Fig. 8.13). Some RFMOs were recently created (SEAFO, SIOFA, SPRFMO), several seamounts were closed to deep-sea fishing temporally or permanently (see Table 8.3), and the use of several fishing gears was restricted at certain depths and in certain areas (e.g., gillnetting >200 m depth), through initiatives of several RFMOs (e.g., NEAFC, SEAFO) and countries (see Table 8.3).

The Conference of the Parties to the Convention on Biological Diversity (COP) in 2004 and 2006 also designated the seamounts and cold water corals as conservation priorities and called for their protection (Probert et al. 2007). In this sequence a working group was established in 2006, to explore the establishment of MPAs in ABNJ (COP VII/28) and ecological criteria for high seas MPAs. In 2008 the COP 9 adopted the scientific criteria for identifying ecologically or biologically significant marine areas (EBSAs) in need of protection in open-ocean waters and deep-sea habitats (annex I, decision IX/20) (CBD 2009).

The FAO took several initiatives to manage deep-water fisheries in the high seas, including the agreement, in 2003, on the application of an Ecosystem Approach for Fisheries (FAO 2003) that seeks sustainability not only of fish stocks but of the entire ecosystem, placing great emphasis on the conservation and integrity of habitats. In addition to the International Guidelines for the Management of Deep-Sea Fisheries in the High Seas (FAO 2009), other relevant initiatives for the high seas fisheries include the Guidelines on Bycatch Management and Reduction of Discards adopted in 2010 (FAO 2011). The by-catch has been identified as a threat to non-target species and it is recognized as particularly problematic in the case of deep sea fisheries (Devine et al. 2006; Rogers 2015).

As previously mentioned the IUU fishing has been recognized as a serious threat to all fisheries, particularly to the high seas deep-sea fisheries, and is being fought by several international initiatives, like the International Plan of Action to Prevent, Deter and Eliminate Illegal, Unreported and Unregulated Fishing (IPOA-IUU 2001) or the Agreement on Port State Measures to Prevent, Deter and Eliminate Illegal, Unreported and Unregulated Fishing (FAO 2009). More recently, in the sequence of the previous 1993 FAO Agreement to Promote Compliance with International Conservation and Management Measures by Fishing Vessels on the High Seas (Rogers 2015), and also relevant to the problem of the IUU fishing, FAO adopted the FAO Voluntary Guidelines for Flag State Performance (2015). This code foresees a range of actions that countries can take to ensure that vessels registered under their flags do not conduct IUU fishing, including Monitoring, Control and Surveillance (MCS) activities, such as Vessel Monitoring Systems (VMS) and onboard observers (FAO 2015).

In the international fora, the EU plays an important role in the discussion and regulation of the deep-water high seas fisheries. Europe has one of the largest and advanced fishing fleets, is one of the largest importers of world fishery products, and among the highest consumers of marine products in the world, with very high rates of fish consumption per capita in some member states (EC 2010). Europe represents about 6% of global fisheries and has one of the most powerful fleets of the world (EC 2010) thus acting in all ocean areas and bearing great international responsibility in the management of high seas marine resources. Since 2008, the EU established legislation following many of the UNGA recommendations, such as rules for European fishing vessels in international waters, catch limits for some deep-sea species in community and international waters of the North Atlantic (ICES areas and NEAFC and NAFO), limitations to the development of new fisheries and deep sea fisheries in highly vulnerable areas (EC 2008a,b; NEAFC 2009; EC 2010). In the specific case of deep-water sharks, and other deep-water species several reductions on the Total Allowable Catch (TAC) have been implemented in recent years as well as limitations on the use of certain gear types, such as deep-water gillnets (Neat et al. 2015). Taking into account that EU flag vessels are responsible for a major amount of the deep-sea fisheries in the high seas, the introduction of zero TAC for several species applicable to EU flag vessels in ICES areas (which includes ABNJ of member states) is expected to have important effects on North Atlantic deep-sea species and habitats. For its part, the OSPAR continues the promotion and creation of a network of Marine Protected Areas on the high seas (HSMPAs), to contribute for the sustainable use, protection and conservation of marine biodiversity and ecosystems in the North-East Atlantic, including the high-seas areas already designated by other organizations (OSPAR 2003; 2008a,b; see also Table 8.3).

During the year 2015, more important steps were taken towards the conservation and sustainable use of resources in the high seas. In January 2015, discussions within the BBJN WG led to the historic decision agreed by the States of opening negotiations, in 2016, for a new international legally binding instrument under UNCLOS (Rochette et al. 2015b). The process is expected to lead to a decision by UNGA on the date of an intergovernmental conference, to elaborate the new legal instrument, by the end of 2018. The issues to be negotiated related with conservation and sustainable use of marine biodiversity in ABNJ, will include marine genetic resources, area-based management tools, assessment of environmental impacts and capacity building and the transfer of marine technology (Rochette et al. 2015b). Such a legal instrument is seen as a long term measure that will provide a global platform for guaranteeing coherence, cooperation and compliance, filling the spatial governance gaps in the high seas, but that should coexist with the regional and sectoral bodies, adopting an hybrid approach to improve the governance of the high seas (Ban et al. 2014; Rochette et al. 2015b).

In September 2015 the UN Member States adopted Resolution 70/1 "Transforming our world: the 2030 Agenda for Sustainable Development", to build on the Millennium Development Goals (UN Res. A/RES/55/2 in 2000) and complete what they did not achieve. Resolution 70/1 includes 17 sustainable development goals (SDG), one of which is a specific goal for the ocean and seas (SDG 14—Conserve and sustainably use the oceans, seas and marine resources). Goal 14 recognizes the importance of the world's oceans in driving global systems, regulating and providing ecosystem services of vital importance for humankind, and acknowledges that protecting and sustainably managing ocean resources is a key feature for sustainable development. In addition to targets aimed at addressing the issues of conservation and sustainable use of oceans and their resources, ocean acidification, some of the targets listed in Goal 14 refer to aspects, such as the call by 2020 to end "overfishing, illegal, unreported and unregulated fishing and destructive fishing practices...." (14.4) and to "prohibit certain forms of fisheries subsidies which contribute to overcapacity and overfishing" (14.6), which are particularly important in the context of deep-sea fisheries in the high seas.

A general characteristic of deep ecosystems, particularly those in the high seas, is the little scientific information available on their structure, functioning and status, so that many fishing activities are still conducted in the absence of any prior knowledge about the genetic and population structure of the stocks, life-histories of both target and by-catch species. Since most deep-sea fisheries started before gathering adequate scientific knowledge for management purposes, it is foreseeable that many areas were already severely affected. As a result of the remote nature of the high seas most of the fishing is done by large vessels with high autonomy and high fishing capacity, using in their majority trawl gears. Albeit other fishing gears are used on the high seas fishing, as longlines or gillnets, trawls are the most frequently used gear and represent a reason for higher concern, not only for the high quantity of resources caught in a short time/trawl event, but also due to their known impacts on the seafloor and benthic communities (Roberts 2002; Clark et al. 2007; Waller et al. 2007; Williams et al. 2010; Pham et al. 2014).

Nowadays it is estimated that the decline of fish populations in the high seas have been significant, and in certain cases the estimated abundances have been reduced up to 90% (Watling et al. 2011). Fish that transverse multiple EEZs and the high seas are more prone to overexploitation than fish contained in a single EEZs (McWhinnie 2009). Despite the establishment of 200 nm EEZs to alleviate pressures on the international commons, the tragedy of commons has not been resolved, and the percentage of over-harvested stocks has increased (McWhinnie 2009). A fish stock has more probability to be overexploited with a higher number of countries sharing it, and international sharing of fish stocks is considered as a detrimental force for determining stock status (McWhinnie 2009).

The subsidies granted by governments worldwide also strongly mask the reality, and make the deep-sea fishing in the high seas even more unsustainable. Estimates put an annual allocation of around US$ 152 million per year of subsidies (which constitutes 25% of the total value of fish landed), and an effective profit of these fleets of no more than 10% of the landed value (Sumaila et al. 2010). This means that most of trawl fleets operating on the high seas, if operated without subsidies would have constant losses and would therefore be unable to continue fishing (Sumaila et al. 2010; Norse et al. 2012). So if these supports cease, the risk to biological resources in deep-sea would be greatly diminished. The combination of features such as low productivity of the target species, the little selectivity of fishing gears, the economy of these fisheries (a fish in the high seas has a higher cost per unit weight, Sumaila et al. 2015), which favors overexploitation of these populations, and a weak and inefficient regulatory and monitoring regimes, cause most offshore deep-sea fisheries to be now considered unsustainable (Watling et al. 2011; Weaver et al. 2011). The exploitation of deep-sea fishing resources, mostly in the high seas areas, is unprofitable and difficult to sustain, resembling more the kind of 'mining' operation, where resources are successively depleted area after area ('boom-and-bust') (Pitcher et al. 2010). Norse et al. (2012) suggest that instead of exploiting fish stocks on the high seas as a 'mine', a right economic and ecological strategy would be the rebuilding and sustainable fishing of the most resilient and more productive populations on shallow waters of coastal ecosystems, closer to the markets.

In the majority of the cases the absence of data, along with the lack of adequacy of monitoring and stock assessments (Watling et al. 2011), precludes a robust evaluation of the real state of these resources. This makes it impossible to use more conventional management measures such as fishing quotas, since they are based on stocks biomass estimates that requires large and good quality data. Therefore, other approaches and initiatives such as the protection of large ocean areas—area closures and MPAs—are being taken and appear to be the ones that may be effective to avoid SAIs on high seas VMEs (Auster et al. 2011; Wright et al. 2015), especially considering that other measures, such as the 'move-on' rules, may be under-utilized and ineffective (Wright et al. 2015). Table 8.3 provides a summary of spatial measures implemented by RFMOs and other bodies in the high seas to date. The most active organizations with this respect are those acting in the South-East Atlantic, North-West and North-East Atlantic, with the first closures implemented in the North-East Atlantic, in 2005, to protect VMEs, and more closures in recent years, and fishing gear regulations to prohibit the use of destructive fishing practices in the deep-sea. Some areas of the ocean—where RFMOs are only very recent or not functioning yet—are still not covered by spatial measures in the high seas, such as the South-West Atlantic, the North and South Pacific, the Indian Ocean. In some ocean regions where RFMOs are absent or only recently constituted, spatial measures have been implemented by Member States (such as New Zealand in the South Pacific) or on a volunteer basis (such as in the South Indian Ocean by a Fishery Industry Association, the SIODFA). One of the shortcomings of the spatial measures implemented by RFMOs identified elsewhere is represented by the fact that closures often cover depth strata that are considered unfishable (reviewed in Wright et al. 2015). The list of closures in Table 8.3 highlights also that there are still gaps between fishing closures and MPAs, as seen in the North-East Atlantic, where NEAFC bottom fishing closures do not fully overlap with high seas MPA established by OSPAR (Ban et al. 2014; Wright et al. 2015), although a cooperative approach and effort between the bodies operating in the North-East Atlantic is ongoing (Wright et al. 2015).

Another protective measure that can be potentially located in the high seas is a 'Particularly Sensitive Sea Area' (PSSA), as established by the Assembly of IMO (International Maritime Organization) (Resolution A.720(17) adopted on 6 November 1991, Resolution A.885(21) adopted on 25 November 1999, and Resolution A.927(22) adopted on 29 November 2001). A PSSA is defined "as an area that needs special protection through action by IMO because of its significance for recognized ecological or socio-economic or scientific reasons and which may be vulnerable to damage by international maritime activities". Measures to protect PSSAs include the control of maritime activities (such as routing, discharge), within the specific competence of IMO. To date, no PSSAs have been designated in ABNJ (Rochette and Wright 2015).

Beyond the bottom fishing closures there are suggestions of creating high seas regeneration zones (GOC 2014), or completely close the high seas to bottom fishing, to address the global challenge of overexploitation of valuable species such as highly migratory species fished over EEZs and high seas (White and Costello 2014). Applying a spatial bio-economic model to simulate the closure of high seas to fishing, White and Costello (2014) showed it would raise fisheries profit and fish stock conservation. Although they recognized the political limitations in putting in act such a measure, they pointed out how the effects of closing the high seas would benefit different aspects of a global issue. Most of the taxa fished in the high seas are straddling between high seas and EEZs, which share ecological and economical resources (Sumaila et al. 2015). Depletion of high-seas stocks can influence coastal fishes, and total closure of high-seas to fishing is projected to generate an increase of 18% in catch of straddling taxa, and to provide socio-ecological benefits, reducing burning of fossil fuels and distributing benefits from fisheries more equally, especially to developing tropical countries, likely more affected by the effects of climate change (Sumaila et al. 2015). Although the study by Sumaila et al. (2015) is mostly based on catches in the high seas represented by pelagic fishes and invertebrates, the projections on the effects of closing the high seas to fishing are even more pertinent in the case of deep-sea fishes, which make up at maximum around 5% of global landed value and catches considered in the study, and are represented by low productive species.

Monitoring and control of these remote areas are however a major flaw of high seas MPAs and area closures. It has been suggested that a combination of remote monitoring and surveillance with port state control would be the best option, and that the costs for enforcing high seas may be obtained by a fraction of fisheries subsidies (GOC 2013). An efficient enforcement should include better use satellite monitoring systems of vessels (VMS—Vessel Monitoring System) with more frequent and detailed communications combined with more effective controls on the discharge ports and the presence of observers on board. Sumaila et al. (2015) also suggest the amendment of the International Maritime Organization's International Convention for the Safety of Life at Sea (SOLAS) to apply to the fishing vessels the mandatory requirement for shipping vessels to carry and use the Automatic Identification System (AIS). Data obtained by VMS, AIS and other geo-referenced data from monitoring equipment are a valuable source of information on the use of marine space, spatial distribution of fishing effort, fishers' behavior, fishing impacts (i.e., Murawski et al. 2005; Bertrand et al. 2007; Marchal et al. 2007; Lee et al. 2010; Jennings et al. 2012). Unfortunately, the access to this information is often limited, hampering the use and analysis of data, and representing a limitation to the assessment and management of high seas fisheries.

Conclusion

Exploitation of the high seas for fishing demersal and deep sea fish occurs mainly on the deep sea-on seamounts, ridges and canyons-by bottom trawls, essentially in a tiny portion of the ocean seafloor. Fishing for deep-sea resources has a documented history of fisheries collapses, dramatic declines of fish populations, and negative impacts on habitats, including VMEs, particularly on seamounts.

The reconstructed catch data (1950–2010) reported in this chapter clearly show an expansion of these fisheries along time, involving a growing number of nations and species, and fluctuating catches-probably a consequence of serial depletions and better fishing technology, both supporting the search for new exploitable areas and species. Despite the large-scale character of high seas deep-sea fisheries, the volumes of fish and corresponding values delivered only account for small fractions of EEZs catches and values. Even considering the possible underestimation of catches due to illegal, unreported and unregulated fishing, these values put accent on the low sustainability of these fisheries, also worsened by high by-catches and the wide use of subsidies from fleets operating in this sector. The impacts on high seas deep-sea fisheries from stressors such as exploitation are possibly going to be exacerbated by changing climate, whose effects were already documented for some deep-water fishes and fisheries; predicted climate change scenarios draw declines of biomass and size of exploitable resources in all areas where this type of fisheries occur.

In recent years the discussions at political and scientific for a undoubtedly have been giving more attention to the conservation and management of high seas. Still, efforts and measures for protecting and managing high seas areas and resources are fragmented, and often rely on limited scientific information on exploited areas and species. In this context, promising conservation and management measures to protect VMEs and associated resources include protecting large ocean areas, or even closing the high seas to fishing. Important steps have been taken, such as the historical consensus for a global legally binding instrument specifically dedicated to the conservation and sustainable use of marine biodiversity in ABNJ, to be negotiated in the next years. Since the principle "out of sight out of mind" well applies to high seas, keeping the interest high is of vital importance for tackling the complex multi-faceted issue related with sustainable use of high seas deep-sea living resources.

Acknowledgments

The authors acknowledge Fundação para a Ciência e Tecnologia (FCT), through the strategic project UID/MAR/04292/2013 granted to MARE.

References

Agnew, D.J., Pearce, J., Pramod, G., Peatman, T., Watson, R., Beddington, J.R. and Pitcher, T.J. 2009. Estimating the worldwide extent of illegal fishing. PLoS ONE 4(2): e4570.

Althaus, F., Williams, A., Schlacher, T.A., Kloser, R.J., Green, M.A., Barker, B.A., Bax, N.J., Brodie, P. and Schlacher-Hoenlinger, M.A. 2009. Impacts of bottom trawling on deep-coral ecosystems of seamounts are long-lasting. Mar. Ecol. Prog. Ser. 397: 279–294.

Auster, P.J., Gjerde, K., Heupel, E., Watling, L., Grehan, A. and Rogers, A.D. 2011. Definition and detection of vulnerable marine ecosystems on the high seas: problems with the "move-on" rule. ICES Journal of Marine Science 68: 254–264.

Baillon, S., Hamel, J.F., Wareham, V.E. and Mercier, A. 2012. Deep cold-water corals as nurseries for fish larvae. Front. Ecol. Environ. 10: 351–356.

Ban, N.C., Bax, N.J., Gjerde, K.M., Devillers, R., Dunn, D.C., Dunstan, P.K., Hobday, A.J., Maxwell, S.M., Kaplan, D.M., Pressey, R.L., Ardron, J.A., Game, E.T. and Halpin, P.N. 2014. Systematic conservation planning: a better recipe for managing the high seas for biodiversity conservation and sustainable use. Conserv. Lett. 7: 41–54.

Basson, M., Gordon, G.D.M., Large, P., Lorance, P., Pope, J. and Rackham, B. 2001. The effects of fishing on deep-water fish species to the West of Britain. JNCC Report No. 324. Available at http://jncc.defra.gov.uk/pdf/jncc324_web.pdf/.

Beaugrand, G., Reid, P.C., Ibanez, F., Lindley, J.A. and Edwards, M. 2002. Reorganization of North Atlantic marine copepod biodiversity and climate. Science 296: 1692–1694.

Bensch, A., Gianni, M., Gréboval, D., Sanders, J.S. and Hjort, A. 2008. Worldwide review of bottom fisheries in the high seas. FAO Fisheries and Aquaculture Technical Paper. No. 522. Rome, FAO.

Bergstad, O.A., Menezes, G. and Høines, Å.S. 2008. Demersal fish on a mid-ocean ridge: Distribution patterns and structuring factors. Deep Sea Research Part II: Topical Studies in Oceanography 55(1–2): 185–202.

Bergstad, O.A. 2013. North Atlantic demersal deep-water fish distribution and biology: present knowledge and challenges for the future. J. Fish Biol. 83: 1489–1507.

Berkes, F., Hughes, T.P., Steneck, R.S., Wilson, J.A., Bellwood, D.R., Crona, B., Folke, C., Gunderson, L.H., Leslie, H.M., Norberg, J., Nyström, M., Olsson, P., Österblom, H., Scheffer, M. and Worm, B. 2006. Globalization, roving bandits, and marine resources. Science 311(5767): 1557–1558.

Bertrand, S., Bertrand, A., Guevara-Carrasco, R. and Gerlotto, F. 2007. Scale-invariant movements of fishermen: the same foraging strategy as natural predators. Ecol. Appl. 17(2): 331–337.

Branch, T.A. 2001. A Review of Orange Roughy (*Hoplostethus atlanticus*) Fisheries, Estimation Methods, Biology and Stock Structure. Payne, A.I.L., Pillar, S.C. and Crawford, R.J.M. (eds.). A Decade of Namibian Fisheries Science. South African Journal of Marine Science. 23: 181–203.

Canals, M., Puig, P., Durrieu de Madron, X., Heussner, S., Palanques, A. and Fabres, J. 2006. Flushing submarine canyons. Nature 444: 354–357.

Catarino, D., Knutsen, H., Veríssimo, A., Olsen, E.M., Jorde, P.E., Menezes, G., Sannæs, H., Stanković, D., Company, J.B., Neat, F., Danovaro, R., Dell'Anno, A., Rochowski, B. and Stefanni, S. 2015. The pillars of hercules as a bathymetric barrier to gene flow promoting isolation in a global deep-sea shark (*Centroscymnus coelolepis*). Mol. Ecol. 24: 6061–6079. doi:10.1111/mec.13453.

CBD. 2009. Azores scientific criteria and guidance for identifying ecologically or biologically significant marine areas and designing representative networks of marine protected areas in open ocean waters and deep sea habitats.

Chan, F., Barth, J., Lubchenco, J., Kirincich, A., Weeks, H., Peterson, W. and Menge, B. 2008. Emergence of anoxia in the California current large marine ecosystem. Science 319: 920.

Cheung, W.W.L., Pitcher, T. and Pauly, D. 2005. Fuzzy logic expert system to estimate intrinsic extinction vulnerabilities of marine fishes to fishing. Biol. Conserv. 124: 97–111.

Cheung, W.W.L., Watson, R., Morato, T. and Pauly, D. 2007. Intrinsic vulnerability in the global fish catch. Mar. Ecol. Prog. Ser. 333: 1–12.

Cheung, W.W.L., Lam, V.W.Y., Sarmiento, J.L., Kearney, K., Watson, R., Zeller, D. and Pauly, D. 2010. Large-scale redistribution of maximum fisheries catch potential in the global ocean under climate change. Glob. Chang. Biol. 16: 24–35.

Cheung, W.W.L., Sarmiento, J.L., Dunne, J., Frolicher, T.L., Lam, V.W.Y., Deng Palomares, M.L., Watson, R. and Pauly, D. 2012. Shrinking of fishes exacerbates impacts of global ocean changes on marine ecosystems. Nat. Clim. Chang. 3: 254–258.

Christiansen, J.S., Mecklenburg, C.W. and Karamushko, O.V. 2014. Arctic marine fishes and their fisheries in light of global change. Global Change 20: 352–359.

Clark, M.R. 1999. Fisheries for orange roughy (*Hoplostethus atlanticus*) on seamounts in New Zealand. Oceanol Acta 22: 593–602.

Clark, M.R., Tittensor, D., Rogers, A.D., Brewin, P., Schlacher, T., Rowden, A. and Stocks, K. and Consalvey, M. 2006. Seamounts, deep-sea corals and fisheries: vulnerability of deep-sea corals to fishing on seamounts beyond areas of national jurisdiction. UNEPWCMC, Cambridge, UK.

Clark, M.R. and Koslow, J.A. 2007. Impacts of fisheries on seamounts. pp. 413–441. *In*: Pitcher, T.J., Morato, T., Hart, P.J.B., Clark, M.R., Haggan, N. and Santos, R.S. (eds.). Seamounts: Ecology, Fisheries, and Conservation. Blackwell Fisheries and Aquatic Resources Series 12, Blackwell Publishing, Oxford.

Clark, M.R., Vinnichenko, V.I., Gordon, J.D.M., Beck-Bulat, G.Z., Kukharev, N.N. and Kakora, A.F. 2007. Large scale distant water trawl fisheries on seamounts. pp. 361–399. *In*: Pitcher, T.J., Morato, T., Hart, P.J.B., Clark, M.R., Haggan, N. and Santos, R.S. (eds.). Seamounts: Ecology, Fisheries, and Conservation. Blackwell Fisheries and Aquatic Resources Series 12, Blackwell Publishing, Oxford.

Clark, M.R. and Rowden, A.A. 2009. Effect of deepwater trawling on the macro-invertebrate assemblages of seamounts on the Chatham Rise, New Zealand. Deep-Sea Res. Pt. I 56: 1540–1554.

Colaço, A., Giacomello, E., Porteiro, F. and Menezes, G.M. 2013. Trophodynamic studies on the Condor seamount (Azores, Portugal, North Atlantic). Deep-Sea Res. Pt I, vol. 98 Part A 178–189.

Company, J.B., Puig, P., Sarda, F., Palanques, A., Latasa, M. and Scharek, R. 2008. Climate influence on deep sea populations. PLoS ONE 1: e1431.

Costello, M.J., Cheung, A. and De Hauwere, N. 2010. Surface area and the seabed area, volume, depth, slope, and topographic variation for the world's seas, oceans, and countries. Environ. Sci. Technol. 44: 8821–28.

Costello, M.J., Smith, M. and Fraczek, W. 2015. Correction to surface area and the seabed area, volume, depth, slope, and topographic variation for the world's Seas, Oceans, and Countries. Environmental Science & Technology 49(11): 7071–7072.

Croxall, J.P. 1992. Southern-Ocean environmental changes effects on seabird, seal and whale populations. Philos. T. Roy. Soc. B 1285: 319–328.

Cunha, R.L., Coscia, I., Madeira, C., Mariani, S. and Stefanni, S. 2012. Ancient divergence in the trans-oceanic deep-sea shark Centroscymnus crepidater. PLoS One 7: e49196.

Danovaro, R., Dell'Anno, A., Fabiano, M., Pusceddu, A. and Tselepides, A. 2001. Deep-sea ecosystem response to climate changes: the eastern Mediterranean case study. Trends Ecol. Evol. 16: 505–510.

Danovaro, R., Dell'Anno, A. and Pusceddu, A. 2004. Biodiversity response to climate change in a warm deep sea. Ecol. Lett. 7: 821–828.

Daufresne, M., Lengfellner, K. and Sommer, U. 2009. Global warming benefits the small in aquatic ecosystems. P. Natl. Acad. Sci. USA 106(31): 12788–12793.

DeLeo, F.C., Drazen, J.C., Vetter, E.W., Rowden, A.A. and Smith, C.R. 2012. The effects of submarine canyons and the oxygen minimum zone on deep-sea fish assemblages off Hawai'i. Deep-Sea Res. Pt I 64: 54–70.

Devine, J.A., Baker, K.D. and Haedrich, R.L. 2006. Deep-sea fishes qualify as endangered. A shift from shelf fisheries to the deep sea is exhausting late-maturing species that recover only slowly. Nature 439: 29.

Dulvy, N.K., Rogers, S.I., Jennings, S., Stelzenmüller, V., Dye, S.R. and Skjoldal, H.R. 2008. Climate change and deepening of the North Sea fish assemblage: a biotic indicator of warming seas. J. Appl. Ecol. 45: 1029–1039.

Dulvy, N.K. and Forrest, R.E. 2010. Life histories, population dynamics, and extinction risks in chondrichthyans. pp. 635–676. *In:* Carrier, J.C., Musick, J.A. and Heithaus, M.R. (eds.). Sharks and their Relatives. Biodiversity, Adaptive Physiology and Conservation, Vol. II, Boca Raton, FL: CRC Press.

EC. 2008a. Council Regulation (EC) No. 40/2008 of 16 January 2008 fixing for 2008 the fishing opportunities and associated conditions for certain fish stocks and groups of fish stocks, applicable in Community waters and, for Community vessels, in waters where catch limitations are required. Brussels. Official Journal of the European Union L 19: 23 January 2008.

EC. 2008b. Council Regulation (EC) No. 734/2008 of 16 July 2008 on the protection of vulnerable marine ecosystems in the high seas from the adverse impacts of bottom fishing gears. Brussels. Official Journal of the European Union L 201: 8–13, 30 July 2008.

EC. 2010. Facts and figures on the Common Fisheries Policy—Basic statistical data—European Commission, 2010 Edition, Luxembourg: Publications Office of the European Union.

FAO. 1995. Precautionary approach to fisheries. Part 1: Guidelines on the precautionary approach to capture fisheries and species introductions. Elaborated by the Technical Consultation on the Precautionary Approach to Capture Fisheries (Including Species Introductions). Lysekil, Sweden, 6–13 June 1995. FAO Fisheries Technical Paper. No. 350, Part 1. Rome.

FAO. 2003. Fisheries Department. The ecosystem approach to fisheries. FAO Technical Guidelines for Responsible Fisheries 4(Suppl. 2).

FAO. 2008. Report of the Technical Consultation on International Guidelines for the Management of Deep-sea Fisheries in the High Seas, Rome, 4–8 February 2008 and 25–29 August 2008. FAO Fisheries and Aquaculture Report.

FAO. 2009. International guidelines for the management of deep-sea fisheries in the High Seas. FAO, Rome.

FAO. 2011. International Guidelines on Bycatch Management and Reduction of Discards. Food and Agriculture Organisation of the United Nations, Rome, Italy, 86 pp.

FAO. 2015. Voluntary Guidelines for Flag State Performance. Food and Agriculture Organisation of the United Nations, Rome, Italy, 53 pp.

Fenton, G.E., Short, S.A. and Ritz, D.A. 1991. Age determination of orange roughy, *Hoplostethus atlanticus* (Pisces: Trachichthyidae) using z °Pb: 226Ra disequilibria. Mar. Biol. 109: 197–202.

Francis, R.I.C.C. and Horn, P.L. 1997. Transition zone in otoliths of orange roughy (*Hoplostethus atlanticus*) and its relationship to the onset of maturity. Mar. Biol. 129: 681–687.

Froese, R. and Pauly, D. Editors. 2015. FishBase. World Wide Web electronic publication. www. fishbase.org, version (08/2015).

Gage, J.D. and Tyler, P.A. 1991. Deep-sea biology: a natural history of organisms at the deep-sea floor. Cambridge University Press, Cambridge.

Gianni, M. 2004. High Seas Bottom Trawl Fisheries and their Impacts on the Biodiversity of Vulnerable Deep-Sea Ecosystems: Options for International Action. IUCN, Gland, Switzerland.

Gilly, W.F., Beman, M., Litvin, S.Y. and Robison, B.H. 2013. Oceanographic and biological effects of shoaling of the Oxygen Minimum Zone. Annu. Rev. Mar. Sci. 5: 393–420.

Gjerde, K.M. 2012. The environmental provisions of the LOSC for the high seas and seabed area beyond national jurisdiction. Int. J. Mar. Coast. Law 27: 839–847.

GOC. 2013. Policy Options Paper # 7: MPAs: Protecting high seas biodiversity. 3rd meeting of the Global Ocean Commission, November 2013.

GOC. 2014. From decline to recovery. A rescue package for the global ocean. Global Ocean Commission Report 2014, 47 pp.

Gordon, J.D.M. 2001. Deep-water fisheries at the Atlantic Frontier. Cont. Shelf Res. 21(8-10): 987–1003.

Gordon, J.D.M., Bergstad, O.A., Figueiredo, I. and Menezes, G. 2003. Deep-water Fisheries in the Northeast Atlantic: I Description and current trends. Journal of Northwest Atlantic Fishery Science 31: 137–150.

Graham, K.J., Andrew, N.L. and Hodgson, K.E. 2001. Changes in relative abundance of sharks and rays on Australian South East Fishery trawl grounds after twenty years of fishing. Marine and Freshwater Research 52: 549–561.

Gregory, J.M., Dixon, K.W., Stouffer, R.J., Weaver, A.J., Driesschaert, E., Eby, M., Fichefet, T., Hasumi, H., Hu, A., Jungclaus, J.H., Kamenkovich, I.V., Levermann, A., Montoya, M., Murakami, S., Nawrath, S., Oka, A., Sokolov, A.P. and Thorpe, R.B. 2005. A model intercomparison of changes in the Atlantic thermohaline circulation in response to increasing atmospheric CO_2 concentration. Geophys. Res. Lett. 32, L12703, doi:10.1029/2005GL023209.

Guinotte, J.M., Orr, J., Cairns, S., Freiwald, A., Morgan, L. and George, R. 2006. Will human-induced changes in seawater chemistry alter the distribution of deep-sea scleractinian corals? Frontiers in Ecology and the Environment 4(3): 141–146.

Haedrich, R.L., Merrett, N.R. and O'Dea, N.R. 2001. Can ecological knowledge catch up with deep-water fishing? A North Atlantic perspective. Fish. Res. 51(2-3): 113–122.

Harris, P.T., Macmillan-Lawler, M., Rupp, J. and Baker, E.K. 2014. Geomorphology of the oceans. Mar. Geol. 352: 4–24.

Henry, L.A., Navas, J.M., Hennige, S.J., Wicks, L.C., Vad, J. and Roberts, J.M. 2013. Cold-water coral reef habitats benefit recreationally valuable sharks. Biol. Conserv. 161: 67–70.

High Seas Task Force. 2006. Closing the net: Stopping illegal fishing on the high seas. Governments of Australia, Canada, Chile, Namibia, New Zealand, and the United Kingdom, WWF, IUCN and the Earth Institute at Columbia University.

Hoffmann, M., Hilton-Taylor, C., Angulo, A., Böhm, M., Brooks, T.M., Butchart, S.H., Carpenter, K.E., Chanson, J., Collen, B., Cox, N.A., Darwall, W.R., Dulvy, N.K., Harrison, L.R., Katariya, V., Pollock, C.M., Quader, S., Richman, N.I., Rodrigues, A.S., Tognelli, M.F., Vié, J.C., Aguiar, J.M., Allen, D.J., Allen, G.R., Amori, G., Ananjeva, N.B., Andreone, F., Andrew, P., Aquino Ortiz, A.L., Baillie, J.E., Baldi, R., Bell, B.D., Biju, S.D., Bird, J.P., Black-Decima, P., Blanc, J.J., Bolaños, F., Bolivar-G, W., Burfield, I.J., Burton, J.A., Capper, D.R., Castro, F., Catullo, G., Cavanagh, R.D., Channing, A., Chao, N.L., Chenery, A.M., Chiozza, F., Clausnitzer, V., Collar, N.J., Collett, L.C., Collette, B.B., Cortez Fernandez, C.F., Craig, M.T., Crosby, M.J., Cumberlidge, N., Cuttelod, A., Derocher, A.E., Diesmos, A.C., Donaldson, J.S., Duckworth, J.W., Dutson, G., Dutta, S.K., Emslie, R.H., Farjon, A., Fowler, S., Freyhof, J., Garshelis, D.L., Gerlach, J., Gower, D.J., Grant, T.D., Hammerson, G.A., Harris, R.B., Heaney, L.R., Hedges, S.B., Hero, J.M., Hughes, B., Hussain, S.A., Icochea, M.J., Inger, R.F., Ishii, N., Iskandar, D.T., Jenkins, R.K., Kaneko, Y., Kottelat, M., Kovacs, K.M., Kuzmin, S.L., La Marca, E., Lamoreux, J.F., Lau, M.W., Lavilla, E.O., Leus, K., Lewison, R.L., Lichtenstein, G., Livingstone, S.R., Lukoschek, V., Mallon, D.P., McGowan, P.J., McIvor, A., Moehlman, P.D., Molur, S., Muñoz Alonso, A., Musick, J.A., Nowell, K., Nussbaum, R.A., Olech, W., Orlov, N.L., Papenfuss, T.J., Parra-Olea, G., Perrin, W.F., Polidoro, B.A., Pourkazemi, M., Racey, P.A., Ragle, J.S., Ram, M., Rathbun, G., Reynolds, R.P., Rhodin, A.G., Richards, S.J., Rodríguez, L.O., Ron, S.R., Rondinini, C., Rylands, A.B., Sadovy de Mitcheson, Y., Sanciangco, J.C., Sanders, K.L., Santos-Barrera, G., Schipper, J., Self-Sullivan, C., Shi, Y., Shoemaker, A., Short, F.T., Sillero-Zubiri, C., Silvano, D.L., Smith, K.G., Smith, A.T., Snoeks, J., Stattersfield, A.J., Symes, A.J., Taber, A.B., Talukdar, B.K., Temple, H.J., Timmins, R., Tobias, J.A., Tsytsulina, K., Tweddle, D., Ubeda, C., Valenti, S.V., van Dijk, P.P., Veiga, L.M., Veloso, A., Wege, D.C., Wilkinson, M., Williamson, E.A., Xie, F., Young, B.E., Akçakaya, H.R., Bennun, L., Blackburn, T.M., Boitani, L., Dublin, H.T., da Fonseca, G.A., Gascon, C., Jr. Lacher, T.E., Mace, G.M., Mainka, S.A., McNeely, J.A.,

Mittermeier, R.A., Reid, G.M., Rodriguez, J.P., Rosenberg, A.A., Samways, M.J., Smart, J., Stein, B.A. and Stuart, S.N. 2010. The impact of conservation on the status of the world's vertebrates. Science 330: 1503–1509.

Horn, P.L., Tracey, D.M. and Clark, M.R. 1998. Between-area differences in age and length at first maturity of the orange roughy *Hoplostethus atlanticus*. Mar. Biol. 132: 187–194.

Hughes, D.J. and Narayanaswamy, B.E. 2013. Impacts of climate change on deep-sea habitats. MCCIP Science Review 2013: 204–210.

International Maritime Organization (IMO). 1992. Resolution A.720(17)—Guidelines for the designation of special areas and the identification of particularly sensitive areas, 9 January 1992. Available at: http://www.imo.org/en/KnowledgeCentre/IndexofIMOResolutions/Pages/Assembly-%28A%29.aspx

International Maritime Organization (IMO). 2000. Resolution A.885 (21)—Procedures for the identification of particularly sensitive sea areas and the adoption of associated protective measures and amendments to the guidelines contained in resolution A.720(17), 4 February 2000. Available at: http://www.imo.org/en/KnowledgeCentre/IndexofIMOResolutions/Pages/Assembly-%28A%29.aspx

International Maritime Organization (IMO). 2002. Resolution A.927 (22)—Guidelines for the designation of special areas under MARPOL 73/78 and guidelines for the identification and designation of particularly sensitive sea areas, 15 January 2002. Available at: http://www.imo.org/en/KnowledgeCentre/IndexofIMOResolutions/Pages/Assembly-%28A%29.aspx

IPCC. 2014. Climate Change 2014: Synthesis Report. Contribution of Working Groups I, II and III to the Fifth Assessment Report of the Intergovernmental Panel on Climate Change [Core Writing Team, R.K. Pachauri and L.A. Meyer (eds.)]. IPCC, Geneva, Switzerland, 151 pp.

Jennings, J., Lee, J. and Hiddink, G. 2012. Assessing fishery footprints and the trade-offs between landings value, habitat sensitivity, and fishing impacts to inform marine spatial planning and an ecosystem approach. ICES J. Mar. Sci. 69(6): 1053–1063.

Jones, D.O.B., Yool, A., We, C.-L., Henson, S.A., Ruhl, H.A., Watson, A. and Gehlen, M. 2014. Global reductions in seafloor biomass in response to climate change. Global Change Biology 20: 1861–1872.

Keeling, R.F., Körtzinger, A. and Gruber, N. 2010. Ocean deoxygenation in a warming world. Annu. Rev. Mar. Sci. 2: 199–229.

Kitchingman, A., Lai, S., Morato, T. and Pauly, D. 2007. How many seamounts are there and where are they located? pp. 26–40. *In*: Pitcher, T.J., Morato, T., Hart, P.J.B., Clark, M.R., Haggan, N. and Santos, R.S. (eds.). Seamounts: Ecology, Fisheries, and Conservation. Blackwell Fisheries and Aquatic Resources Series 12, Blackwell Publishing, Oxford.

Koslow, J.A. 1997. Seamounts and the ecology of deep-sea fisheries. Am. Sci. 85: 168–176.

Koslow, J.A., Boehlert, G.W., Gordon, D.M., Haedrich, R.L., Lorance, P. and Parin, N. 2000. Continental slope and deep-sea fisheries: implications for a fragile ecosystem. ICES J. Mar. Sci. 57: 548–557.

Koslow, J.A., Gowlett-Holmes, K., Lowry, J.K., O'Hara, T., Poore, G.C.B. and Williams, A. 2001. Seamount benthic macrofauna off southern Tasmania: Community structure and impacts of trawling. Mar. Ecol. Prog. Ser. 213: 111–125.

Koslow, J., Goericke, R., Lara-Lopez, A. and Watson, W. 2011. Impact of declining intermediate-water oxygen on deepwater fishes in the California Current. Mar. Ecol. Prog Ser. 436: 207–218.

Kroeker, K.J., Kordas, R.L., Crim, R.N. and Singh, G.G. 2010. Meta-analysis reveals negative yet variable effects of ocean acidification on marine organisms. Ecol. Lett. 13: 1419–1434.

Kroeker, K.J., Kordas, R.L., Crim, R., Hendriks, I.E., Ramajo, L., Singh, G.S., Duarte, C.M. and Gattuso, J.P. 2013. Impacts of ocean acidification on marine organisms: quantifying ensitivities and interaction with warming. Glob. Chang. Biol. 19: 1884–96.

Lee, J., South, A.B. and Jennings, S. 2010. Developing reliable, repeatable, and accessible methods to provide high-resolution estimates of fishing-effort distributions from vessel monitoring system (VMS) data. ICES J. Mar. Sci. 67: 1260–1271.

Lehodey, P. and Grandperrin, R. 1996. Influence of temperature and ENS0 events on the growth of the deep demersal fish alfonsino, *Bevyx splendens*, off New Caledonia in the western tropical South Pacific Ocean. Deep-sea Res. Pt. I 43(1): 49–57.

Marchal, P., Poos, J.-J. and Quirijns, F. 2007. Linkage between fishers' foraging, market and fish stocks density: Examples from some North Sea fisheries. Fish. Res. 83: 33–43.

Matear, R.J. and Hirst, A.C. 2003. Long-term changes in dissolved oxygen concentrations in the ocean caused by protracted global warming. Glob. Biogeochem. Cycles 17: 1125 doi:10.1029/2002GB001997.

McAllister, D.E., Baquero, J., Spiller, G. and Campbell, R. 1999. A Global Trawling Ground Survey. A study carried out for the World Resources Institute, Marine Conservation Biology Institute and Ocean Voice International.

McClain, C.R., Allen, A.P., Tittensor, D.P. and Rex, M.A. 2012. Energetics of life on the deep seafloor. P. Natl. Acad. Sci. USA 109: 15366–15371.

McWhinnie, S.F. 2009. The tragedy of the commons in international fisheries: An empirical examination. J. Environ. Econ. Manage. 57: 321–333.

Melzner, F., Gutowska, M.A., Langenbuch, M., Dupont, S., Lucassen, M., Thorndyke, M.C., Bleich, M. and Pörtner, H.-O. 2009. Physiological basis for high CO_2 tolerance in marine ectothermic animals: pre-adaptation through lifestyle and ontogeny? Biogeosciences 6: 2313–2331.

Merret, N.R. and Haedrich, R.L. 1997. Deep-sea demersal fish and fisheries. London: Chapman and Hall.

Möllmann, C., Diekmann, R., Müller-Karulis, B., Kornilovs, G., Plikshs, M. and Axe, P. 2009. Reorganization of a large marine ecosystem due to atmospheric and anthropogenic pressure: a discontinuous regime shift in the Central Baltic Sea. Global Change Biology 15: 1377–1393.

Mora, C., Wei, C.-L., Rollo, A., Amaro, T., Baco, A.R., Billett, D., Bopp, L., Chen, Q., Collier, M., Danovaro, R., Gooday, A.J., Grupe, B.M., Halloran, P.R., Ingels, J., Jones, D.O.B., Levin, L.A., Nakano, H., Norling1, K., Ramirez-Llodra, E., Rex, M., Ruh, H.A., Smith, C.R., Sweetman, A.K., Thurber, A.R., Tjiputra, J.F., Usseglio, P., Watling, L., Wu, T. and Yasuhara, M. 2014. Biotic and human vulnerability to projected changes in ocean biogeochemistry over the 21st century. PLOS Biology 11(10): e1001682.

Morato, T., Watson, R., Pitcher, T. and Pauly, D. 2006. Fishing down the deep. Fish and Fisheries 7: 24–34.

Morgan, L.E., Norse, E.A., Rogers, A.D., Haedrich, R.L. and Maxwell, S.M. 2005. Why the world need a time-out on high-seas bottom trawling. The Deep Sea Conservation Coalition: 1–14.

Mortensen, P.B., Buhl-Mortensen, L. and Gordon, D.C. 2005. Evidence of fisheries damage to deep-water gorgonians in the Northeast Channel, Nova Scotia. *In*: Thomas, J. and Barnes, P. (eds.). Proceeding from the Symposium on the Effects of Fishing Activities on Benthic Habitats: Linking Geology, Biology, Socioeconomics and Management, American Fisheries Society, November 12–14, 2002, Florida, USA.

Moura, T., Jones, E., Clarke, M.W., Cotton, C.F., Crozier, P., Daley, R.K., Diez, G., Dobby, H., Dyb, J.E., Fossen, I., Irvine, S.B., Jakobsdottir, K., Lopez-Abellan, L.J., Lorance, P., Pascual-Alayon, P., Severino, R.B. and Figueiredo, I. 2014. Large-scale distribution of three deep-water squaloid sharks: integrating data on sex, maturity and environment. Fish. Res. 157: 47–61.

MRAG. 2005. Review of Impacts of Illegal, Unreported and Unregulated Fishing on Developing Countries. London: MRAG, Available at: http://www.dfid. gov.uk/pubs/files/illegal-fishing-mrag-report.pdf.

Murawski, S.A., Wigley, S.E., Fogarty, M.J., Rago, P.J. and Mountain, D.G. 2005. Effort distribution and catch patterns adjacent to temperate MPAs. ICES J. Mar. Sci. 62: 1150–1167.

NEAFC. 2009. Recommendation XIV: 2009. Recommendation by the North-east Atlantic Fisheries Commission in accordance with article 5 of the convention on future multilateral cooperation in North-east Atlantic Fisheries at its annual meeting on 10–14 November-2008 to adopt the following recommendation for the protection of vulnerable deep-water habitats in the NEAFC regulatory Area. London. North East Atlantic Fisheries Commission. NEAFC.

Neat, F.C., Burns, F., Jones, E. and Blasdale, T. 2015. The diversity, distribution and status of deep-water elasmobranchs in the Rockall Trough, north-east Atlantic Ocean. J. Fish Biol. 87: 1469–1488.

Nellemann, C., Hain, S. and Alder, J. (eds.). 2008. *In*: Dead Water—Merging of climate change with pollution, over-harvest, and infestations in the world's fishing grounds. United Nations Environment Programme, GRID-Arendal, Norway.

Netburn, A.N. and Koslow, J.A. 2015. Dissolved oxygen as a constraint on daytime deep scattering layer depth in the southern California Current Ecosystem. Deep-Sea Res. Pt I 06/2015; doi:10.1016/j.dsr.2015.06.006.

Norse, E.A., Brooke, S., Cheung, W.W.L., Clark, R.M., Ekeland, I., Froese, R., Gjerde, K.M., Haedrich, R.L., Heppell, S.S., Morato, T., Morgan, L.E., Pauly, D., Sumaila, R. and Watson, R. 2012. Sustainability of deep-sea fisheries. Mar. Policy 36(2): 307–320.

OSPAR. 2003. Guidelines for the Identification and Selection of Marine Protected Areas in the OSPAR Maritime Area. OSPAR Convention for the Protection of the Marine Environment of the North-east Atlantic. Meeting of the OSPAR Commission, Bremen, 23–27 June 2003. Reference number 2003–17, A-4.44b(i), Annex 10. 8 pp.

OSPAR. 2008a. Memorandum of Understanding between the North East Atlantic Fisheries Commission (NEAFC) and the OSPAR Commission. Agreement 2008-4. London. OSPAR Commission.

OSPAR. 2008b. General outline of roadmap for further work on the Charlie Gibbs Fracture Zone (CGFZ)/Mid Atlantic Ridge proposal 2008/2009. Meeting of the OSPAR Commission. 23–27 June 2008. Brest. France. Annex 10. Ref. 7.24d. 1 pp.

Pauly, D. and Zeller, D. (eds.). 2015. Sea Around Us Concepts, Design and Data (seaaroundus.org).

Perry, A.L., Low, P.J., Ellis, J.R. and Reynolds, J.D. 2005. Climate change and distribution shifts in marine fishes. Science 308: 1912–1915.

Pham, C.K., Diogo, H., Menezes, G., Porteiro, F., Braga-Henriques, A., Vandeperre, F. and Morato, T. 2014. Deep-water longline fishing has reduced impact on Vulnerable Marine Ecosystems. Scientific Reports 4: 4837. doi:10.1038/srep04837.

Pham, C.K., Vandeperre, F., Menezes, G., Porteiro, F., Isidro, E. and Morato, T. 2015. The importance of deep-sea vulnerable marine ecosystems for demersal fish in the Azores. Deep-Sea Res. Pt. I 96: 80–88.

Pitcher, T.J., Clark, M.R., Morato, T. and Watson, R. 2010. Seamount fisheries: Do they have a future? Oceanography 23(1): 134–144.

Poloczanska, E.S., Brown, C.J., Sydeman, W.J., Kiessling, W., Schoeman, D.S., Moore, P.J., Brander, K., Bruno, J.F., Buckley, L.B., Burrows, M.T., Duarte, C.M., Halpern, B.S., Holding, J., Kappel, C.V., O'Connor, M.I., Pandolfi, J.M., Parmesan, C., Schwing, F., Thompson, S.A. and Richardson, A.J. 2013. Global imprint of climate change on marine life. Nature Clim. Change 3: 919–925.

Pörtner, H.-O. 2008. Ecosystem effects of ocean acidification in times of ocean warming: a physiologist's view. Mar. Ecol. Prog. Ser. 373: 203–217.

Pörtner, H.O. and Peck, M.A. 2010. Climate change effects on fishes and fisheries: towards a cause-and-effect understanding. J. Fish Biol. 77: 1745–1779.

Probert, P.K., Christiansen, S., Gjerde, K.M., Gubbay, S. and Santos, R.S. 2007. Management and conservation of seamounts. pp. 442–475. *In*: Pitcher, T.J., Morato, T., Hart, P.J.B., Clark, M.R., Haggan, N. and Santos, R.S. (eds.). Seamounts: Ecology, Fisheries, and Conservation. Blackwell Fisheries and Aquatic Resources Series 12, Blackwell Publishing, Oxford.

Ramirez-Llodra, E., Brandt, A., Danovaro, R., De Mol, B., Escobar, E., German, C.R., Levin, L.A., Martinez Arbizu, P., Menot, L., Buhl-Mortensen, P., Narayanaswamy, B.E., Smith, C.R., Tittensor, D.P., Tyler, P.A., Vanreusel, A. and Vecchione, M. 2010. Deep, diverse and definitely different: unique attributes of the world's largest ecosystem. Biogeosciences 7: 2851–2899.

Ramirez-Llodra, E., Tyler, P.A., Baker, M.C., Bergstad, O.A., Clark, M.R., Escobar, E., Levin, L.A., Menot, L., Rowden, A.A., Smith, C.R. and Van Dover, C.L. 2011. Man and the last great wilderness: Human impact on the deep sea. PLoS ONE 6: e22588.

Rex, M.A., Etter, R.J., Morris, J.S., Crouse, J., McClain, C.R., Johnson, N.A., Stuart, C.T., Deming, J.W., Thies, R. and Avery, R. 2006. Global bathymetric patterns of standing stock and body size in the deep-sea benthos. Mar. Ecol. Prog. Ser. 317: 1–8.

Roark, E.B., Guilderson, T.P., Dunbar, R.B. and Ingram, B.L. 2006. Radiocarbon-based ages and growth rates of Hawaiian deep-sea corals. Mar. Ecol. Prog. Ser. 327: 1–14.

Roberts, C.M. 2002. Deep impact: the rising toll of fishing in the deep sea. Trends Ecol. Evol. 17(5): 242–245.

Roberts, J.M., Wheeler, A.J., Freiwald, A. and Cairns, S.D. 2009. Cold-Water Corals: The Biology and Geology of Deep-Sea Coral Habitats. Cambridge University Press.

Rochette, J., Billé, R., Molenaar, E.J., Drankier, P. and Chabason, L. 2015a. Regional oceans governance mechanisms: A review. Mar. Policy 60: 9–19.

Rochette, J. and Wright, G. 2015. Developing area-based management tools in areas beyond national jurisdiction: possible options for the Western Indian Ocean. IDDRI, Working Paper no. 6, Paris, 2015.

Rochette, J., Wright, G., Gjerde, K.M., Greiber, T., Unger, S. and Spadone, A. 2015b. A new chapter for the high seas? Historic decision to negotiate an international legally binding instrument on the conservation and sustainable use of marine biodiversity in areas beyond national jurisdiction. IDDRI, Working Paper No. 2, Paris, 2015.

Rogers, A.D. 1994. The biology of seamounts. Adv. Mar. Biol. 30: 305–350.

Rogers, A.D. and Gianni, M. 2010. The Implementation of UNGA Resolutions 61/105 and 64/72 in the Management of Deep-Sea Fisheries on the High Seas. Report prepared for the Deep-Sea Conservation Coalition. International Programme on the State of the Ocean, London, United Kingdom.

Rogers, A.D. 2015. Environmental change in the deep ocean. Annu. Rev. Env. Resour. 40: 1–38.

Rowden, A.A., Dower, J.F., Schlacher, T.A., Consalvey, M. and Clark, M.R. 2010. Paradigms in seamount ecology: fact, fiction and future. Mar. Ecol. 31: 226–241.

Ruhl, H., Ellena, J. and Smith, K. 2008. Connections between climate, food limitation, and carbon cycling in abyssal sediment communities. P. Natl. Acad. Sci. USA 105: 17006–17011.

Ruhl, H.A. and Smith, K.L. 2004. Shifts in deep-sea community structure linked to climate and Food Supply. Science 305: 513–515.

Sheridan, J.A. and Bickford, D. 2011. Shrinking body size as an ecological response to climate change. Nat. Clim. Chang. 1: 401–406.

Simpfendorfer, C.A. and Kyne, P.M. 2009. Limited potential to recover from overfishing raises concerns for deep-sea sharks, rays and chimaeras. Environ. Conserv. 36: 97–103.

Smith, C.R., De Leo, F.C., Bernardino, A.F., Sweetman, A.K. and Arbizu, P.M. 2008. Abyssal food limitation, ecosystem structure and climate change. Trends Ecol. Evol. 23: 518–528.

Stauffer, B., Gellene, A., Schnetzer, A., Seubert, E., Oberg, C., Sukhatme, G. and Caron, D. 2012. An oceanographic, meteorological, and biological "perfect storm" yields a massive fish kill. Mar. Ecol. Prog. Ser. 468: 231–243.

Stone, G.S., Madin, L.P., Stocks, K., Hovermale, G., Hoagland, P., Schumacher, M., Etnoyer, P., Sotka, C. and Tausig, H. 2004. Seamount biodiversity, exploitation and conservation. pp. 41–70. In: Glover, L.K. and Earle, S.A. (eds.). Defying Ocean's End: An Agenda for Action. Island Press, Washington, DC.

Stramma, L., Schmidtko, S., Levin, L.A. and Johnson, G.C. 2010. Ocean oxygen minima expansions and their biological impacts. Deep-Sea Res. Pt. I 57: 587–595.

Stramma, L., Prince, E.D., Schmidtko, S., Luo, J., Hoolihan, J.P., Visbeck, M., Wallace, D.W.R., Brandt, P. and Körtzinger, A. 2011. Expansion of oxygen minimum zones may reduce available habitat for tropical pelagic fishes. Nat. Clim. Chang. 1–5.

Sumaila, U.R., Khan, A., Teh, L., Watson, R., Tyedmers, P. and Pauly, D. 2010. Subsidies to high seas bottom trawl fleets and the sustainability of deep-sea demersal fish stocks. Mar. Policy 34(3): 495–497.

Sumaila, U.R., Lam, V.W.Y., Miller, D.D., The, L., Watson, R.A., Zeller, D., Cheung, W.W.L., Côté, I.M., Rogers, A.D., Roberts, C., Sala, E. and Pauly, D. 2015. Winners and losers in a world where the high seas is closed to fishing. Scientific Reports 5: 8481.

ter Hofstede, R. and Rijnsdorp, A.D. 2011. Comparing demersal fish assemblages between periods of contrasting climate and fishing pressure. ICES J. Mar. Sci. 68(6): 1189–1198.

Thresher, R.E., Koslow, J.A., Morison, A.K. and Smith, D.C. 2007. Depth-mediated reversal of the effects of climate change on long-term growth rates of exploited marine fish. P. Natl. Acad. Sci. USA 104(18): 7461–7465.

Thurber, A.R., Sweetman, A.K., Narayanaswamy, B.E., Jones, D.O.B., Ingels, J. and Hansman, R.L. 2014. Ecosystem function and services provided by the deep sea. Biogeosciences 11: 3941–3963.

Tittensor, D.P., Baco, A.R., Hall-Spencer, J.M., Orr, J.C. and Rogers, A.D. 2010. Seamounts as refugia from ocean acidification for cold-water stony corals. Marine Ecology Special Issue: Recent Advances in Seamount Ecology 31(Supplement s1): 212–225.

Toropova, C., Meliane, I., Laffoley, D., Matthews, E. and Spalding, M. (eds.). 2010. Global Ocean Protection: Present Status and Future Possibilities. Brest, France: Agence des aires marines protégées, Gland, Switzerland, Washington, DC and New York, USA: IUCN WCPA, Cambridge, UK: UNEP-WCMC, Arlington, USA: TNC, Tokyo, Japan: UNU, New York, USA: WCS.

Troyanovsky, F.M. and Lisovsky, S.F. 1995. Russian (USSR) fisheries research in deep waters (below 500 m) in the North Atlantic. pp. 357–366. In: Hopper, A.G. (ed.). Deep-water Fisheries of the North-Atlantic Oceanic Slope. NATO ASI Series, Series E. Applied Sciences, vol. 296.

UNGA. 2000. United Nations Millennium Declaration, Resolution Adopted by the General Assembly, 18 September 2000, A/RES/55/2, Available at: http://daccess-dds-ny.un.org/doc/UNDOC/GEN/N00/559/51/PDF/N0055951.pdf?OpenElement.

UNGA. 2004. A/RES/59/25 - Sustainable fisheries, including through the 1995 Agreement for the Implementation of the Provisions of the United Nations Convention on the Law of the Sea of 10 December 1982 relating to the Conservation and Management of Straddling Fish Stocks and Highly Migratory Fish Stocks, and related instruments. Available at: http://daccess-dds-ny.un.org/doc/UNDOC/GEN/N04/477/70/PDF/N0447770.pdf?OpenElement.

UNGA. 2006. A/RES/61/105 - Sustainable fisheries, including through the 1995 Agreement for the Implementation of the Provisions of the United Nations Convention on the Law of the Sea of 10 December 1982 relating to the Conservation and Management of Straddling Fish Stocks and Highly Migratory Fish Stocks, and related instruments. Available at: http://daccess-dds-ny.un.org/doc/UNDOC/GEN/N06/500/73/PDF/N0650073.pdf?OpenElement.

UNGA. 2009. A/RES/64/72 - Sustainable fisheries, including through the 1995 Agreement for the Implementation of the Provisions of the United Nations Convention on the Law of the Sea of 10 December 1982 relating to the Conservation and Management of Straddling Fish Stocks and Highly Migratory Fish Stocks, and related instruments. Available at: http://daccess-dds-ny.un.org/doc/UNDOC/GEN/N09/466/15/PDF/N0946615.pdf?OpenElement.

UNGA. 2015. Transforming our world: the 2030 Agenda for Sustainable Development, Resolution adopted by the General Assembly on 25 September 2015, A/RES/70/1, Available at: http://www.un.org/en/ga/search/view_doc.asp?symbol=A/RES/70/1.

Vecchione, M., Bergstad, O.A., Byrkjedal, I., Falkenhaug, T., Gebruk, A.V., Godo, O.R., Gislason, A., Heino, M., Hoines, A.S., Menezes, G.M.M., Piatkowski, U., Priede, I.G., Skov, H., Søiland, H., Sutton, T. and de Lange Wenneck, T. 2010. Biodiversity patterns and processes on the Mid-Atlantic Ridge. pp. 103–121. *In*: McIntyre, A.D. (ed.). Life in the World's Oceans: Diversity, Distribution and Abundance. Oxford, Blackwell Publishing Ltd.

Waller, R., Watling, L., Auster, P. and Shank, T. 2007. Anthropogenic impacts on the Corner Rise seamounts, north-west Atlantic Ocean. J. Mar. Biol. Assoc. UK 87: 1075–1076.

Walther, G.R., Post, E., Convey, P., Menzel, A., Parmesan, C., Beebee, T.J.C., Fromentin, J.-M., Hoegh-Guldberg, O. and Bairlein, F. 2002. Ecological responses to recent climate change. Nature 416: 389–395.

Warner, R., Verlaan, P. and Lugten, G. 2012. An Ecosystem Approach to Management of Seamounts in the Southern Indian Ocean. Volume 3—Legal and Institutional Gap Analysis. Gland, Switzerland: IUCN.

Watling, L., Haedrich, R.L., Devine, J., Drazen, J., Dunn, M.R., Gianni, M., Baker, K., Cailliet, G., Figueiredo, I., Kyne, P.M., Menezes, G., Neat, F., Orlov, A., Duran, P., Perez, J.A., Ardron, J.A., Bezaury, J., Revenga, C. and Nouvian, C. 2011. Can Ecosystem-Based Deep-Sea Fishing Be Sustained? Report of a Workshop held 31 August—3 September 2010. Darling Marine Center Special Publication 11-1, University of Maine.

Weaver, P.P.E., Benn, A., Arana, P.M., Ardron, J.A., Bailey, D.M., Baker, K., Billett, D.S.M., Clark, M.R., Davies, A.J., Durán Muñoz, P., Fuller, S.D., Gianni, M., Grehan, A.J., Guinotte, J., Kenny, A., Koslow, J.A., Morato, T., Penney, A.J., Perez, J.A.A., Priede, I.G., Rogers, A.D., Santos, R.S. and Watling, L. 2011. The impact of deep-sea fisheries and implementation of the UNGA Resolutions 61/105 and 64/72. Report of an international scientific workshop, National Oceanography Centre, Southampton. http://hdl.handle.net/10013/epic.37995.

Wessel, P. 2007. Seamount characteristics. pp. 3–25. *In*: Pitcher, T.J., Morato, T., Hart, P.J.B., Clark, M.R., Haggan, N. and Santos, R.S. (eds.). Seamounts: Ecology, Fisheries, and Conservation. Blackwell Fisheries and Aquatic Resources Series 12, Blackwell Publishing, Oxford.

Wessel, P., Sandwell, D.T. and Kim, S.S. 2010. The global seamount census. Oceanography 23(1): 24–33.

White, B. 1987. Anoxic events and allopatric speciation in the deep sea. Biol. Oceanogr. 5: 243–259.

White, C. and Costello, C. 2014. Close the High Seas to Fishing? PLoS Biol 12: e1001826.

Wigham, B., Tyler, P. and Billett, D. 2003. Reproductive biology of the abyssal holothurian Amperima rosea: an opportunistic response to variable flux of surface derived organic matter? J. Mar. Biol. Assoc. UK 83: 175–188.

Williams, A., Schlacher, T.A., Rowden, A.A., Althaus, F., Clark, M.R., Bowden, D.A., Stewart, R., Bax, N.J., Consalvey, M. and Kloser, R.J. 2010. Seamount megabenthic assemblages fail to recover from trawling impacts. Marine Ecology Special Issue: Recent Advances in Seamount Ecology 31(Supplement s1): 183–199.

Wright, G., Ardron, J., Gjerde, K., Currie, D. and Rochette, J. 2015. Advancing marine biodiversity protection through regional fisheries management: A review of bottom fisheries closures in areas beyond national jurisdiction. Mar. Policy: 134–148.

Globalization of the Antarctic Seas
Pollution and Climate Change Perspectives

Rosalinda Carmela Montone,[1,*] *César de Castro Martins,*[2]
Marcos Henrique Maruch Tonelli,[1,a] *Tailisi Hoppe Trevizani,*[1,b]
Marcia Caruso Bícego,[1,c] *Rubens Cesar Lopes Figueira,*[1,d]
Ilana Elazari Klein Coaracy Wainer[1,e] and
Jorge E. Marcovecchio[3]

Introduction

The Antarctic continent is entirely surrounded by the Antarctic Ocean also named Southern Ocean, forming a barrier to the movement of organisms into and out of Antarctica. Therefore, the Antarctic Continent has unique characteristics because of its geographic isolation.

It is regarded as one of the remaining environments on the planet where human activities have little direct impact. Thus, this region has been considered an ideal observatory for research on global change and environmental impacts caused by man (Xuebin et al. 2006).

[1] Instituto Oceanográfico – Universidade de São Paulo (IOUSP – São Paulo – Brazil), Pça do Oceanográfico 191 – Cidade Universitária São Paulo – SP – Brasil – 05508-120.
[a] E-mail: mtonelli@usp.br
[b] E-mail: tailisi@usp.br
[c] E-mail: marciabicego@usp.br
[d] E-mail: rfigueira@usp.br
[e] E-mail: wainer@usp.br
[2] Centro de Estudos do Mar – UFPR – Caixa Postal: 61, AV. Beira Mar, S/N, Pontal do Paraná, PR, Brasil – CEP: 83255-976.
E-mail: ccmart@ufpr.br
[3] Instituto Argentino de Oceanografía (IADO – Bahía Blanca – Argentina).
E-mail: jorgemar@criba.edu.ar
* Corresponding author: rmontone@usp.br

The Southern Ocean and the Climate System

Presenting a straightforward definition for the Southern Ocean (SO) is somewhat difficult, since such a definition will be built on processes and mechanisms that vary from each field of research. Although finding a common ground regarding SO's southern limit can be fairly simple (since it is south–limited by the Antarctic Continent), there is no singular definition of a northern boundary. From a broader perspective the SO may be referred as the whole oceanic area surrounding Antarctica, but again this would not be suitable for scientific purposes. The Antarctic Treaty suggested the 60°S latitude to be this northern limit, but if one considers the oceanographic processes, this boundary would easily reach 30°S as that is the northernmost extent of the Subtropical Front (STF, Sokolov and Rintoul 2009).

All in all, should one bear in mind the circumpolar geometry south of 55°S, the global influence of the SO may be basically accredited to one aspect of its geomorphology: the Drake Passage sits in the only continuous zonal oceanic band where ocean waters circle the Earth, connecting the ocean basins and transferring climate anomalies between them (Rintoul et al. 2012b). The lack of continental boundaries allows the establishment of a dynamical barrier for the north–south exchange of heat above the height of the shallowest bathymetry that supports the present glacial climate of Antarctica: the Antarctic Circumpolar Current (ACC, Sokolov and Rintoul 2009). The ACC is the most important dynamic feature of the SO and consists of a westerly–driven clockwise geostrophic–balanced flux found approximately between 45°–55°S around Antarctica (Fig. 9.1; Trenberth et al. 1990). Carrying more water than any other current (147 Sv south of Australia (Rintoul and Sokolov 2001) and 137 Sv close to South Africa and South America (Cunningham et al. 2003)), the ACC has a complex structure consisting of multiple narrow frontal jets: maximum sea level height and current speed gradients can be found in specific lines, as the jets keep the water masses properties that define the fronts. In places where the ACC main flow extends far enough from the Antarctic Continent, large subpolar cyclonic cells of recirculating waters are established, where the two most prominent are the Weddell Gyre and the Ross Gyre (Fig. 9.1; Deacon 1979; Rodman and Gordon 1982; Reid 1997).

The cryosphere, featured in some of the most up-to-date scientific investigations, is one component of the climate system that plays a major part on modulating the Southern Ocean dynamics and the Meridional Overturning Circulation (MOC). The cryosphere components found in the Antarctic environment are largely responsible for most of the regional dynamics. Throughout almost the entire year, the Antarctic Continent is surrounded by a zone of frozen seawater 1 or 2 m thick (Lythe et al. 2001). By September, Antarctic sea ice covers an area of 19–20 x 10^6 km² spreading

Figure 9.1. A schematic representation of the current systems in the Southern Ocean. The Antarctic Circumpolar Current (ACC) flows from west to east around Antarctica in two major branches, the polar front and sub Antarctic front. The Weddell and Ross gyres fill the deep basins between the Antarctic Continent and the ACC. This clockwise gyres act as warm water source for the Weddell and Ross seas as they branch off waters from ACC and advect them into the continental shelf. Adapted from Tonelli (2014).

the first thin sea ice of the year close to 60°S around most of the continent and even farther 55°S at the Weddell Sea section of the Southern Ocean.

A huge amount of this ice will eventually melt during the austral summer shrinking to approximately 3 x 106 km² in March, i.e., less than 20% of its maximum annual extent (Fig. 9.2). But even the thin ice can impact the Antarctic climate. A 10 cm thick layer will strongly reduce the air-sea exchange of moisture and gases, lessening the air-sea heat transfer by 90% (Rintoul et al. 2012a). The widespread thin ice will also affect the albedo of the ocean. While the ice–free surface of the ocean can absorb about 90% of the incident solar radiation, depending on the thickness and snow cover, the ice surface can reflect a similar amount of this radiation, insulating cold waters underneath. However, the positive feedback takes place when the sea ice retreats; an ice–free ocean with reduced albedo absorbs radiation that warms the water and leads to further sea ice melting (Rintoul et al. 2012a).

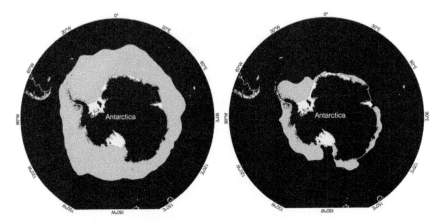

Figure 9.2. Maps of sea ice extent in the Antarctic winter (September – left panel) and summer (February – right panel). They represent averaged sea ice extent from 1979 to 2002/2003, based on ESMR–nimbus–5 satellite observations. Adapted from Tonelli (2014).

Due to much energy and momentum exchange with the lower atmospheric layers (Wallace and Hobbs 1977), the ocean redistributes incoming solar radiation from the tropics to the poles through a system known as the Global Conveyor Belt (Broecker 1987). Broecker suggested that the heat flux around the globe was performed by a large scale oceanic current system with two limbs; an upper limb, that absorbs heat around the tropics and releases it back to the atmosphere at the poles, and a lower limb resulting from the cold water downwelling at the North Atlantic (Fig. 9.3). Due to mass conservation, water from the lower limb flows along the ocean bottom to upwell again at the northern section of the Pacific and Indian Oceans. The Global Conveyor Belt was actually a representation of the resulting effect from the interaction of many currents with different temperatures, which created a vertical circulation pattern that ventilates the ocean bottom also known as the Thermohaline Circulation (THC). Although this representation was a first approach to explain the global oceanic heat transport, if one takes into account that 50% of the solar radiation at the top of the atmosphere ends up stored in the ocean, it gets clear that the ocean is indeed Earth's great climate modulator.

Schmitz (1996) presented an update for the THC scheme, highlighting not only the water masses interaction to create the heat transport distribution pattern, but also the Antarctic Continent and the Southern Ocean (SO), key roles in the process, since unique exchange processes and water transformations over the Antarctic continental shelf are responsible for the formation of the Antarctic Bottom Water (AABW); one of the main components of the THC's lower limb (Orsi et al. 2002). Finally, since the THC is driven not only by temperature and salinity gradients, but also by ocean-

Figure 9.3. Representation of the global conveyor belt as suggested by Broecker (1987) with a warm surface limb (red) that sinks at the northern Atlantic Ocean connecting to the deeper cold limb (blue) to ventilate the global ocean bottom to later upwell at the northern Indian and Pacific Oceans. Adapted from Tonelli (2014).

atmosphere-cryosphere momentum exchange, the more comprehensive term Meridional Overturning Circulation was ultimately adopted by the scientific community.

MOC impacts on the climate system go beyond Earth's energy budget. It also affects the global cycle of carbon and nutrients. Besides heat, the sinking of Southern Ocean surface waters carries oxygen and carbon dioxide into the ocean interior, being spread throughout the globe, renewing oxygen levels in the deep layers and boosting the capacity of the southern hemisphere oceans to store heat and carbon. Thanks to the MOC's conveyor–like mechanism, the ocean has stored more than 85% of the total increase in heat accumulated by the Earth system over the second half of the 20th century (most of this heat being trapped at the SO), preventing the energy excess to build up atmosphere warming or intensify ice melting (Levitus et al. 2005). Rintoul et al. (2012b) discuss that integrated around the globe, the Southern Ocean stores more of the excess heat trapped by the Earth system than any other latitude band.

As for the carbon dioxide, Sabine et al. (2004) suggested that about 40% of the total ocean inventory of anthropogenic CO_2 is found south of 30°S. This shows the efficiency with which the upper cell of the MOC transfers CO_2-full water from the surface into the deep ocean. The sinking of intermediate and bottom waters removes anthropogenic carbon dioxide from the atmosphere and traps it in the ocean. Reciprocally, deep carbon–rich water upwelling at high latitudes tends to release CO_2 back to the atmosphere so that wind-driven fluctuations of the MOC will eventually lead to changes in ocean uptake of CO_2 (Butler et al. 2007; Le Quéré et al. 2007; Verdy et al. 2007).

Should one look at the biology and chemistry of the global ocean, deep SO overturning influence will be found. The upwelling of deep water returns nutrients to the surface ocean, where, associated with incoming solar radiation, they become available to be used by phytoplankton. However, the upwelled nutrients are not fully consumed at the SO, and get transported to lower latitudes within the intermediate and mode waters. Numerical investigations by Sarmiento et al. (2004) showed that setting the MOC's export of nutrients to zero leads to a 75% reduction of the primary production in the rest of the ocean. This primary production also plays an important role on the carbon dioxide sequestration as organic material sinks from the surface ocean and decomposes in the deep sea (Tréguer and Jacques 1993). Thus, by means of either biological or chemical mechanisms, the balance between the upwelling and downwelling limbs of the MOC will ultimately determine how much carbon dioxide is absorbed and stored by the ocean.

Although it might seem evident that the cryosphere affects the SO, and therefore, the rest of the world, assessing this climate component within its own variability, its interactions with the Southern Ocean and further impacts, like sea level rise, is not an easy task. The freezing of the ocean surface during winter forms sea ice covering about 16 million km², which is actually larger than the area of the Antarctic Continent (Fig. 9.2; Convey et al. 2009). Associated changes in sea ice extent or volume can lead to changes in many aspects: deep water mass formation rates, air-sea exchange of gases such as carbon dioxide, the trophic web from phytoplankton to superior animals and even the Earth's albedo on a planetary scale (Rintoul et al. 2012a). Warm waters advected to the inner continental shelves induce basal melting of ice shelves modifying high-latitude freshwater budget and stratification, finally altering the mass balance of the Antarctic ice sheet and the rate at which glacial ice flows into the sea (Rignot et al. 2004).

Given the unmistakable worldwide influence of the Southern Ocean, changes in this peculiar environment must be traced, assessed and uncovered. Changes in the ocean–atmosphere–cryosphere, which are already underway, coupling mechanisms like MOC's heat transport, sea ice extent, ice shelf melting, freshwater balance will produce positive feedbacks that will enhance the climate changing scenario (Turner 2010).

Evidences for the critical role played by the Southern Ocean in global budgets of heat, freshwater, carbon and nutrients cannot be neglected and, as the indications that the Southern Ocean is changing seem to pile up, we can only hope that the science–politics understanding will thrive to enhance the quality of Antarctic research.

Contaminants in the Antarctic Marine Environment

The Antarctic ecosystem is particularly sensitive to anthropic modifications, due to the latter being concentrated in ice free areas, which represents less than 2% of the continent (Santos et al. 2005). Furthermore, Antarctic marine biota has characteristics that leave it highly vulnerable to environmental contamination, such as low fertility, low larval dispersion capacity, short reproductive season and low growth rates during critical life stages (King and Riddle 2001; Stark et al. 2003b; Santos et al. 2005). All of these characteristics reflect the high degree of endemism and ecophysiological adaptations that Antarctic species developed over an isolated evolutive process, leading to the creation of unique communitary structures, with greater efficiency and facilitating the comprehension of contaminants transfer rates between trophic levels (Nygard et al. 2001; Santos et al. 2006).

There was an increase of human activities on the continent since the International Geophysical Year (1957), and the ratification of the Antarctic Treaty (1961). The Council of Managers of National Antarctic Programs (COMNAP) currently lists 103 national research facilities (41 opened all year round and 62 seasonally only, typically opened every summer or most summers) operated by 30 countries. The total capacity of the accommodation established for national Antarctic research programs is at least 4000, and much of this is occupied during the summer period. In winter there are typically approximately 1000 people in Antarctica maintaining station facilities and undertaking the science activity that occurs during this time of year (Poland et al. 2003).

It was the consolidation of political and scientific interests and a substantial increase in the number of permanent research stations, particularly on King George Island (Table 9.1), leading to the creation of environmental protocols, with the aim of preventing pollution (Vodopivez et al. 2001; Santos et al. 2005).

Thus, the Antarctic environment has been affected by the development of research and increased tourism in recent decades, introducing fuel oil, sewage and metals as a result of local activities, but also by the presence of Persistent Organic Pollutants (POPs) predominantly from global anthropogenic activities.

Hydrocarbons in Antarctica

Hydrocarbons such as *n*-alkanes, isoprenoid hydrocarbons and Polycyclic Aromatic Hydrocarbons (PAHs) are the main components of petroleum and its derivatives. PAHs are an important group of marine environmental contaminants, having anthropogenic sources including atmospheric deposition of compounds from incomplete combustion of fossil fuels (NRC 1985). The carcinogenic properties of some PAHs have raised an interest in the study of their sources, distribution, transport mechanisms, environmental impact and destination (Neff 2002; Kasiotis and Emmanouil 2015).

Table 9.1. Bases and Research Stations in the King George Island - Antarctica.

	Base/Research station	Para Latitude	Para Longitude	Country	Established
1	Carlini Base (Ex-Jubany)	62°14'16.7"S	58°40'0.2"W	Argentina	1953
2	Bellingshausen Station	62°11'47"S	58°57'39"W	Russia	1968
3	Base Presidente Eduardo Frei Montalva and Villa Las Estrellas	62°11.7'S	58°58.7'W	Chile	1969
4	Henryk Arctowski Polish Antarctic Station	62°09'0.14"S	058°28'2.1"W	Poland	1977
5	Comandante Ferraz Antarctic Station	62°05'00"S	58°23'28.2"W	Brazil	1984
6	Artigas Base	62°11'3.4"S	58°54'11.9"W	Uruguay	1984
7	King Sejong Station	62°13'23.2"S	58°47'13.4"W	South Korea	1988
8	Machu Picchu Research Station	62°05'29.9"S	58°28'15.4"W	Peru	1989
9	Professor Julio Escudero Base	62°12'4.2"S	58°57'45.3"W	Chile	1994

https://en.wikipedia.org/wiki/Research_stations_in_Antarctica#Research_stations.

Hydrocarbons can be introduced to Antarctic environment recurrently or accidentally during scientific support activities, ship operations, incineration and tourism (Cripps 1991, 1992; Kennicutt et al. 1995; UNEP 2002; Bicego et al. 2009; Klein et al. 2012). According to COMNAP (2016), there are about 4462 people working at research stations in the summer at Antarctica. During winter, 38 stations still operate and provide space for about 1000 people. In the summer season 2014/15, tourism brought to the Antarctic region 36,702 people (including paying passengers, crew and staff) (IAATO 2016). All kinds of transportation and energy in these activities need oil derivatives, which makes the region susceptible to localized oil contamination (Priddle 2002).

Accidents with oil spills have occurred in the area, such as the Bahia Paraiso shipwreck in 1989 (Cripps 1992), the fuel spill at the Faraday Research Station in 1992 (Cripps and Shears 1997), the fire at the Brazilian Antarctic Station in 2012 (Colabuono et al. 2015) and others (Ruopollo et al. 2013). Chemical contamination from abandoned sites and past fuel spills are yet another source of contamination in Antarctica (Deprez et al. 1999; Tin et al. 2009).

Studies about hydrocarbons in the marine environment have been carried out in the Antarctic region and have showed the influence of anthropic activities. Some of these studies have revealed the existence of hydrocarbon concentrations in the vicinities of research stations (e.g.,

Kennicutt et al. 1991; Cripps 1992; Kennicutt et al. 1992; 1995; Kim et al. 2006; Cincinelli et al. 2007; Bícego et al. 2009; Prus et al. 2015). A study conducted near the McMurdo station, which is the largest Antarctic research facility, reported hydrocarbon concentrations in sediments comparable to the most polluted harbors in temperate regions (Lenihan 1992). The study also found drastic changes in the biota due to contamination. Polycyclic aromatic hydrocarbons (PAHs) and Aliphatic Hydrocarbons (AHs) were detected in seawater, sediment and biota at Admiralty Bay (Bícego et al. 1996; 1998; 2003; 2009; Martins et al. 2004; 2010; Colabuono et al. 2015; Montone et al. 2016) as a result from fossil-fuel consumption, combustion of organic matter and petroleum derivatives mainly from the research stations in the area. Cripps (1992) reported a low level of hydrocarbon contaminated seawater and sediment at a British research station at Signy station, South Orkney Islands. In Clark Peninsula a monitoring of the contamination extent showed that hydrocarbons in sediment and water were below quantitation limits in most samples (Fryirs et al. 2015). Finally, recent studies developed in sediments of Potter Cove, located in Maxwell Bay, King George Island, at the vicinities of the Argentine Carlini research station, presented hydrocarbons levels less than or within the range of those found near other research stations (Dauner et al. 2015).

Local data and laboratory experiments showed biological effects of anthropic hydrocarbons in marine Antarctic organisms (e.g., Lenihan 1992; Regoli et al. 2005; Lister et al. 2015; Mearns et al. 2015; Polmear et al. 2015).

The increase of research and tourism activities in Antarctica have been exposing the region to hydrocarbons, thus continuous attention to the levels of anthropic hydrocarbons regarding the potential influence on the local biota is necessary.

Sewage Contribution to Antarctica

In Antarctica, waste can be released into the sea both by ships and research stations, representing an exception of the Protocol on Environmental Protection to the Antarctic Treaty, which ascertains the removal of all residue generated in Antarctica. Waste disposal is usually discharged below sea level and after maceration, and the disposal in ice-free or freshwater areas being strictly forbidden. The prohibition of effluent discharge on ice-free areas is due to the fact that such areas are poor in nutrients, thus would significantly alter the biogeochemical balance of elements by the addition of high nutrient effluent.

The extension of waste pollution in Antarctica is usually influenced by the number of people at the station, the efficiency of the sewage treatment plant and the physiographic features that affect the effluent dispersion and dilution of the effluent. In general, the impact of sewage input in Antarctic regions tends to be more often located near the station's sewage outfalls (Aronson et al. 2011).

Grondahl et al. (2009) presented an inventory of sewage treatment systems accounting 71 Antarctic stations and verifying that 37% of the permanent stations and 69% of the summer stations lack any type of wastewater treatment. Nevertheless, some larger stations have been implementing wastewater management procedures following their national standards for water quality. The processing commonly includes maceration, screening, followed by aerobic biological treatment and in some cases ultraviolet (UV) disinfection. Hence, the raw sewage load is reduced before released into the sea.

Fecal pollution indicators used in Antarctica usually include microorganisms (Bruni et al. 1997; Edwards et al. 1998; Hughes 2003; Martins et al. 2005) and organic markers, such as fecal steroids and linear alkylbenzenes (LABs) (Green and Nichols 1995; Martins et al. 2002, 2012; Hughes and Thompson 2004; Dauner et al. 2015). The earliest studies on fecal pollution in Antarctica were performed on Ross Sea, in the proximities of US station McMurdo. Howington et al. (1992) observed that McMurdo Station's sewage plume reached 200–300 m seawards. High densities of coliforms were also found within 1 km from the station. Venkatesan and Mirsadeghi (1992) found fecal sterols (coprostanol and epicoprostanol) in sediments near the sewage outfall with values comparable to urban centers. Indeed, the US station McMurdo houses over 1000 people during summer and consequently generates a significant amount of effluent and waste. The coprostanol data along with current measurements corroborates (endorses) with the coliform densities mapped near McMurdo Station in the following year by McFeters et al. (1993) and Edwards et al. (1998).

Near the Italian station sewage treatment plant in Terra Nova Bay the occurrence of fecal coliforms and streptococci was observed in seawater samples (Bruni et al. 1997). The highest bacterian densities were found near the sewage release, and particularly when the population at the Italian station was approximately 80 people.

These pioneering works demonstrated the importance of monitoring sewage introduction from scientific research stations in Antarctica. Therefore, more adequate sewage treatment plants were implemented by many research stations in Antarctica, in order to reduce sewage effect in Antarctic ecosystems, particularly marine environments, since most scientific stations are located in coastal zones.

Although monitoring activity impacts developed by nations is mandatory according to the Protocol on Environmental Protection to the Antarctic Treaty (Madrid Protocol), there are few studies available about the extension and impact of sewage disposal in Antarctic environment.

Sewage contribution to sediments of the Admiralty Bay (King George Island) has been monitored since 1997/98 using fecal steroids and linear alkylbenzenes (LABs) as sewage contribution indicators. Martins et al. (2002) showed that the critical point was the sewage outfall of the Brazilian Station Comandante Ferraz. Although LABs concentrations are low, these results have shown that detergent may reach up to 1000 m away from the sewage outfall.

A later study conducted in the austral summer of 1999–2000 by Martins et al. (2005) comparing fecal sterols and microbiological indicators also observed that the residual contamination was restricted to the vicinities of Ferraz Station decreasing with distance from the sewage outfall. In the following austral summers (2003 and 2004) sewage contribution gradually increased as a result of doubling human occupation in the Brazilian station. Conversely, especially due to tidal effects, local hydrodynamic conditions have favored the dispersion of sewage effluent in the shallow coastal zone (Montone et al. 2010).

After implementing the new effluent treatment system in 2005 a reduction of sewage influence was observed around Ferraz Station. Martins et al. (2012) found low fecal sterols concentrations (<0.01–0.17 µg g^{-1} dry weight) and linear alkylbenzenes (<1.0 to 46.5 ng g^{-1} dry weight) in marine sediments. The maximum concentration of fecal sterol was similar to the value set as background for this region (0.19 µg g^{-1}) (Montone et al. 2010), being lower than the concentrations observed in previous studies (1997–2008), while LABs concentrations remained virtually constant (35 ng g^{-1}).

Hughes (2004) also demonstrated better environmental conditions after the implementation of a sewage treatment in British Rothera station located in the Antarctic Peninsula. From 2004, sewage influence has not exceeded 50 meters from the station, whereas prior to such a treatment plant started the operation, the influence of untreated sewage released reached up to 800 m away from the station.

Since 2003, a sewage treatment plant similar to the one in Rothera station was installed in the US station McMurdo. Several studies were conducted to verify the influence of sewage in the region. Conlan et al. (2004) observed that after implementing of sewage treatment on site there was a decrease in the abundance of tolerant to organic enrichment polychaete and the recolonization of polychaete intolerant to sewage. There are signs of detoxification and limited recovery of benthic communities. In this location as in other places however, the contamination and its associated impacts upon benthic communities are located within 1 km from the pollution sources (Conlan et al. 2010).

The disposal of untreated sewage (population of 20–50 people) during many decades resulted in similar alterations, meaning the reduction of the abundance of benthic invertebrates around sewage outfalls in the

Australian station Casey (Stark et al. 2003a). Recent analysis of residual water discharged from Australian station Davis (population of 70 people) evidence the necessity of a more efficient treatment. The microbiological load was typical of untreated residual water and the ecotoxicological tests show risks for the biota (Stark et al. 2015). Such a study contributed in developing a more advanced wastewater treatment system for the station.

In addition to high concentrations of nutrients, and the abundance of pathogens, sewage effluent may also contain high concentrations of other contaminants such as metals, hydrocarbons and Persistent Organic Pollutants (POPs). Studies already reveal the presence of more recent pollutants as flame retardants (polybrominated diphenyl ethers (PBDEs), hexabromocyclododecane (HBCD)) and other Emerging Pollutants (EPs) such as perfluoroalkylated substances (PFASs) and active ingredients from medications and personal care products in effluent disposed through outfalls from Antarctic stations (Hale et al. 2008; Chen et al. 2015; Emnet et al. 2015; Wild et al. 2015).

Therefore, it should be noted that despite the improvement of sewage treatment facilities (Table 9.2), the number of people in Antarctica is increasing, which implies the need for continued environmental monitoring in order to minimize anthropogenic impacts.

Metals

Currently in the Antarctic environment there is a tendency of increase in some contaminant levels, among which metals and metalloids, very reactive chemical elements present in the seawater in small quantities, which are quickly removed to the sediment or assimilated by sea organisms, representing a problem due to their toxicity, accumulation capacity in the biota and persistence and stability in the environment (Islam and Tanaka 2004). The Antarctic Continent is vulnerable to metal contamination through atmospheric circulation and sea currents from lower latitudes. Direct release of theses contaminants can also occur in areas where research stations are present, as through burning fossil fuels and residues, from ships, aircrafts, small vehicles, used for research data collection, fishing and tourism, as well as paints and sewage from research stations (Hong et al. 1999; Montone et al. 2001b; Vodopivez et al. 2001; Santos et al. 2006; Ribeiro et al. 2011).

Metals naturally occur in the Earth's crust, however anthropogenic activities artificially introduce them into the environment (Santos et al. 2006). Representing a fundamental role in the functioning of the planet's life, because they are essential for the body's metabolic activities, and exist in the limit between essentiality and toxicity some of these elements are considered essential micronutrients such as zinc (Zn), copper (Cu), nickel (Ni), cobalt (Co), iron (Fe) and manganese (Mn). Other elements like

Table 9.2. Treatment systems and sewage markers in the marine environment around the Antarctic stations.

Station	Location	Sampling	Treatment system	Sewage markers	Distance (m)	References
Ferraz–Brazil	King George Is. Antarctic Peninsula	1997/1998	Primary and secondary treatment	Fecal sterols/LABs C. perfringens	700	Martins et al. (2002)
		1999/2000				Martins et al. (2005)
		2002/2003		Fecal sterols/LABs		Montone et al. (2010)
		2009/2010		Fecal sterols/LABs		Martins et al. (2012)
Carlini–Argentina	King George Is. Antarctic Peninsula	2010/2011	Primary treatment	Fecal sterols/LABs		Dauner et al. (2015)
Davis–Australia	Princess Elizabeth Land	No information available	Secondary treatment (2005) Maceration (after 2005)	Faecal sterols	2000	Leeming et al. (2015)
	East Antarctica	October 2009–March 2010		Fecal sterols	2000	Stark et al. (2016)
Rothera–UK	Adelaide Island, Antarctic Peninsula	February 2000	Aerated biological filter sewage treatment plant	Fecal sterols C. perfringens	200	Hughes and Thompson (2004)
McMurdo EUA	Ross Island East Antarctica	October–November 1999 and 2000	Sewage treatment plant	C. perfringens	900	Lisle et al. (2004)

cadmium (Cd), mercury (Hg), chrome (Cr) and lead (Pb) do not possess a known biological function and present extreme toxicity, making it necessary to monitor sensitive to contamination marine environments. Arsenic (As) and selenium (Se) are metalloids, elements presenting metal and non-metal characteristics, however they are generally considered as metals due to their toxicity (Karadede-Akin and Unlu 2007). Mercury is given special attention since it has high mobility and affinity with organic matter, besides its biomagnification capacity, being one of the most harmful metals for the biota (Santos et al. 2006).

Assessing metal concentration in sediments and in the biota has special importance in isolated coastal areas such as Antarctica, for it allows the detection of base levels for these elements to be adopted as world references, tracking their fluctuation and detecting the impact of local human activities and their damaging effects to the marine ecosystem (Berkman and Nigro 1992; Jerez et al. 2011).

Considering the high concentration of metals in sediments obtained from studies conducted in several Antarctic regions, elevated levels of Cu and Zn are observed in this environment (Table 9.3).

The high concentration of Cu in the sediments from Admiralty Bay, Ross Sea and McMurdo Sound, reported in these studies, may be associated to chalcopyrite mineralization (Fe and Cu sulfide) in the area and also due to these sediments mineralogy, originated from the glacial erosion of volcanic rocks, such as basaltic andesite (Fourcade 1960; Machado et al. 2001). During the magmatic differentiation, Cu and other metals, Zn for example, are incorporated in olivine, pyroxene and plagioclase, which make up rocks like basaltic andesite (Salomons and Forstner 1984). Therefore, the high concentrations of Cu and Zn in the sediments may be associated with natural/basal levels of these elements in Antarctica.

The presence of metals in sediments in Antarctica usually attributed to a natural origin, can also be ascribed to local human activities and to atmospheric transport of contaminants from lower latitudes, namely from other continents of the southern hemisphere, Cr, Cu, Ni, Pb and Zn in particular, which are released in coal, oil and gas combustion and in non-ferrous metal production (Bargagli 2008).

Taking into account that aquatic environment contamination has been a growing concern, and organisms are capable of integrating pollutant concentration variation over time through bioaccumulation and following the increase caused by biomagnification, the determination of heavy metals in organisms must be part of monitoring and assessment of marine and coastal environment programs (Marcovecchio 2004).

The comprehension of metal bioaccumulation and biomagnification in the biota of different regions of Antarctica is rather complex. Despite being a characteristic of Antarctic pelagic trophic webs to present few levels and

Table 9.3. Metal concentrations in the Antarctic sediments (mg kg^{-1}, dry weight).

Local	As	Cd	Cr	Cu	Ni	Pb	Zn	References
Ferraz Station	7.8		11.9	66.6	9.2	5.5	61	Trevizani et al. (2016)
Punta Ullman	6		11.1	60.8	8.6	5.3	59	Trevizani et al. (2016)
Arctowski Station	6.8		8.4	50.4	5	3.9	65.8	Trevizani et al. (2016)
Botany Point	6.5		15.4	73	12.1	5.4	64.6	Trevizani et al. (2016)
Ferraz Station	8.5	0.6		60	4.3	6	63	Majer et al. (2014)
Admiralty Bay	2–12	0.4–0.9	7–12	47–84	3–10	3–11	44–89	Ribeiro et al. (2011)
Ross Sea		0.1–1.6	12–97	10–46	10–46	4–20	52–144	Ianni et al. (2010)
Ferraz Station	8–33		25–52				87–134	Santos et al. (2007)
Botany Point	4–6		36–45				81–95	Santos et al. (2007)
McMurdo Station	4–5	0.2–0.5		31–100		8–66	114–156	Negri et al. (2006)
Ferraz Station			31	92	10.1	10.5	89	Santos et al. (2005)
Potter Cove			4–8	73–156		2–6	46–63	Andrade et al. (2001)

For sediment cores were used results of fraction 0–2 cm.

have Krill as base, the biomass of these euphausiaceae is smaller. Hence, the coastal trophic webs have a longer and more complex structure, in which the base is composed of benthic invertebrates, which can transfer contaminants from phytoplankton and superficial sediments to fishes, birds and mammals (Bargagli et al. 1998; 2000).

Particularly at Antarctic coastal regions high levels of heavy metals from activities developed in research stations, land, sediments, seawater and lichen are recorded (Hong et al. 1999; Ahn et al. 2002; 2004). Taking into consideration that the intertidal zone has great ecological importance, since it serves as habitat for diatoms, macroalgae and small invertebrates, and as food source for microphytobenthos (Ahn et al. 1996). Efficient monitoring of these zones and their biota becomes necessary, thus early detection of harmful effects to the Antarctic marine ecosystem becomes possible (Ahn et al. 2002).

Arsenic bioaccumulation was verified along all of the Admiralty Bay in 2003 (Trevizani et al. 2016), it was also detected at Potter Cove in the year of 2002 (Farías et al. 2007) and at around Ferraz station in 2005 (Majer et al. 2014), having greater concentration in the macroalgae *Himantothallus grandifolius*. The highest levels of As have been recorded especially in macroalgae, an expected behavior, since As concentration is more intense in base levels of the trophic chain (Jakimska et al. 2011). This procedure has been attributed to a natural process in algae in Antarctica, due to the region's geological composition (Farías et al. 2002; 2007). The algae would assimilate arsenate as analogous to phosphor during the metabolic process, so that As would accumulate in its tissues (Murray et al. 2003).

Cadmium is a toxic element for aquatic organisms even in low concentrations (Farías et al. 2002). Cd bioaccumulation in Antarctic biota was verified in *Laternula elliptica* at Maxwell Bay (Ahn et al. 1996), in echinoderms, sponges and algae in Antarctic islands (de Moreno et al. 1997), in *L. elliptica* at King George island (Ahn et al. 2001), in several benthic species at Terra Nova Bay (Bargagli et al. 1996; Vodopivez et al. 2001; Dalla Riva et al. 2004; Grotti et al. 2008) and in sponges and mollusks at McMurdo Sound (Negri et al. 2006). Cadmium presents a particular distribution tendency in Terra Nova Bay's trophic web, accumulating with greater intensity in invertebrate benthic detritivores, omnivores and suspension feeders (Dalla Riva et al. 2004). A tendency of Cd biomagnification around Brazilian Ferraz station benthic trophic chain was verified in 2005 (Majer et al. 2014) and Terra Nova Bay's biota (Bargagli et al. 1996).

The bioaccumulation of Cd in the Antarctic biota has been associated to this metal's high bioavailability in the Antarctic marine environment, a fact that is possibly connected to rich deep waters outbreak, algae proliferation, high concentrations of Cd in the water column in determined regions and to volcanism (Bargagli et al. 1996; 2008; Rodrigues et al. 2007).

Although Cr concentration in Antarctic sediments has increased over time, the Cr levels are basal in the biota, with similar concentration to Admiralty Bay, Terra Nova Bay and Evans Cape (Trevizani et al. 2016; Grotti et al. 2008).

Different regions of Antarctica have similar Cu concentration, in several species in Admiralty Bay (Trevizani et al. 2016), in the mollusk *Nacella concinna* collected at King George Island (Ahn et al. 2004), in the mollusk *L. elliptica* and in the starfish *Odontaster validus* at Terra Nova Bay and Cape Evans (Grotti et al. 2008). Despite presenting elevated levels in sediments, Cu bioaccumulation does not occur in trophic webs at Admiralty Bay and Ferraz Station region (Trevizani et al. 2016; Majer et al. 2014).

Studies performed at Maxwell Bay, Terra Nova Bay and Cape Evans verified Ni accumulation in mollusks and echinoderms (Ahn et al. 1996; Grotti et al. 2008). In 2003, Ni bioaccumulation was confirmed in macroalgae at Brazilian Ferraz station (Trevizani et al. 2016), such a process was not found at the same place in 2005 (Majer et al. 2014).

Admiralty Bay's Pb level resembles the one found in McMurdo Sound (Trevizani et al. 2016; Negri et al. 2006), and higher than the one detected in Terra Nova Bay and Cape Evans (Grotti et al. 2008). Bioaccumulation of this element occurs in microphytobenthos at Brazilian Ferraz station (Trevizani et al. 2016).

Zn presented high concentration in Admiralty Bay's biota, with bioaccumulation for most of the species at Ferraz, Punta Ullman and Botany Point, that is, in localized points at Martel Inlet (Trevizani et al. 2016). Zn bioaccumulation in the biota was also registered by Grotti et al. (2008), at Terra Nova Bay and Cape Evans. Results higher than those obtained at Admiralty Bay (Trevizani et al. 2016) were attained for the biota of Maxwell Bay, Deception Island, McMurdo Sound, King George Island, Terra Nova Bay and Cape Evans (Ahn et al. 1996; Deheyn et al. 2005; Negri et al. 2006; Santos et al. 2006; Grotti et al. 2008). The high levels of Zn in the biota are justified by this element's presence in large quantities in the Antarctic environment and in sediments, and also because this metal is essential, exhibiting biological functions in marine organisms (Majer et al. 2014).

Mercury, an element with proven possibility of biomagnification, has been found in the Antarctic biota, especially in birds and marine mammals. Dalla Riva et al. (2004) noticed an increase in Hg concentration in Terra Nova Bay trophic chain. Bargagli (1998) observed variable Hg concentrations in Terra Nova Bay biota, even though it was not found in sediments, as in the study by Trevizani et al. (2016) at Admiralty Bay. The Hg presence in the biota may occur due to atmospheric deposition or to natural presence in sediments associated with past volcanism, organic matter or attached to sulfides (Bargagli 1998; 2008; Dallas Riva et al. 2004; Santos et al. 2005).

In a study by Sanchez-Hernandez (2000) there is a compilation of studies' results in which metals were investigated in biotic and abiotic components of the Antarctic ecosystem. Most of the Pb found in ice is of anthropic origin and atmospheric source, while the environmental impact of metals like Cd, Zn and Hg is restricted to areas of few hundreds of meters of anthropogenic origin.

Generally, metals concentration in abiotic matrices is around trace level, but metal concentrations in the biota are comparable to polar and temperate areas of the northern hemisphere, mainly Cd and Hg in particular. Biological factors favor a greater metal accumulation throughout the Antarctic biota, such as the growth rate of organisms and detoxification mechanisms which are affected by extreme environmental conditions and that interfere with captation, storage and excretion of trace elements by organisms. Moreover, environmental factors, like volcanism and upwelling, increase metal availability in the environment. In this context, fishes, mollusks, lichens and macroalgae have been suggested as adequate biomonitors, and their vestigial concentrations of elements as baseline (Sanchez-Hernandez 2000).

In the Antarctic sediments high concentrations of Cu and Zn were found, attributed to the region's geological composition. Bioaccumulation of As occurs in macroalgae and bioaccumulation of Zn occurs in the Antarctic biota. These processes have been reported as natural, with the local geological composition as source. Furthermore, Cd, Pb and Hg, important pollutants, have been registered in the Antarctic biota. Their origin has been discussed by many authors and accredited to the presence of Cd in water and volcanism, in addition to the presence of Pb and Hg derived from atmospheric transport and local pollution. Such information is useful for monitoring and comprehending metal contamination in Antarctica and also for present and future studies comparisons.

Bargagli (2000) revised studies which investigated metal concentrations in abiotic matrices such as atmospheric particulate matter, snow, superficial soils, superficial sediments, freshwater, and seawater in Antarctica, and found values that can be considered on a global basis. Moreover, Bargagli verified that Pb biogeochemical cycle is probably the only cycle significantly altered by anthropogenic emissions in Antarctica and other southern hemisphere continents. Under these circumstances, metal pollution by anthropogenic activities in Antarctica may compromise studies for evaluating the biogeochemical cycle of these elements and the global climate change effects (Bargagli 2000).

Taking into account the impact of metal pollution and its relation to climate change, we could consider that the Antarctic biota, adapted to the Antarctic Ocean's low temperatures, would be severely affected by climate change. The increase in global temperature could affect metabolic processes involved in the uptake and detoxification from environmental pollutants. Despite not having records of significant changes of temperature in continental Antarctica, in the Antarctic Peninsula ice loss have been reported, the latter affecting atmospheric precipitation, environmental biogeochemistry along with distribution and composition of biotic communities. Thus, global warming might reinforce Antarctica's role of cold trap for Hg, through ice sheet alterations and the increase in atmospheric input of metals (Bargagli 2000; 2008).

Persistent Organic Pollutants (POPs) in the Marine Environment

Persistent organic pollutants (POPs) include various synthetic chemicals that comprise a group of semivolatile, bioaccumulative, persistent and toxic pollutants (Jones and Voogt 1999). The Stockholm Convention signed on May 2001 established a list of 12 priority compounds, including eight organochlorine pesticides (Aldrin, Dieldrin, Endrin, Chlordane, Heptachlor, DDT, Toxaphene and Mirex), two industrial products (polychlorinated biphenyls (PCBs) and hexachlorobenzene (HCB), which also have insecticidal properties) and two unintended by-products (dioxins and furans). In August 2009, nine new chemicals (α-hexachlorocyclohexane; β-hexachlorocyclohexane; Lindane; chlordecone; hexabromobiphenyl; tetrabromodiphenyl ether, pentabromodiphenyl ether, hexabromodiphenyl ether and heptabromodiphenyl ether; pentachlorobenzene; perfluorooctane sulfonic acid and its salts and perfluorooctane sulfonyl fluoride) were added in an amendment that was implemented one year later. During the 5th meeting held in 2011, Endosulfan became the 22nd POP (Xu et al. 2013). These compounds are distributed globally and reach remote areas like Antarctica.

The presence of POPs in the Antarctic marine environment has been observed since the 1960 with the pioneering work of Sladen et al. (1966) and George and Frear (1966) who found DDT residues in marine biota (penguins and seals) on Ross Island (77° 27'S –169° 14'E) in the vicinity of the US Antarctic station at McMurdo Sound. Tatton and Ruzicka (1967) have also detected DDT and other pesticides (Lindane, Dieldrin and Heptachlor epoxide) in Antarctic wildlife from the British Antarctic Survey station at Signy lsland in the South Orkneys. The occurrence of PCBs in Antarctica was also reported in the 1960s and 1970s by Risebrough et al. (1968; 1976).

Since the organochlorine pesticides (OCPs) and industrial compounds (mainly PCBs) are neither manufactured nor applied in the region, the major source of POPs is attributed to the long-range transport. However, local sources (waste burning and dumping sites) from Antarctic stations might also introduce POPs into the region (Risebrough et al. 1990). The presence of anthropogenic radionuclides such as ^{90}Sr and ^{137}Cs (e.g., Desideri et al. 2003; Ferreira et al. 2013) have also demonstrated that long-range atmospheric transport of emitted particles from tropical and temperate regions to the south polar areas occurs, and provided a clear and well-publicized indication of the global impacts of human activities.

Studies of POP contamination and its environmental impact in Antarctic ecosystems have been conducted by various Antarctic programs. The POP data for marine environment include several sections, such as atmosphere, seawater, sediment and biota.

POPs data in Antarctic marine atmosphere are scarce compared to the Arctic data. The first data concerning organochlorines (OCs) in marine air samples from low-mid to high latitudes were reported by Tanabe et al. (1982a,b), who undertook air sampling on an Antarctic supply voyage between Japan and the Syowa Research Station in 1980–1981. This study showed a north–south gradient for DDTs and HCHs. Concentrations of OCs in the air were again measured by Tanabe et al. (1983) along two extensive cruise tracks (Australia to Antarctica and Mauritius Island to Syowa Research Station) in 1980–1981 and 1982, respectively. Kawano et al. (1985) collected samples of marine air in the eastern Indian Ocean and in the Southern Ocean by Australia during 1983–1984. Bidleman et al. (1993) measured chlorinated pesticides in air samples during a cruise between New Zealand and Ross Island, Antarctica in 1990. Iwata et al. (1993) made an important study of OCs distribution in the oceanic air and surface seawater from various oceans, including the Southern Ocean during 1989–1990. Organochlorine data from the Southwest Atlantic and Antarctic Oceans was reported by Weber and Montone (1990) during a cruise between Brazil and the Antarctic Peninsula in the austral summer of 1987. Subsequent studies by Montone et al. (2005) for similar transects confirmed the presence of these compounds, although at levels that were several orders of magnitude lower than previously reported. A decline in organochlorine pesticide levels with an increase in latitude was observed in both studies.

Atmospheric concentrations of PCBs in the vicinity of a Brazilian Ferraz station (Comandante Ferraz) were measured in the austral summer of 1994–1995 by Montone et al. (2003). PCB levels ranged from 12.1 to 92.6 pg m^{-3} (mean 37.4 pg m^{-3}), with a predominance of low chlorinated congeners. These atmospheric levels of PCBs were correlated with meteorological conditions; higher concentrations were associated with the passage of frontal systems arriving from South America. These results are similar to those reported by Larsson et al. (1992) (Ross Island), Kallenborn et al. (1998) (Signy Island) and Gambaro (2005) (Terra Nova Bay).

Kallenborn et al. (1998) found concentrations of individual PCB congeners in ambient air at Signy Island comparable to those in Arctic air. The atmospheric long-range transport events from South America were also addressed by the authors. Evidences of atmospheric transport of PCBs in the Antarctic environment were also provided by Fuoco et al. (1996) who observed high contributions of atmospheric particulate matter, primary vehicle of transport and diffusion of PCB in the environment, in Antarctic lake sediment.

Gambaro (2005) detected very low levels of PCBs in the air of Terra Nova Bay (from below detected levels to 0.25 pg m^{-3} for individual PCB congeners), indicating that the Antarctic environment still has its pristine characteristics.

The levels and patterns of PCBs and OCPs were determined by Choi et al. (2008) using air samplers for one year at Korean polar research stations at King George Island, Antarctica (2004–2005). These measured patterns also suggest that the Antarctic area is affected by long-range transport originating from regions located in the southern tip of South America. These results are consistent with a hypothesis of global fractionation and long-range transport, which was confirmed by measurements of black carbon (BC) and insoluble particulates respectively attributed to sources in Brazil and Africa (Pereira et al. 2006) and Chilean Patagonia (Pereira et al. 2004) and by measurements of spheroidal carbonaceous particles (SCPs) from fossil fuel combustion related to industrial activities in South America (Martins et al. 2010).

Few investigations have been carried out to determine POPs content in Antarctic seawater. Tanabe et al. (1982a,b) detected HCH (260–920 pg L^{-1}), DDTs (5–58 pg L^{-1}) and PCBs (40–80 pg L^{-1}) in surface seawater samples during an Antarctic supply voyage between Japan and the Syowa Research Station in 1980–1981. Tanabe et al. (1983) investigated POPs close to Syowa station. The levels of POPs in seawater varied from 210 to 930 pg L^{-1} (HCHs), 1.3 to 21 pg L^{-1} (DDTs) and 35 to 69 pg L^{-1} (PCBs). Concentrations of HCHs and PCBs were similar in water samples under fast ice and at outer margins of pack ice, while DDTs were lower under fast ice.

Montone et al. (2001a) measured PCBs in surface seawater from Admiralty Bay (King George Island) during the summer of 1994–1995. Low levels of total PCBs were observed in the dissolved (0.08–0.30 ng L^{-1}) and particulate fractions (0.08–0.51 ng L^{-1}). Low chlorinated PCB congeners were predominant in both fractions, but small quantities of the heavier congeners (138, 180 and 187) were observed near Brazilian Ferraz station (EACF). These concentrations are of the same order of magnitude as the PCB concentrations from superficial seawater collected near Syowa Station (Tanabe et al. 1983).

Studies carried out in a large area of the Ross Sea and Victoria Land during summer campaign between 1988–1989 to 1991–1992, showed differences between PCB levels before pack ice melting (90–180 pg L^{-1}) and after pack ice melting (150–230 pg L^{-1}). The increase of higher PCB concentration can be explained considering that ice acts as an accumulator that traps atmospheric particulates during its formation and transfers them and the pollutants adsorbed by the ice to seawater during its melting as an indirect consequence of global climate change (Fuoco et al. 1996).

Dickhut et al. (2005) detected HCB, HCHs, Heptachlor and Heptachlor epoxide in seawater during the austral winter (September–October 2001) and summer (January–February 2002) along a transect in the Western Antarctic Peninsula. These pesticides were also investigated in air, sea ice, and snow to estimate the exchange at the air/water interface and to assess the accumulation on the pack ice.

Vertical distribution of PCBs in the coastal area of the Ross Sea was conducted by Fuoco et al. (2005) during the Antarctic summer (1997–1998) at the Gerlache Inlet, Terra Nova Bay. PCB levels in the water column ranged between 30–120 pg L^{-1} and these values were strongly dependent on the suspended matter content. Further studies conducted by Fuoco et al. (2009) also detected PCBs (mean values 50 pg L^{-1}) and the data was comparable with the previous campaigns.

Coastal sediments can serve as temporary or long-term sinks for many classes of anthropogenic contaminants and, consequently, as sources of these substances in oceans and biota (Guzzella et al. 2005). Furthermore, sediments accumulate POPs over time, storing valuable information concerning past environmental and climatic events (Fuoco et al. 1996).

Risebrough et al. (1990) and Kennicutt et al. (1995) reported high concentrations of PCBs in sediments from Winter Quarters Bay, mainly due to a single point source as a result of anthropogenic introduction of a commercial mixture of PCB (Aroclor 1260). Other Antarctic regions present low contamination of POPs in sediment as noted by Pu et al. (1996) for the Great Wall Bay, King George Island.

The distribution of PCBs in the sediment from Terra Nova Bay and Ross Sea was studied by Fuoco et al. (1996) during the 1990–1991 expedition. The variability of PCB concentrations (0.03–0.16 ng g^{-1}) was related to the particle size distribution. The depth profiles of PCB content in marine sediment samples collected in a few stations clearly show that PCBs are confined in a surface layer of about 10 cm.

Montone et al. (2001b) measured the PCB levels in superficial marine sediments collected from Admiralty Bay in the 1993–1994 austral summer. The PCB levels were also low, ranging from 0.85 to 2.47 ng g^{-1} dry weight (dw) and were comparable to those determined in Antarctic sediments collected from Great Wall Bay (Pu et al. 1996), Terra Nova Bay (Fuoco et al. 1996), James Ross Island (Klanova et al. 2008) and a remote site in McMurdo Sound (Kennicutt et al. 1995). The PCB profiles were generally similar for all samples, with a predominance of low chlorinated congeners. The presence of heavier congeners at low concentrations (for 138 and 153) and below detection limits (for 128, 180 and 187) indicates that there are no significant local sources of PCBs. Recent studies have detected low and uniform POP levels, 0.4 to 1.0 ng.g^{-1} dw (unpubl. data), indicating a slight decrease and a lack of a significant contribution in the region. The levels of OCPs were also low and uniform over the last decade, with measurable amounts of HCB (<0.1 to 0.3 ng g^{-1} dw), HCHs (<0.1 to 0.8 ng g^{-1} dw) and DDTs (<0.03 to 0.1 ng g^{-1} dw).

Klanova et al. (2008) detected PCBs and OC pesticides in sediment samples from James Ross Island. Sediment PCB concentrations ranged from 0.32 to 0.83 ng g^{-1}, while they ranged from 0.14 to 0.76 ng g^{-1} for HCHs, and from 0.19 to 1.15 ng g^{-1} for DDTs. The predominance of less chlorinated PCBs and more volatile chemicals was explained by the long-range atmospheric transport of POPs from populated areas of Africa, South America and Australia.

Another study conducted by Zhang et al. (2013) also reported similar POP levels in sediments collected from the Western Antarctic Peninsula (WAP) in July, 2008. Hexachlorinated PCBs dominated the PCB profiles in those sediments (40–100%), suggesting that local activities may be the dominant source of POPs in the WAP. HCB was the most frequently detected OCP in those sediments, at concentrations ranging from 0.002 to 0.130 ng g^{-1} dw.

Recently, Combi et al. (2017) study the temporal patterns, fluxes and inventories of polychlorinated biphenyls (PCBs) in nine sediment cores collected from selected areas of Admiralty Bay off the Antarctic Peninsula. Concentrations of total PCBs were low, but slightly higher in comparison to low-impacted, remote environments in the world, ranging from below the detection limit to 11.9 ng g^{-1} in dry weight.

The food web has been the most studied subject for environmental contamination in Antarctica. Antarctic ecosystems are typically dominated by short trophic webs due to the extreme climate. The organisms even have peculiar features: they grow very slowly, may present gigantism when adults, are often long-lived, and may use lipids as an energy reserve or to protect the body against cold. These features make Antarctic organisms prone to accumulate chemicals and vulnerable to toxic effects (Corsolini 2012). Primary production is transferred through the food chain to top predators with relatively high efficiency (Priddle et al. 1998), which should tend to reduce the amplification of pollutant load resulting from biomagnification compared to more complex food webs. Thus, Antarctic ecosystems are relatively simple compared to other ecosystems due to the limited number of trophic levels compared with, for example, tropical ecosystems (Hidaka et al. 1984).

In the pelagic system, producers are microscopic phytoplankton with very short turnover times. Zooplankton, however, tend to be large and are often long-lived. Predators include cephalopods, fish, birds and mammals. With the exception of cephalopods, these predators are typically long-lived. Many species of birds and mammals migrate to the higher latitudes of the southern hemisphere or even into the northern hemisphere during the austral winter. Consequently, their pollutant loads reflect inputs from both outside and inside the Antarctic region in the case of POPs, which have long turnover times. The mobility of pelagic organisms also means that they will integrate environmental loads of POPs over large spatial and temporal scales (UNEP 2002).

Concentrations of POPs in the Antarctic biota have been reported for organisms of various trophic levels, from plankton to top predators like birds (penguins and flying birds) and pinnipeds (seals).

Joiris and Overloop (1991) analyzed POP residues in samples of particulate matter (mainly phytoplankton) from the Indian sector of the Southern Ocean. PCB concentration (0.7 µg g^{-1} dw) was similar to that of the temperate zones. Organochlorine pesticides have been detected at very low concentrations. They observed an increase in the ratio DDT/DDE, indicating recent Antarctic organochlorines originating from the southern hemisphere, and suggested a possible increase in contamination of Antarctica by POPs. Phytoplankton samples from Ross Sea presented PCB concentration of 1–4,2 ng g^{-1} wet weight (ww) with predominance of heavier congeners as PCB 138, 153, 180 and 195 (Corsolini and Focardi 2000).

Some studies have been conducted to determine the concentration of POPs in krill, a key species of the Antarctic food chain. Risebrough et al. (1976) detected 3 ng g^{-1} lipid weight (lw) for PCBs, 14 ng g^{-1} lw for *pp'*-DDE and 19 ng g^{-1} lw for *pp'*-DDT in a sample collected in 1975 along the Antarctic Peninsula. In January 2000, krill samples collected in the Ross Sea presented *pp'*-DDE concentrations of 0.86 ± 0.98 and PCB 167 ± 85 ng g^{-1} (ww). Tetra-PCB were the predominant congeners (Corsolini et al. 2002). Levels of PBDEs in krill (5.6 ng g^{-1} lw) presented a lower order of magnitude than PCBs (Corsolini et al. 2006).

Analysis of POPs in Antarctic fish such as *Trematomus bernacchii*, which has been used as a bioindicator, showed similar values of DDTs and PCBs to krill (Larsson et al. 1992; Focardi et al. 1992; Bargagli 2005). Moreover, HCB concentrations (20 ng g^{-1}, ww) were elevated and associated with the cold condensation process (Weber and Goerke 1996). An increase in concentration of POPs was observed in two species of benthic fish (*Gobionotothen gibberifrons, Chaenocephalus aceratus*) between 1987–1996 by Goerke et al. (2004). However, there was no such increase in the *Champsocephalus gunnari* species that feeds on krill. Analysis on fish collected near sewage outfall showed high levels of PBDEs (1840 ng.g^{-1} lw) in *T. bernacchii* (Hale et al. 2008).

A large number of studies have been conducted to determine POPs on penguins and other Antarctic birds. The first POP studies were performed by Risebrough et al. (1976) on penguin eggs from different areas of the Antarctic Peninsula and compared with data from penguins in the sub-Antarctic islands. The results showed that the atmosphere was the main transport of PCBs and DDT to Antarctica.

Donnewald et al. (1979) detected HCB in accumulated subcutaneous fat ranging between 87 to 600 ng.g^{-1} (lw) in penguins from the Antarctic Peninsula. Lukowski (1983a,b) measured concentrations of DDT and its derivatives in penguins and others seabirds from King George Island in the late 1970s. The derivative *pp'*-DDE was detected in adult penguins, while trace amounts of *pp'*-DDT and *pp'*-DDE were present in penguin eggs. The concentrations of DDTs (*pp'*-DDE + *pp'*-DDT) in samples of penguin fat ranged from 340 to 550 ng g^{-1} (ww). Wide-ranging migrants such as Wilson's storm petrel and skuas had the highest levels of DDT residues among the flying birds, and *pp'*-DDE predominated and reached levels of 400 to 8,000 ng g^{-1} ww in fat tissues. The low levels of DDT in relation to DDE in these samples indicated relatively little new input of DDT. Lukowski et al. (1987) measured the total PCB concentrations in subcutaneous fat and liver in three pygoscelid penguin species and four flying seabirds species from the breeding colony. The highest concentrations of total PCBs were found in Wilson's storm petrel, which forages extensively outside the Antarctic region. The concentrations in fat were 15,000 ng g^{-1} (ww), and in livers were 1,800 ng g^{-1} (ww). Penguins, which are mostly restricted to the Southern Ocean, contained the lowest levels of PCBs; the concentrations in the fat and liver were 150 to 720 ng g^{-1} and 60 to 290 ng g^{-1} (ww), respectively.

High levels of DDTs (30.8–972.3 ng g^{-1} ww) were measured in the fat tissues of Gentoo penguins (*P. papua*) sampled in 1991–1993 by Inomata et al. (1996). Although *pp'*-DDE predominated in all samples, the occurrence of *pp'*-DDT suggests that the use of DDT in the southern hemisphere was significant during the late 1980s and early 1990s (Barra et al. 2005). The PCB levels varied from 43.2 to 1,583 ng g^{-1}, and heavier congeners (138, 153 and 180) were detected. The PCB profiles matched most closely to that of Aroclor 1260. HCH (<0.1–24.9 ng g^{-1}), HCB (56.7–1,160 ng g^{-1}), Dieldrin (6.9–106.7 ng g^{-1}) and Endrin (0.8–49.3 ng g^{-1}) were also detected in these fat samples. Analyses of fat samples from penguins and skuas collected in Admiralty Bay in the austral summer of 1997 by Taniguchi et al. (2009) confirmed the presence of polychlorinated biphenyls (PCBs) and chlorinated pesticides. Penguin fat contained high levels of *pp'*-DDE (177 ± 98 ng g^{-1}), PCBs (256 ± 125 ng g^{-1}) and Mirex (90.6 ± 70.6 ng g^{-1}) in lipid weight (lw). There were no significant differences among penguin species in this study, and the highest concentrations were measured in skuas, most likely because of their migratory and feeding habits. However, the detection of these pollutants in Adélie penguins (*P. adeliae*), Chinstrap penguins (*P. antarctica*) and Gentoo penguins (*P. papua*), which are more restricted to the Antarctic environment, confirms the existence of these compounds in this Antarctic ecosystem and their incorporation into its trophic web. Similar values of PCBs were measured in the same penguin species collected during the Brazilian Antarctic expeditions in the austral summers of 2005/2006 and 2006/2007. Chicks of all three species showed similar profiles of PCB congeners, with predominance of low chlorinated compounds (Montone et al. 2016).

Corsolini et al. (2007) observed different chemical accumulation patterns in relation to penguin species and gender in blood samples collected at Admiralty Bay in February 2004. POP levels were higher in Adélie penguins than in Chinstrap and Gentoo penguins. The authors associated the differences among species with ecological factors or metabolic features. Moreover, Adélie penguins consume more krill (a fatty resource) during their rearing period compared to the other two species.

Cipro et al. (2010) reported measurements of POPs in unhatched penguin eggs collected from penguins at Admiralty Bay during the austral summers from 2005 to 2007. The predominant compounds were PCBs (2.53–78.7 ng g^{-1} ww), DDTs (2.07–38.0 ng g^{-1} ww) and HCB (4.99–39.1 ng g^{-1} ww), and their occurrence seemed to be species-specific for the Pygoscelis genus.

Similar accumulations of POPs (ΣPCB > ΣDDT > HCB > ΣHCH) were observed by Zhang et al. (2007) in the sampled penguin eggs at Fildes Peninsula adjacent to the Great Wall Station, China, in 2001/2002. The lower accumulation of HCHs in seabird eggs compared to other POPs was associated with the reduced use of HCHs in agriculture and its more rapid degradation relative to PCBs, DDTs and HCB.

Corsolini et al. (2011) evaluated the presence of POPs in unhatched eggs of Adélie penguins (*Pygoscelis adeliae*) and Brown skuas (*Catharacta lonnbergi*) collected in the Brainsfield Strait near the U.S. field camp, Lenie Field Station, Admiralty Bay, King George Island during the 2003/2004 and 2004/2005 breeding seasons and in eggs collected from the Ross Sea (East Antarctica) for comparison. Higher contaminant concentrations were detected in migrating seabirds, respectively: (South Polar skua and Brown skua), sub-Antarctic species (snow petrel), Antarctic species (penguins) at both sampling sites, suggesting that contamination events occurred at lower latitudes for those birds migrating northward. PCBs were the most abundant contaminants, and HCHs were present at the lowest concentrations in all species. The detection of *pp'*-DDT in Adélie penguins and in both species of skua confirm its use in certain malaria-endemic countries, from which it was transferred via long-range transport to the polar regions.

PBDEs were detected in bird eggs and livers at concentrations as high as 39.1 and 8.0 ng g^{-1} ww, respectively, by Cipro et al. (2013). BDEs 47 and 99 were the prevalent congeners among brominated homologues in Antarctic penguins, similar to the patterns observed in vegetation samples collected at Admiralty Bay (Cipro et al. 2011). In blood samples collected from three species of Antarctic pygoscelids at Admiralty Bay in February 2004, PBDEs were also detected with concentrations ranging from 107 to 291 pg.g^{-1} ww (Corsolini et al. 2007). The most abundant congeners were BDE-47 in Adélie and Chinstrap penguins and BDE-17 in Gentoo penguins. Other POPs were detected, such as polychlorinated dibenzo-dioxins (PCDDs) at mean concentrations of 22 ± 32, 6.5 ± 7.4 and 18 ± 23 pg.g^{-1} ww in Adélie, Chinstrap and Gentoo penguins, respectively. Polychlorinated dibenzo–furans (PCDFs) were in greater concentration in Adélie penguins and lower in Chinstrap penguins. PCDDs/Fs and PBDEs were more concentrated in male than female Gentoo and Chinstrap penguins; these differences in concentrations were likely related to the partial detoxification that occurs in females during egg formation.

There are few data on POP levels in pinniped tissues from Antarctica. Karolewski et al. (1987) analyzed adipose tissues from five Antarctic seal species collected in 1979–1981. Concentrations of *pp'*-DDE varied from species to species, ranging from 5 to 36 ng g^{-1} (ww) in crabeater seals, from 54 to 69 ng g^{-1} (ww) in Weddell seals. DDT was undetectable in elephant, Weddell, crabeater and fur seals. The highest mean concentration was measured in leopard seals, which contained from 158 to 164 ng g^{-1} (ww) and 267 to 456 ng g^{-1} (ww) of *pp'*-DDT and *pp'*-DDE, respectively. The differences among species may be related to their feeding habits. Weddell seals feed on fish and squid, while crabeater seals feed on krill. The leopard seal is a top predator that feeds mainly on penguins and other seals. By contrast, high levels of HCHs were observed in Weddell seals (52.1 ng g^{-1}), while low levels were detected in young elephant seals (8.0 ng g^{-1}). Leopard seals contained levels similar to (or lower than) those reported for other species.

Studies conducted by Cipro et al. (2012) in the same area from November 2004 to February 2005 measured POP levels of the same order of magnitude in fat tissues (Table 9.4); for example, PBDEs were detected at low levels –2.0 ng g^{-1} lipid weight (lw). The POP levels in Weddell seals were similar to those reported by Miranda-Filho et al. (2007) for the blubber of Southern elephant seal (*Mirounga leonina*) pups. However, the PCB and DDT levels were an order of magnitude lower than those detected in samples collected from Antarctic fur seal (*Arctocephalus gazelle*) pups at Livingston Island. The transfer of DDTs through the placenta may explain the high values of *pp'*-DDE in stillborn seal pups (Schiavone et al. 2009).

The variability in the bioaccumulation of POPs may be related to differences in diet among seals, as was observed by Karolewski et al. (1987). Overall, PCBs and DDTs are the predominant POPs in pinnipeds from Antarctica, while the levels of HCHs and HCB are more variable.

A common profile having PCBs > DDTs > HCB was observed in Antarctic organisms, with the exception of birds, particularly pygoscelids,

Table 9.4. Mean concentrations in Antarctic marine organisms (ng g⁻¹).

Local	Species	Sampling	Tissue	HCB	DDT	PCBs	Unit	References
King George Island	*Euphausia superba*	2004–05	whole	0.1	0.4	8	ww	Cipro et al. (2010)
Peninsula Antarctica	*Nototlienia* spp.	2006–07	muscle	1	3	7	ww	Cipro et al. (2013)
	Pygoscelis spp.	1997–98	fat	373	177	256	ww	Taniguchi et al. (2009)
	Leptonychotes weddelli	2004–05	fat	5.8	131	300	ww	Cipro et al. (2012)
Ross Sea	*Euphausia superba*	1999–00	whole	0.2	0.2	1.7	ww	Corsolini et al. (2006)
Terra Nova Bay	*Trematomus bernacchii*	1999–00	muscle	1.4	8.6	6.4	ww	
	Pygoscelis adeliae	1995–96	eggs	18.7	23	25	ww	
Weddell Sea	*Euphausia superba*	1986	whole	1	0.7[a]	0.1[b]	EOM	Goercke et al. (2004)
	Pleourogramma antarcticum	2000	muscle	1	3[a]	0.5[b]	EOM	
	Pygoscelis adeliae	1995	preen gland oil	25	3.5[a]	2[b]	EOM	
	Leptonychotes weddelli	1995	blubber	7.5	110[a]	10[b]	EOM	
McMurdo Sound	*Nacella concinna*	1991	whole	na	na	18.3	dw	Kennicutt et al. (1995)
	Orcinus orca	2005–06	blubber	740	4300	1600	lw	Kran et al. (2008)

ww = wet weight; lw = lipidic weight; dw = dry weight; EOM = extractable organic matter
a = *pp'*-DDE
b = PCB-153

which contained higher levels of HCB than DDTs. This pattern was similar to that observed in Gentoo, Chinstrap and Adélie penguins in the Antarctic Peninsula, while Emperor penguins mainly accumulated HCB and DDTs (in comparison to PCBs) in the Ross Sea and Antarctic Peninsula (Corsolini 2012). Van den Brink (1997) also observed higher levels of HCB than PCB in several species of seabirds from the East Antarctic coast. Hence, the predominance of HCB in seabirds may be the result of a higher bioavailability of HCB compared to other POPs in cold regions. Moreover, the species-specific metabolism of the organisms as well as the preferential bioconcentration and/or biomagnification of POPs must also be considered (Corsolini 2012).

The HCB levels in Antarctic trophic webs can be influenced by the global distribution of HCB, its condensation in cold environments and trapping by polar ice. The contribution of HCB to POP residues is relatively high in organisms in Antarctica, when compared to other areas of the world (Corsolini et al. 2011). The concentrations of HCB in Antarctic seals are comparable to those detected in other Antarctic pinnipeds. Goerke et al. (2004) measured considerably lower HCB levels in Weddell seals and in southern elephant seals than those in their dominant prey, pelagic fish species. Lower HCB levels were also detected in ringed seals (*Phoca hispida*) and harbor seals (*Phoca vitulina*) in the northern hemisphere and seals from highly industrial regions in Canada (Hobbs et al. 2002) compared to their dominant prey. Because the physical and chemical properties of HCB require that pinnipeds absorb this compound from their food, an exceptional capacity to eliminate HCB by biotransformation must be postulated for these mammals (Goerke et al. 2004). Low HCB levels were also observed by Yogui et al. (2003) for cetaceans sampled from the Brazilian coast. These results suggest that, in general despite its global dispersion, HCB can be eliminated by mammals. Mirex has been mainly accumulated in migratory birds such as skuas. However, Mirex was also present in endemic birds (penguins). Taken together, these findings may indicate both an input of this pollutant to the region and a low rate of elimination by these birds.

Studies on the presence and trend of POPs in polar regions have been mainly carried out for the Arctic (Bidleman et al. 2010; Muir et al. 2010). An important review about contamination profile and temporal trend of POPs in the Antarctic marine biota was conducted by Corsolini (2012) that presents a critical assessment and compares it with Arctic data. Although there was no recognizable overall temporal trend for POPs in Adélie penguin eggs from the Ross Sea and the Antarctic Peninsula areas, the data indicate that concentrations of DDTs declined from the end of the 1980s until the mid-1990s, but then began increasing again at the end of the 1990s. The same trend was observed for HCB and PCBs. By contrast, POP concentrations in Arctic biota have exhibited significantly decreasing trends for the past two to three decades.

POP data for marine biota from Admiralty Bay (Taniguchi et al. 2009; Cipro et al. 2012; Montone et al. 2016) indicated that PCB concentrations have remained constant over time, despite the decrease in environmental levels. The concentrations of pesticides in tissues from marine birds and mammals remain low compared to the levels observed in similar species from regions with greater environmental pollution, and concentrations are higher in migrant species than predominantly endemic species.

A recent study conducted by Cincinelli et al. (2016) evaluated the trends of POPs in Antarctic benthic seawaters using *Trematomus bernacchii* as the indicator. A slight decreasing PCB trend was detected during a 30 year time span (from early 1980s to 2010) in Antarctic seawaters. Two higher concentration peaks were reported in 2001 and 2005 in Ross Sea and may reflect the melting of icebergs.

In general, despite the dramatic decrease in input of POP worldwide due to prohibitions, the occurrence of this contaminants in the Antarctic marine sediments and biota may occur for the next decades based on the chemical resistance and low degradation rates of this compounds and the frequently input of POPs previously retained in the continental environment and ice sheets as consequence of global warming and melting processes.

Climate Change and Contaminants in Antarctica

Climate change has been the subject of several studies and concern. However, investigations to estimate the impact of climate change on the levels of environmental contaminants are still incipient.

A recent review conducted by Nadal et al. (2015) show the state of the art of climate change impact on environmental concentrations of some organic pollutants such as PAHs, PCBs and organochlorine pesticides. It was estimated that POP concentrations can change within a factor from 2 to 3, but there are few data available on PAHs and emerging contaminants such as PBDEs and perfluoroalkyl substances (PFASs). Bargagli (2008) conducted a major revision of the distribution of persistent contaminants, like metals and POPs, in several sections of Antarctic ecosystems, and has discussed their possible biological effects and the likely environmental contamination trend related to climate changes and increasing human activity in Antarctica and the southern hemisphere. This author points out that the volatilization and transport of POPs may be affected by heating the surface water and in turn affect the Antarctic organisms. In contrast to organisms in temperate and tropical seas, those in the Southern Ocean are well adapted to narrow ranges of water temperature close to the freezing point. Slight increases in temperature may have a disproportionate influence on the properties of cell membranes and biological processes involved in the uptake and detoxification of environmental pollutants (Bargagli 2008).

A decade before, Van den Brink (1997) had noted that the process of cold condensation could cause polar regions to act as a sink for the most volatile pollutants and that these pollutants may then be present at higher environmental levels in these regions than at lower latitudes. On this basis, animals occupying a specific trophic niche in a polar ecosystem may have higher POP loads than the corresponding species in temperate or tropical ecosystems. Furthermore, the mean annual temperature of the Antarctic Peninsula has increased by 6°C over the last 30 years, increasing the volume and frequency of glacial retreat and meltwater (Geisz et al. 2008) and thereby contributing to the release of POPs into Antarctic environments.

More recent studies by Cabrerizo et al. (2013) show the influence of climate change on PCB remobilization and reservoirs in Antarctica. It was estimated that an increase of 1°C in temperature may increase 21–45% in atmospheric PCB burden. However, a concomitant increase of 0.5% solid organic material can minimize the influence of heating, reducing POP fugacity in the soil.

Conclusion

These results suggest the importance to continue studying the dynamics of organic pollutants, which have persisted even after their prohibition or restricted use, and their consequences for Antarctic environments. Furthermore, the long-term monitoring of POPs in this polar region is necessary because Antarctica may become the major POP sink and source in the future.

Monitoring programs on a continental scale have also been suggested to assess metal pollution long term effects and climate change in terrestrial and marine Antarctic ecosystems (Bargagli 2000; 2008; Sanchez-Hernandez 2000).

Although the Antarctic Treaty provides guidelines for protection of the Antarctic environment and regulations upon human activities in the continent and in the Antarctic Ocean, global warming, population growth and industrial development in countries from the southern hemisphere will probably increase the impact of anthropogenic contaminants in the Antarctic ecosystem in the future.

Acknowledgments

This work is a contribution to the Brazilian Antarctic Program (PROANTAR) and National Science and Technology Institute on Antarctic Environmental Research (INCT-APA)—(CNPq 574018/2008–5 and FAPERJ E-16/170023/2008).

References

Ahn, I.Y., Lee, S.H., Kim, K.T., Shim, J.H. and Kim, D.Y. 1996. Baseline heavy metal concentrations in the Antarctic clam, *Laternula elliptica* in Maxwell Bay, King George Island, Antarctica. Mar. Pollut. Bull. 32: 592–598.

Ahn, I.Y., Kang, J. and Kim, K.W. 2001. The effect of body size on metal accumulations in the bivalve *Laternula elliptica*. Antarctic Sci. 13(4): 355–362.

Ahn, I.Y., Kim, K.W. and Choi, H.J. 2002. A baseline study on metal concentrations in the Antarctic limpet *Nacella concinna* (Gastropod: *Patellidae*) on King George Island: variations with sex and body parts. Mar. Pollut. Bull. 44: 421–431.

Ahn, I.Y., Chung, K.H. and Choi, H.J. 2004. Influence of glacial runoff on baseline metal accumulation in the Antarctic limpet *Nacella concinna* from King George Island. Mar. Pollut. Bull. 49: 119–141.

Andrade, S., Poblet, A., Scagliola, M., Vodopivez, C., Curtosi, A., Pucci, A. and Marcovecchio, J. 2001. Distribution of heavy metals in surface sediments from an Antarctic marine ecosystem. Environ. Monit. Assess. 66: 147–158.

Aronson, R.B., Thatje, S., McClintock, J.B. and Huges, K.A. 2011. Anthropogenic impacts on marine ecosystems in Antarctica. Annals of the New York Academy of Sciences 1223: 82–107.

Bargagli, R., Nelli, L., Ancora, S. and Focardi, S. 1996. Elevated cadmium accumulation in marine organisms from Terra Nova Bay (Antarctica). Polar Biol. 16: 513–520.

Bargagli, R., Monaci, F., Sanchez-Hernandez, J.C. and Cateni, D. 1998. Biomagnification of mercury in an Antarctic marine coastal food web. Mar. Ecol. Prog. Ser. 169: 65–76.

Bargagli, R. 2000. Trace metals in Antarctica related to climate change and increasing human impact. Rev. Envion. Contam. Toxicol. 166: 129–173.

Bargagli, R. 2005. Ecological Studies 175: Antarctic Ecosystems – Environmental Contamination, Climate Change and Human Impact. Springer-Verlag, Berlin 395 pp.

Bargagli, R. 2008. Environmental contamination in Antarctic ecosystems. Sci. Total Environ. 400(1-3): 212–226.

Barra, R., Colombo, C.J., Eguren, G., Jardim, W.F., Gamboa, N. and Endonza, G. 2005. Persistent Organic pollutants (POPs) in Eastern and Westearn South American countries. Rev. Environ. Contam. T. 185: 1–33.

Berkman, P.A. and Nigro, M. 1992. Trace metal concentrations in Scallop around Antarctica: extending the mussel watch Programme to the Southern Ocean. Mar. Pollut. Bull. 24: 322–323.

Bícego, M.C., Weber, R.R. and Ito, R.G. 1996. Hydrocarbons on surface waters of Admiralty Bay, King George Island, Antarctica. Mar. Pollut. Bull. 32: 549–553.

Bícego, M.C., Zanardi, E., Ito, R.G. and Weber, R.R. 1998. Hydrocarbons in surface sediments of Admiralty Bay, King George Island, Antarctica, Peninsula. Pesq. Antar Bras. 3: 15–21.

Bícego, M.C., Zanardi-Lamardo, E. and Weber, R.R. 2003. Four-year survey of dissolved/ dispersed petroleum hydrocarbons on surface waters of Admiralty Bay, King George Island, Antarctica. Braz. J. Oceanog. 51: 33–38.

Bícego, M.C., Zanardi-Lamardo, E., Taniguchi, S., Martins, C.C., Silva, D.A.M., Sasaki, S.T., Albergaria-Barbosa, A.C., Paolo, F.S., Weber, R.R. and Montone, R.C. 2009. Results from a 15-year study on hydrocarbon concentrations in water and sediment from Admiralty Bay, King George Island, Antarctica. Antarct. Sci. 21: 209–220.

Bidleman, T.F., Walla, M.D., Roura, R., Carr, E. and Schmidt, S. 1993. Organochlorine pesticides in the atmosphere of the Southern Ocean and Antarctica, January–March, 1990. Mar. Pollut. Bull. 26: 258–262.

Bidleman, T.F., Helm, P.A., Braune, B.M. and Gabrielsen, G.W. 2010. Polychlorinated naphthalenes in polar environments—A review. Sci. Total Environ. 408(15): 2919–2935.

Broecker, W.S. 1987. Unpleasant surprises in the greenhouse? Nature. 328: 123–126.

Bruni, V., Maugeri, T.L. and Monticelli, L. 1997. Faecal pollution indicators in the Terra Nova Bay (Ross Sea, Antarctica). Mar. Pollut. Bull. 34: 908–912.

Butler, J.H., King, D.B., Lobert, J.M., Montzka, S.A., Yvon-Lewis, S.A., Hall, B.D., Warwick, N.J., Mondeel, D.J., Aydin, M. and Elkins, J.W. 2007. Oceanic distributions and emissions of short-lived halocarbons, Global biogeochemical cycles 21.

Cabrerizo, A., Dachs, J., Barceló, D. and Jones, K.C. 2013. Climatic and biogeochemical controls on the remobilization and reservoirs of persistent organic pollutants in Antarctica. Environ. Sci. Technol. 47: 4299–4306.

Chen, D., Hale, R.C., La Guardia, M.J., Luellen, D., Kim, S. and Geisz, H.N. 2015. Hexabromocyclododecane flame retardant in Antarctica: Research stations as sources. Environ. Pollut. 206: 611–618.

Choi, S.D., Baek, S.Y., Chang, Y.S., Wania, F., Ikonomou, M.G., Yoon, Y.J., Park, B.K. and Hong, S. 2008. Passive air sampling of polychlorinated biphenyls and organochlorine pesticides at the Korean Arctic and Antarctic research stations: Implications for long-range transport and local pollution. Environ. Sci. Technol. 42(19): 7125–7131.

Cincinelli, A., Martellini, T., Bittoni, L., Russo, A., Gambaro, A. and Lepri, L. 2007. Natural and anthropogenic hydrocarbons in the water column of the Ross Sea (Antarctica). J. Mar. Syst. 73: 208–220.

Cincinelli, A., Martellin, T., Pozo, K., Kukucka, P., Audy, O. and Corsolini, S. 2016. *Trematomus bernacchii* as an indicator of POP temporal trend in the Antarctic seawaters. Environ. Pollut. (in press).

Cipro, C.V.Z., Taniguchi, S. and Montone, R.C. 2010. Occurrence of organochlorine compounds in *Euphausia superba* and unhatched eggs of *Pygoscelis* genus penguins from Admiralty Bay (King George Island, Antarctica) and estimation of biomagnification factors. Chemosphere 78: 767–771.

Cipro, C.V.Z., Yogui, G.T., Bustamante, P., Taniguchi, S., Sericano, J.L. and Montone, R.C. 2011. Organic pollutants and their correlation with stable isotopes in vegetation from King George Island, Antarctica. Chemosphere. 85(3): 393–398.

Cipro, C.V.Z., Bustamante, P., Taniguchi, S. and Montone, R.C. 2012. Persistent organic pollutants and stable isotopes in pinnipeds from King George Island, Antarctica. Mar. Pollut. Bull. 64: 2650–2655.

Cipro, C.V.Z., Colabuono, F.I., Taniguchi, S. and Montone, R.C. 2013. Persistent organic pollutants in bird, fish and invertebrate samples from King George Island, Antarctica. Antarct. Sci. 25(4): 545–552.

Colabuono, S., Taniguchi, C.V.Z., Cipro, J. da Silva, Bícego, M.C. and Montone, R.C. 2015. Persistent organic pollutants and polycyclic aromatic hydrocarbons in mosses after fire at the Brazilian Antarctic Station. Mar. Pollut. Bull. 93: 266–269.

Combi, T., Martins, C.C., Taniguchi, S., Leonel, J., Lourenço, R.A. and Montone, R.C. 2017. Depositional history and inventories of polychlorinated biphenyls (PCBs) in sediment cores from an Antarctic Specially Managed Area (Admiralty Bay, King George Island). Marine Pollution Bulletin, http://dx.doi.org/10.1016/j.marpolbul.2017.03.031.

COMNAP (Council of Managers of National Antarctic Programs). 2015. Main Antarctic facilities operated by National Programs in the Antarctic Treaty area (South of 60°S). Available at https://www.comnap.aq/.../Antarctic_Facilities_List_24Feb2015.

Conlan, K.E., Kim, S.L., Lenihan, H.S. and Oliver, J.S. 2004. Benthic changes during 10 years of organic enrichment by McMurdo Station, Antarctica. Mar. Pollut. Bull. 49: 43–60.

Conlan, K.E., Kim, S.L., Thurber, A.R. and Hendrycks, E. 2010. Benthic changes at McMurdo Station, Antarctica following local sewage treatment and regional iceberg-mediated productivity decline. Mar. Pollut. Bull. 60: 49–432.

Convey, P., Stevens, M.I., Hodgson, D.A., Smellie, J.L., Hillenbrand, C.D., Barnes, D.K., Clarke, A., Pugh, P.J., Linse, K. and Cary, S.C. 2009. Exploring biological constraints on the glacial history of Antarctica. Quaternary Sci. Rev. 28: 3035–3048.

Corsolini, S. and Focardi, S. 2000. Bioconcentration of polychlorinated biphenyls in the pelagic food chain of the Ross Sea. pp. 575–584. *In*: Faranda, F.M., Guglielmo, L. and Ianora, A. (eds.). Ross Sea Ecology. Springer, Berlin, Heidelberg, New York.

Corsolini, S., Romeo, T., Ademollo, N., Greco, S. and Focardi, S. 2002. POPs in key species of marine Antarctic ecosystem. Microchem. J. 73: 187–193.

Corsolini, S., Covaci, A., Ademollo, N., Focardi, S. and Schepens, P. 2006. Occurrence of organochlorine pesticides (OCPs) and their enantiomeric signatures, and concentrations of polybrominated diphenyl ethers (PBDEs) in the Adelie penguin food web. Antarctica. Environ. Pollut. 140: 371–382.

Corsolini, S., Borghesi, N., Schiamone, A. and Focardi, S. 2007. Polybrominated diphenyl ethers, polychlorinated dibenzo-dioxins, -furans, and -biphenyls in three species of Antarctic Penguins. Env. Sci. Pollut. Res. 14(6): 421–429.

Corsolini, S., Borghesi, N., Ademollo, N. and Focardi, S. 2011. Chlorinated biphenyls and pesticides in migrating and resident seabirds from East and West Antarctica. Environ. Int. 37: 1329–1335.

Corsolini, S. 2012. Contamination profile and temporal trend of POPs in Antarctic biota. Chapter 25. pp. 571–591. *In:* Loganathan, B.G. and Lam, P.K.S. (eds.). Global Contamination Trends of Persistent Organic Chemicals. CRC Press Taylor & Francis Group.

Cripps, G.C. and Priddle, J. 1991. Hydrocarbons in the Antarctic marine environment. Antarct. Sci. 3: 233–250.

Cripps, G.C. 1992. Natural and anthropogenic hydrocarbons in the Antarctic marine environment. Mar. Pollut. Bull. 25: 266–273.

Cripps, G.C. and Shears, J. 1997. The fate in the marine environment of a minor diesel fuel spill from an Antarctic research station. Environ. Monit. Assess. 46: 221–232.

Cunningham, S.A., Alderson, S.G., King, B.A. and Brandon, M.A. 2003. Transport and variability of the Antarctic Circumpolar Current in Drake Passage. J. Geophys. Res. (Oceans). 108 (C5): 8084. doi:10.1029/2001JC001147.

Dalla Riva, S., Abelmoschi, M.L., Magi, E. and Soggia, F. 2004. The utilization of the Antarctic environmental specimen bank (BCAA) in monitoring Cd and Hg in an Antarctic coastal area in Terra Nova Bay (Ross Sea—Northern Victoria Land). Chemosphere 56: 59–69.

Dauner, A.L.L., Maccormack, W.P., Hernandez, E.A. and Martins, C.C. 2015. Molecular characterisation of anthropogenic sources of sedimentary organic matter from Potter Cove, King George Island, Antarctica. Sci. Total Environ. 502: 408–416.

De Moreno, J.E.A., Gerpe, M.S., Moreno, V.J. and Vodopivez, C. 1997. Heavy metals in Antarctic organisms. Polar Biol. 17: 131–140.

Deacon, G. 1979. The Weddell Gyre. Deep-Sea Res. Part A. Oceanographic Research Papers 26: 981–995.

Deheyn, D.D., Gendreau, P., Baldwin, R.J. and Latz, M.I. 2005. Evidence for enhanced bioavailability of trace elements in the marine ecosystem of Deception Island, a volcano in Antarctica. Mar. Environ. Res. 60: 17–24.

Deprez, P.P., Arens, M. and Locher, H. 1999. Identification and assessment of contaminated sites at Casey. Polar Rec. 195: 299–316.

Desideri, D., Giuliani, S., Testa, C. and Triulzi, C. 2003. ^{90}Sr, ^{137}Cs, ^{238}Pu and ^{241}Am levels in terrestrial and marine ecosystems around the Italian base in Antarctica. J. Radioanal. Nucl. Chem. 258(2): 221–5.

Dickhut, R.M., Cincinelli, A., Cochran, M. and Ducklow, H.W. 2005. Atmospheric concentrations and air-water flux of organochlorine pesticides along the western Antarctic Peninsula. Environ. Sci. Technol. 39: 465–470.

Donnewald, H., Astolfi, E., Belifera, J.C. and Fernandez, J.C.G. 1979. Contaminacion del medio en el sector Antarctico Argentino: 1 Plagiucidas organoclorados en grasa del pinguinos papua. Contrib. Inst. Antart. Argent. 239: 111–9.

Edwards, D.D., McFeters, G.A. and Venkatesan, M.I. 1998. Distribution of *Clostridium perfringens* and fecal sterols in a benthic coastal marine environment influenced by the sewage outfall from McMurdo station, Antarctica. Appl. Environ. Microb. 64: 2596–2600.

Emnet, P., Gaw, S., Northcott, G., Storey, B. and Graham, L. 2015. Personal care products and steroid hormones in the Antarctic coastal environment associated with two Antarctic research stations, McMurdo Station and Scott Base. Environ. Res. 136: 331–342.

Farías, S.S., Pérez Arisnabarreta, S., Vodopívez, C. and Smichowski, P. 2002. Levels of essential and potentially toxic trace metals in Antarctic macroalgae. Spectrochim. Acta part. B. 57: 2133–2140.

Farías, S.S., Smichowski, P., Vélez, D., Montoro, R., Curtosi, A. and Vodopívez, C. 2007. Total and inorganic arsenic in Antarctic macroalgae. Chemosphere 69: 1017–1024.

Ferreira, P.A.L., Ribeiro, A.P., Nascimento, M.G., Martins, C.C., Mahiques, M.M., Montone, R.C. and Figueira, R.C.L. 2013. ^{137}Cs in marine sediments of Admiralty Bay, King George Island, Antarctica. Sci. Total Environ. 443: 505–510.

Focardi, S., Fossi, M.C., Leonzio, C., Lari, L., Marsili, L., Court, G.S. and Davis, L.S. 1992. Mixed function oxidase activity and chlorinated hydrocarbon residues in Antarctic sea birds: South polar skua (*Catharacta maccormicki*) and adélie penguin (*Pygoscelis adeliae*). Mar. Environ. Res. 34(1-4): 201–205.

Fourcade, N.H. 1960. Estudio geológico y petrográfico de Caleta Potter, isla 25 de Mayo, Islas Shetland del Sur. Instituto Antártico Argentino. Publicación No. 8, 115 p.

Fryirs, K.A., Hafsteinsdóttir, E.G., Stark, S.C. and Gore, D.B. 2015. Metal and petroleum hydrocarbon contamination at Wilkes Station, East Antarctica Antarct. Sci. 27: 118–133.

Fuoco, R., Colombini, M.P., Ceccarini, A. and Abete, C. 1996. Polychlorobiphenyls in Antarctica. Microchem. J. 54: 384–390.

Fuoco, R., Giannarelli, S., Wei, Y., Abete, C., Francesconi, S. and Termine, M. 2005. Polychlorobiphenyls and polycyclic aromatic hydrocarbons in the sea-surface micro-layer and the water column at Gerlache Inlet, Antarctica. J. Environ. Monit. 7: 1313–1319.

Fuoco, R., Giannarelli, S., Wei, Y., Ceccarini, A., Abete, C., Francesconi, S. and Termine, M. 2009. Persistent organic pollutants (POPs) at Ross Sea (Antarctica). Microchem. J. 92(1): 44–48.

Gambarro, A., Manodori, L., Zangrando, R., Cincinelli, A., Capodaglio, G. and Cescon, P. 2005. Atmospheric PCB concentrations at Terra Nova Bay, Antarctica. Environ. Sci. Technol. 39: 9406–9411.

Geisz, H.N., Dickhut, R.M., Cochran, M.A., Fraser, W.R. and Ducklow, H.W. 2008. Melting glaciers: A probable source of DDT to the antarctic marine ecosystem. Environ. Sci. Technol. 42(11): 3958–3962.

George, J.L. and Frear, D.L.H. 1966. Pesticides in the environment and their effects on wildlife. J. Appl. Ecol. 3(suppl): 155–167.

Goerke, H., Weber, K., Bornemann, H., Ramdohr, S. and Plotz, J. 2004. Increasing levels and biomagnification of persistent organic pollutants (POPs) in Antarctic biota. Mar. Pollut. Bull. 48: 295–302.

Green, G. and Nichols, P.D. 1995. Hydrocarbons and sterols in marine sediments and soils at Davis station, Antarctica: a survey for human-derived contaminants. Antarct. Sci. 7: 137–144.

Gröndahl, F., Sidenmark, J. and Thomsen, A. 2009. Survey of waste water disposal practices at Antarctic research stations. Polar Res. 28: 298–306.

Grotti, M., Soggia, F., Lagomarsino, C., Dalla Riva, S., Goessler, W. and Francesconi, K.A. 2008. Natural variability and distribution of trace elements in marine organisms from Antarctic coastal environments. Antarct. Sci. 20: 39–51.

Guzzella, L., Roscioli, C., Vigano, L., Sahab, M., Sarkarb, S.K. and Bhattachary, A. 2005. Evaluation of the concentration of HCH, DDT, HCB, PCB and PAH in the sediments along the lower stretch of Hugli estuary, West Bengal, northeast India. Environ. Int. 31: 523–534.

Hale, C., Kim, S.L., Harvey, E., La Guardia, M.J., Mainor, T.M., Bush, E.O. and Jacobs, E.M. 2008. Antarctic research bases: Local sources of polybrominated diphenlyl ether (PBDE) flame retardants. Environ. Sci. Technol. 42: 1452–1457.

Hidaka, H., Tanabe, S., Kawano, M. and Tatsukawa, R. 1984. Fate of DDTs, PCBs and chlordane compounds in the Antarctic marine ecosystem. Mem.natn. Inst. Pol. Res. (32 special issue): 151–161.

Hobbs, K.E., Lebeuf, M. and Hsmmill, M.O. 2002. PCBs and OCPs in male harbour, grey, harp and hooded seals from the Estuary and Gulf of St. Lawrence, Canada. Sci. Total Environ. 296: 1–18.

Hong, S., Kang, C.Y. and Kang, J. 1999. Lichen biomonitoring for the detection of local heavy metal pollution aroun King Sejong Station, King George Island, Antarctica. Korean J. Polar Res. 10: 17–24.

Howington, J.P., McFeters, G.A., Barry, J.P. and Smith, J.J. 1992. Distribution of the McMurdo Station sewage plume. Mar. Pollut. Bull. 25: 324–327.

Hughes, K.A. 2003. Influence of seasonal environmental variables on the distribution of presumptive fecal coliforms around an Antarctic research station. Appl. Environ. Microb. 69: 4884–4891.

Hughes, K.A. 2004. Reducing sewage pollution in the Antarctic marine environment using a sewage treatment plant. Mar. Pollut. Bull. 49: 850–853.

Hughes, K.A. and Thompson, A. 2004. Distribution of sewage pollution around a maritime Antarctic research station indicated by faecal coliforms, *Clostridium perfringens* and faecal sterol markers. Environ. Pollut. 127: 315–321.

IAATO. 2016. http://iaato.org/documents/10157/1017626/20142015+Tourism+Summary+by+Expedition/f79481bc-71ec-4095-a174-d6d6beb22b5d.

Ianni, C., Magi, E., Soggia, F., Rivaro, P. and Frache, R. 2010. Trace metal speciation in coastal and off-shore sediments from Ross Sea (Antarctica). Microchem. J. 96: 203–212.

Inomata, O.N.K., Montone, R.C., Lara, W.H., Weber, R.R. and Toledo, H.H.B. 1996. Tissue distribution of organochlorine residues—PCBs and pesticides—in Antarctic penguins. Antarct. Sci. 8: 253–255.

Islam, M.S. and Tanaka, M. 2004. Impacts of pollution on coastal and marine ecosystems including coastal and marine fisheries and approach for management: a review and synthesis. Mar. Pollut. Bul. 48: 624–649.

Iwata, I., Tanabe, S., Sakai, N. and Tatsukawa, R. 1993. Distribution of persistent organochlorines in the oceanic air and surface seawater and the role of ocean on their global transport and fate. Environ. Sci. and Technol. 27: 1080–1098.

Jakimska, A., Konieczka, P., Skóra, K. and Namieśnik, J. 2011. Bioaccumulation of metals in tissues of marine animals, part I: The role and impact of heavy metals on organisms. Pol. J. Environ. Stud. 20: 1117–1125.

Jerez, S., Motas, M., Palacios, M.J., Valera, F., Cuervo, J.J. and Barbosa, A. 2011. Concentration of trace elements in feathers of three Antarctic penguins and interspecific differences. Environ. Pollut. 159: 2412–2419.

Joiris, C. and Overloop, W. 1991. PCBs and organochlorine pesticides in phytoplankton and zooplankton in the Indian sector of the Southern ocean. Antarct. Sci. 3: 371–377.

Jones, K.C. and de Voogt, P. 1999. Persistent organic pollutants (POPs): state of the science. Environ. Pollut. 100: 209–221.

Kallenborn, R., Oehme, M., Wynn-Williams, D.D., Schlabach, M. and Harris, J. 1998. Ambient air levels and atmospheric long-range transport of persistent organochlorines to Signy Island, Antarctica. Sci. Total Environ. 220: 167–180.

Karadede-Akin, H. and Unlu, E. 2007. Heavy Metal Concentrations in Water, Sediment, Fish and Some Benthic Organisms from Tigris River, Turkey. Environ. Monit. Assess. 131: 323–337.

Karolewski, M.A., Lukowski, A.B. and Halba, R. 1987. Residues of chlorinated hydrocarbons in the adipose tissue of the Antarctic pinnipeds. Polish. Polar Res. 8: 189–197.

Kasiotis, K.M. and Emmanouil, C. 2015. Advanced PAH pollution monitoring by bivalves. Environ. Chem. Letters. 13: 395–411.

Kawano, M., Tanabe, S., Inoue, T. and Tatsukawa, R. 1985. Chlordane compounds found in the marine atmosphere from the southern hemisphere. Trans. Tokyo Univ. Fish. 6: 59–66.

Kennicutt, M.C., Sweet, S.T., Fraser, W.R., Stockton, W.L. and Culver, M. 1991. Grounding of the Bahia Paraiso at Arthur Harbor, Antarctica. 1. Distribution and fate of oil spill related hydrocarbons. Environ. Sci. Technol. 25: 509–518.

Kennicutt II, M.C., McDonald, T.J., Denoux, G.J. and McDonald, S.J. 1992. Hydrocarbon contamination on the Antarctic Peninsula. Mar. Pollut. Bull. 24: 499–506.

Kennicutt, M.C., McDonald, S.J., Sericano, J.L., Boothe, P., Oliver, J., Safe, S., Presley, B.J., Liu, H., Wolfe, D., Wade, T.L., Crockett, A. and Bockus, D. 1995. Human Contamination of the Marine Environment-Arthur Harbor and McMurdo Sound, Antarctica. Environ. Sci. Technol. 29: 1279–1287.

Kim, M., Kennicutt II, M.C. and Qian, Y. 2006. Molecular and stable carbon isotopic characterization of PAH contaminants at McMurdo Station, Antarctica. Mar. Pollut. Bull. 52: 1585–1590.

King, C.K. and Riddle, M.J. 2001. Effects of metal contaminants on the embryonic and larval development of the common Antarctic sea urchin *Sterechinus neumayeri* (Meissner). Mar. Ecol-Prog. Ser. 215: 143–154.

Klanova, J., Matykiewiczova, N., Macka, Z., Prosek, P., Laska, K. and Klan, P. 2008. Persistent organic pollutants in soils and sediments from James Ross Island, Antarctica. Environ. Pollut. 152: 416–423.

Klein, A.G., Sweet, S.T., Wade, T.L., Sericano, J.L. and Kennicutt, M.C. 2012. Spatial patterns of total petroleum hydrocarbons in the terrestrial environment at McMurdo Station. Antarctica Antarct. Sci. 24: 450–466.

Krahn, M.M., Pitman, R.L., Burrows, D.G., Herman, D.P. and Pearce, R.W. 2008. Use of chemical tracers to assess diet and persistent organic pollutants in Antarctic Type C killer whales. Mar. Mammal Sci. 24(3): 643–663.

Larsson, P., Järnmark, C. and Södergren, A. 1992. PCBs and chlorinated pesticides in the atmosphere and aquatic organisms of Ross Island, Antarctica. Mar. Pollut. Bull. 25(9-12): 281–287.

Le Quéré, C., Rödenbeck, C., Buitenhuis, E.T., Conway, T.J., Langenfelds, R., Gomez, A., Labuschagne, C., Ramonet, M., Nakazawa, T., Metzl, N., Gillett, N. and Heimann, M. 2007. Saturation of the Southern Ocean CO_2 sink due to recent climate change. Science 316: 1735–1738.

Leeming, R., Stark, J.S. and Smith, J.J. 2015. Novel use of faecal sterols to assess human faecal contamination in Antarctica: a likelihood assessment matrix for environmental monitoring. Antarct. Sci. 27: 31–43.

Lenihan, H.S. 1992. Benthic marine pollution around McMurdo Station, Antarctica: A summary of Findings. Mar. Pollut. Bull. 25: 318–323.

Levitus, S., Antonov, J. and Boyer, T. 2005. Warming of the world ocean, 1955–2003. Geophys. Res. Lett. 32: L02604. doi:10.1029/2004GL021592.

Lister, K.N., Lamare, M.D. and Burrit, D.J. 2015. Oxidative damage and antioxidant defense parameters in the Antarctic bivalve *Laternula elliptica* as biomarkers for pollution impacts. Polar Biol. 38: 1741–1752.

Lukowski, A.B. 1983a. DDT residues in the tissues and eggs of three species of penguins from breeding colonies at Admiralty Bay (King George Island, South Shetland Islands). Pol. Polar Res. 4: 129–134.

Lukowski, A.B. 1983b. DDT and its metabolites in the tissues and eggs of migrating Antarctic seabirds from the regions of the South Shetland Islands. Pol. Polar Res. 4: 135–141.

Lukowski, A.B., Karolewski, M.A. and Gorski, T. 1987. Polychlorinated biphenyls in the tissues from the breeding colony on King George Island (South Shetland Islands). Pol. Polar Res. 8(2): 179–187.

Lythe, M., Vaughan, D. and the BEDMAP Consortium. 2001. BEDMAP: A new ice thickness and subglacial topographic model of Antarctica. J. Geophys. Res. 106(B6): 11,335–11,351.

Machado, A., Lima, E.F., Chemale Jr., F., Liz, J.D. and Ávila, J.N. 2001. Química mineral de rochas vulcânicas da Península Fildes (Ilha Rei George), Antártica. Rev. Bras. Geoc. 31: 299–306.

Majer, A.P., Petti, M.A., Corbisier, T.N., Ribeiro, A.P., Theophilo, C.Y., Ferreira, P.A. and Figueira, R.C. 2014. Bioaccumulation of potentially toxic trace elements in benthic organisms of Admiralty Bay (King George Island, Antarctica). Mar. Pollut. Bull. 79: 321–325.

Marcovecchio, J.E. 2004. The use of *Micropogonias furnieri* and *Mugil liza* as bioindicators of heavy metals pollution in La Plata river estuary, Argentina. Sci. Total. Environ. 323: 219–226.

Martins, C.C., Venkatesan, M.I. and Montone, R.C. 2002. Sterols and linear alkylbenzenes in marine sediments from Admiralty Bay, King George Island, South Shetland Islands. Antarct. Sci. 14: 244–252.

Martins, C.C., Bícego, M.C., Taniguchi, S. and Montone, R.C. 2004. Aliphatic (AHs) and aromatic hydrocarbons (PAHs) in surface sediments in Admiralty Bay, King George Island, Antarctica. Antarct. Sci. 16: 117–122.

Martins, C.C., Montone, R.C., Gamba, R.C. and Pellizari, V.H. 2005. Sterols and fecal indicator microorganisms in sediments from Admiralty Bay, Antarctica. Braz. J. Oceanogr. 53: 1–12.

Martins, C.C., Bícego, M.C., Rose, N.L., Taniguchi, S., Lourenço, R.A., Figueira, R.C.L., Mahiques, M.M. and Montone, R.C. 2010. Historical record of polycyclic aromatic hydrocarbons (PAHs) and spheroidal carbonaceous particles (SCPs) in marine sediment cores from Admiralty Bay, King George Island, Antarctica. Environ. Pollut. 158: 192–200.

Martins, C.C., Aguiar, S.N., Bícego, M.C. and Montone, R.C. 2012. Sewage organic markers in surface sediments around the Brazilian Antarctic station: Results from the 2009/10 austral summer and historical tendencies. Mar. Pollut. Bull. 64: 2867–2870. http://dx.doi.org/10.1016/j.marpolbul.2012.08.019.

McFeters, G.A., Barry, J.P. and Howington, J.P. 1993. Distribution of enteric bacteria in Antarctic seawater surrounding a sewage outfall. Water Res. 27: 645–650.

Mearns, A.J., Reish, D.J., Oshida, P.S., Ginn, T., Rempel-Hester, M.A., Arthur, C. and Ritheford, N. 2015. Effects of pollution on marine organisms. Water Environ. Res. 87: 1718–1816.

Miranda-Filho, K.C., Metcalfe, T.L., Metcalfe, C.D., Robaldo, R.B., Muelbert, M.M.C., Colares, E.P., Martinez, P.E. and Bianchini, A. 2007. Residues of persistent organochlorine contaminants in Southern Elephant seals (*Mirounga leonina*) from Elephant Island, Antarctica. Environ. Sci. Technol. 41: 3829–3835.

Montone, R.C., Taniguchi, S., Sericano, J.L., Weber, R.R. and Lara, W.H. 2001a. Determination of polychlororinated biphenyls in Antarctic macroalgae "*Desmarestia* sp." Sci. Total. Enviro. 277: 181–186.

Montone, R.C., Taniguchi, S. and Weber, R.R. 2001b. Polychlorinated biphenyls in marine sediments of Admiralty Bay, King George Island, Antarctica. Mar. Pollut. Bull. 42(7): 611–614.

Montone, R.C., Taniguchi, S. and Weber, R.R. 2003. PCBs in the atmosphere of King George Island, Antarctica. Sci. Total Environ. 308: 167–173.

Montone, R.C., Taniguchi, S., Boian, C. and Weber, R.R. 2005. PCBs and chlorinated pesticides (DDTs, HCHs and HCB) in the atmosphere of the southwest Atlantic and Antarctic oceans: Mar. Pollut. Bull. 50: 778–782.

Montone, R.C., Martins, C.C., Bícego, M.C., Taniguchi, S., Silva, D.A.M., Campos, L.S. and Weber, R.R. 2010. Distribution of sewage input in marine sediments around a maritime Antarctic research station indicated by molecular geochemical indicators. Sci. Total Environ. 408: 4665–4671.

Montone, R.C., Taniguchi, S., Colabuono, F.I., Martins, C.C., Cipro, C.V.Z., Barroso, H.S., Silva, J., Bícego, M.C. and Weber, R.R. 2016. Persistent organic pollutants and polycyclic aromatic hydrocarbons in penguins of the genus *Pygoscelis* in Admiralty Bay—An Antarctic specially managed area. Mar. Pollut. Bull. 106: 377–382.

Muir, D.C.G. and de Wit, C.A. 2010. Trends of legacy and new persistent organic pollutants in circumpolar Arctic: overview, conclusion, and recommendations. Sci. Total Environ. 408(15): 3044–3051.

Murray, L.A., Raab, A., Marr, I.L. and Feldmann, J. 2003. Biotransformation of arsenate to arsenosugars by *Chlorella vulgaris*. Appl. Organomet. Chem. 17: 669–674.

Nadal, M., Marquès, M., Mari, M. and Domingo, J.L. 2015. Climate change and environmental concentrations of POPs: A review. Environ. Res. 143: 177–185.

Neff, J.M. 2002. Polycyclic aromatic hydrocarbons in the ocean. pp. 241–318. *In:* Neff, J.M. (ed.). Bioaccumulation in Marine Organisms, Effects of Contaminants from Oil Well Produced Water. Elsevier Science Publishers, Amsterdam, The Netherlands.

Negri, A., Burns, K., Boyle, S., Brinkman, D. and Webster, N. 2006. Contamination in sediments, bivalves and sponges of McMurdo Sound, Antarctica. Environ. Pollut. 143: 456–467.

NRC. 1985. Oil in the sea: Inputs, fates and effects. National Research Council, Washington, DC: National Academy Press, 602 pp.

Nygard, T., Lie, E., Rov, N. and Steinnes, E. 2001. Metal dynamics in an Antarctic food chain. Mar. Pollut. Bull. 42: 598–602.

Orsi, A., Smethie Jr, W. and Bullister, J. 2002. On the total input of Antarctic waters to the deep ocean: A preliminary estimate from chlorofluorocarbon measurements. J. Geophys. Res. 107: 3122.

Pereira, E.B., Evangelista, H., Pereira, K.C.D., Cavalcanti, I.F.A. and Setzer, A.W. 2006. Apportionment of black carbon in the South Shetland islands, Antarctic Peninsula. J. Geophy. Res. 111: D03303.

Pereira, K.C.D., Evangelista, H., Pereira, E.B., Simões, J.C., Johnson, E. and Melo, L.R. 2004. Transport of crustal microparticles from Chilean Patagonia to the Antarctic Peninsula by SEM-EDS analysis. Tellus, Serie B 56: 262–275.

Poland, J.S., Riddle, M.J. and Zeeb, B.A. 2003. Contaminants in the Arctic and the Antarctic: a comparison of sources, impacts, and remediation options. Polar Rec. 39(211): 369–383.

Polmear, R., Stark, J.S., Roberts, D. and McMinn, A. 2015. The effects of oil pollution on Antarctic benthic diatom communities over 5 years. Mar. Pollut. Bull. 90: 33–40.

Priddle, J., Boyd, I.L., Whitehouse, M.J., Murphy, E.J. and Croxall, J.P. 1998. Estimates of Southern Ocean primary production—constraints from predator carbon demand and nutrient drawdown. J. Marine Syst. 17: 275–288.

Priddle, J. 2002. Regionally based assessment of persistent toxic substances—Antarctica. United nations Environment Program—Global Environment Facility. Chatelaine. Switzerland: UNEP Chemicals 86 pp.

Pu, J., Li, Z., Shang, L. and Li, H. 1996. Organic contamination in the Great Wall Bay, Antarctica in Austral summer. J. Environ. Sci. (China) 8: 397–401.

Prus, W., Fabianaska, M.J. and Labno, R. 2015. Geochemical markers of soil anthropogenic contaminants in polar scientific stations nearby (Antarctica, King George Island) Sci. Total Environ. 518-519: 266–279.

Regoli, F., Nigro, M., Benedetti, M., Fattorini, D. and Gorbi, S. 2005. Antioxidant efficiency in early life stages of the Antarctic silverfish, *Pleuragramma antarcticum*: Responsiveness to pro-oxidant conditions of platelet ice and chemical exposure. Aquat. Toxicol. 24: 1475–1482.

Reid, J.L. 1997. On the total geostrophic circulation of the Pacific Ocean: Flow patterns, tracers, and transports. Progr. Oceanogr. 39: 263–352.

Ribeiro, A.P., Figueira, R.C.L., Martins, C.C., Silva, C.R.A., França, E.J., Bícego, M.C., Mahiques, M.M. and Montone, R.C. 2011. Arsenic and trace metal contents in sediment profiles from the Admiralty Bay, King George Island, Antarctica. Mar. Pollut. Bull. 62: 192–196.

Rignot, E., Casassa, G., Gogineni, P., Krabill, W., Rivera, A.U. and Thomas, R. 2004. Accelerated ice discharge from the Antarctic Peninsula following the collapse of Larsen B ice shelf. Geophys. Res. Lett. 31.

Rintoul, S.R. and Sokolov, S. 2001. Baroclinic transport variability of the Antarctic Circumpolar Current south of Australia (WOCE repeat section SR3). J. Geophys. Res. 106(C2): 2815–2832. doi:10.1029/2000JC900107.

Rintoul, S., Sparrow, M., Meredith, M., Wadley, V., Speer, K., Hofmann, E., Summerhayes, C., Urban, E. and Bellerby, R. 2012a. The Southern Ocean Observing System. Oceanography 25: 74.

Rintoul, S.R., Sparrow, M., Meredith, M.P., Wadley, V., Speer, K., Hofmann, E., Summerhayes, C., Urban, E., Bellerby, R. and Ackley, S. 2012b. The Southern Ocean observing system: initial science and implementation strategy, SCAR and SCOR 76 p.

Risebrough, R.W., Reiche, P., Peakall, D.B., Herman, S.G. and Kirven, M.N. 1968. Polychlorinated biphenyls in the global ecosystem. Nature (London) 220: 1098–1102.

Risebrough, R.W., Walker, H.W., Schmidt, T.T., de Lappe, B.W. and Connors, C.W. 1976. Transfer of chlorinated biphenyls to Antarctica. Nature 264: 738–739.

Risebrough, R.W., de Lappe, B.W. and Younghans-Haug, C. 1990. PCB and PCT contamination in Winter Quarters Bay, Antarctica. Mar. Pollut. Bull. 21: 523–529.

Rodman, M. and Gordon, A. 1982. Southern Ocean bottom water of the Australian-New Zealand sector. J. Geophys. Res. 87(C8): 5771–5778.

Rodrigues, E., Vani, G.S. and Lavrado, H.P. 2007. Nitrogen metabolism of the Antarctic bivalve *Laternula elliptica* (King & Broderip) and its potential use as biomarker. Oecol. Bras. 11: 37–49.

Ruoppolo, V., Woehler, E.J., Morgan, K. and Clumpner, C.J. 2013. Wildlife and oil in the Antarctic: a recipe for cold disaster. Polar Rec. 49: 97–109.

Sabine, C.L., Feely, R.A., Gruber, N., Key, R.M., Lee, K., Bullister, J.L., Wanninkhof, R., Wong, C.S., Wallace, D.W.R., Tilbrook, B., Millero, F.J., Peng, T.H., Kozyr, A., Ono, T. and Rios, A.F. 2004. The oceanic sink for anthropogenic CO_2. Science 305(5682): 367–371.

Salomons, W. and Förstner, U. 1984. Metals in the Hydrocycle. Springer-Verlag, Berlin 349 p.

Sanchez-Hernandez, J.C. 2000. Trace element contamination in Antarctic ecosystems. Rev. Environ. Contam. Toxicol. 166: 83–127.

Santos, I.R., Silva-Filho, E.V., Schaefer, C.E., Albuquerque Filho, M.R. and Campos, L.S. 2005. Heavy metals contamination in coastal sediments and soils near the Brazilian Antarctic Station, King George Island. Mar. Pollut. Bull. 50: 185–194.

Santos, I.R., Silva Filho, E.V., Schaefer, C.E., Sella, S.M., Silva, C.A., Gomes, V., Passos, M.J. and Ngan, P.V. 2006. Baseline mercury and zinc concentrations in terrestrial and coastal organisms of Admiralty Bay, Antarctica. Environ. Pollut. 140: 304–311.

Santos, I.R., Fávaro, D.I.T., Schaefer, C.E.G.R. and Silva Filho, E.V. 2007. Sediment geochemistry in coastal maritime Antarctica (Admiralty Bay, King George Island): Evidence from rare earths and others elements. Mar. Chem. 107: 464–474.

Sarmiento, J.L., Slater, R., Barber, R., Bopp, L., Doney, S.C., Hirst, A.C., Kleypas, J., Matear, R., Mikolajewicz, U., Monfray, P., Soldatov, V., Spall, S.A. and Stouffer, R. 2004. Response of ocean ecosystems to climate warming. Global Biogeochemical Cycles. 18 GB3003. doi:10.1029/2003GB002134.

Schiavone, A., Corsolini, S., Borghesi, N. and Focardi, S. 2009. Contamination profiles of selected PCB congeners, chlorinated pesticides, PCDD/Fs in Antarctic fur seal pups and penguin eggs. Chemosphere 76: 264–269.

Schmitz, W.J. 1996. On the eddy field in the Agulhas Retroflection, with some global considerations, J. Geophys. Res. 101: 16259–16272. doi:10.1029/96JC01143.

Sladen, W.J.L., Menzie, C.M. and Reichel, W.L. 1966. DDT residues in Adelie penguins and a crabeater seal from Antarctica. Nature 210: 670–673.

Sokolov, S. and Rintoul, S.R. 2009. Circumpolar structure and distribution of the Antarctic Circumpolar Current fronts: 2. Variability and relationship to sea surface height. J. Geophys. Res. Oceans (1978–2012), 114(C11). doi:10.1029/2008JC005248.

Stark, J.S., Riddle, M.J. and Simpson, R.D. 2003a. Human impacts in soft sediment assemblages at Casey Station, East Antarctica: spatial variation, taxonomic resolution and data transformation. Austral Ecol. 28: 287–304.

Stark, J.S., Riddle, M.J., Snape, I. and Scouller, R.C. 2003b. Human impacts in Antarctic marine soft-sediment assemblages: correlation between multivariate biological patterns and environmental variables as Casey Station. Estuar. Coast. Shelf. S. 56: 717–734.

Stark, J.S., Smith, J., King, C.K., Lindsay, M., Stark, S., Palmer, A.S., Snape, I., Bridgen, P. and Riddle, M. 2015. Physical, chemical, biological and ecotoxicological properties of wastewater discharged from Davis Station, Antarctica. Cold Reg. Sci. Technol. 113: 52–62.

Tanabe, S., Kawano, M. and Tatsukawa, R. 1982a. Chlorinated hydrocarbons in the Antarctic, Western Pacific and Eastern Indian Oceans. Trans. Tokyo Univ. Fish. 5: 97–109.

Tanabe, S., Tatsukawa, R., Kawano, M. and Hidaka, H. 1982b. Global distribution and atmospheric transport of chlorinated hydrocarbons: HCH (BHC) isomers and DDT compounds in the Western Pacific, Eastern Indian and Antarctic Oceans. J. Oceanogr. Soc. Japan 38: 137–148.

Tanabe, S., Hidaka, H. and Tatsukawa, R. 1983. PCBs and chlorinated biphenyls in Antarctic atmosphere and hydrosphere. Chemosphere 12: 277–288.

Taniguchi, S., Montone, R.C., Bícego, M.C., Colabuono, F.I., Weber, R.R. and Sericano, J.L. 2009. Chlorinated pesticides, polychlorinated biphenyls and polycyclic aromatic hydrocarbons in the fat tissue of seabirds from King George Island, Antarctica. Mar. Pollut. Bull. 58(1): 129–133.

Tatton, J.O.G. and Ruzicka, J.H.A. 1967. Organochlorine pesticides in Antarctica. Nature 215: 346–348.

Tin, T., Fleming, Z.L., Hughes, K.A., Ainley, D.G., Convey, P., Moreno, C.A., Pfeiffer, S., Scott, J. and Snape, I. 2009. Impacts of local human activities on the Antarctic environment. Antarct. Sci. 21(1): 3–33.

Tonelli, M.H.M. 2014. Numerical Investigation of the Ross Sea water masses using the Regional Ocean Modeling System – ROMS. Doctoral Thesis, Instituto Oceanográfico, University of São Paulo, São Paulo. Retrieved 2016-04-29. From http://www.teses.usp.br/teses/disponiveis/21/21135/tde-21012015-090549.

Tréguer, P. and Jacques, G. 1993. Review Dynamics of nutrients and phytoplankton, and fluxes of carbon, nitrogen and silicon in the Antarctic Ocean. pp. 149–162. *In:* Hempel, G. (ed.). Weddell Sea Ecology Results of EPOS European "Polarstern" Study. Springer.

Trenberth, K.E., Large, W.G. and Olson, J.G. 1990. The mean annual cycle in global ocean wind stress. J. Phys. Oceanogr. 20: 1742–1760.

Trevizani, T.H., Figueira, R.C.L., Ribeiro, A.P., Theophilo, C.Y.S., Majer, A.P., Petti, M.A.V., Corbisier, T.N. and Montone, R.C. 2016. Bioaccumulation of heavy metals in marine organisms and sediments from Admiralty Bay, King George Island, Antarctica. Mar. Pollut. Bull. 106: 366–371.

Turner, J. 2010. The Melting of ice in the Arctic Ocean: The Influence of Double-Diffusive Transport of Heat from Below. J. Phys. Oceanogr. 40: 249–256.

UNEP. 2002. Regionally based assessment of persistent toxic substances. Antarctica regional report 82 p.

Van den Brink, N.W. 1997. Directed transport of volatile organochlorine pollutants to polar regions: the effect on the contamination pattern of Antarctic seabirds. Sci. Total Environ. 198: 43–50.

Venkatesan, M.I. and Mirsadeghi, F.H. 1992. Coprostanol as sewage tracer in McMurdo Sound, Antarctica. Mar. Pollut. Bull. 25: 9–12.

Verdy, A., Dutkiewicz, S., Follows, M., Marshall, J. and Czaja, A. 2007. Carbon dioxide and oxygen fluxes in the Southern Ocean: Mechanisms of interannual variability. Global Biogeochemical Cycles, 21 GB2020. doi:10.1029/2006GB002916.

Vodopivez, C., Smichowski, P. and Marcovecchio, J. 2001. Trace metals monitoring as a tool for characterization of Antarctic ecosystems and environmental management. The Argentine programme at Jubany Station. Chap. 6. pp. 155–180. *In:* Caroli, S., Cescon, P. and Walton, D. (eds.). Environmental Contamination in Antarctica: A Challenge to Analytical Chemistry. Elsevier Science. New York.

Wallace, J.M. and Hobbs, P.V. 1977. Atmosphere science—an introductory survey. Academic Press (New York) 467 p.

Weber, K. and Goerke, H. 1996. Organochlorine compounds in fish off the Antarctic Peninsula. Chemosphere 33: 377–392.

Weber, R.R. and Montone, R.C. 1990. Distribution of organochlorines in the atmosphere of the South Atlantic and Antarctic Oceans. pp. 185–197. *In:* Kurtz, D.A. (ed.). Long range transport of pesticides. Lewis Publisher.

Wild, S., McLagan, D., Schlabach, M., Bossi, R., Hawker, D., Cropp, R., King, C.K., Stark, J.S., Mondon, J. and Nash, S.B. 2015. An Antarctic research station as a source of brominated and perfluorinated persistent organic pollutants to the Local Environment. Environ. Sci. Technol. 49: 103–112.

Xu, W., Wanga, X. and Cai, Z. 2013. Analytical chemistry of the persistent organic pollutants identified in the Stockholm Convention: A review. Anal. Chim. Acta 790: 1–13.

Xuebin, Y., Xiaodong, L., Linguand, S., Renbin, Z., Zhouqing, X. and Yuhong, W. 2006. A 1500-year record of lead, copper, arsenic, cadmium, zinc level in Antarctic seal hairs and sediments. Sci. Total Environ. 371: 252–257.

Yogui, G.T., Santos, M.C.O. and Montone, R.C. 2003. Chlorinated pesticides and polychlorinated biphenyls in marine tucuxi dolphins (*Sotalia fluviatilis*) from the Cananéia estuary, Southeastern Brazil. Sci. Total Environ. 312: 67–78.

Zhang, H.S., Wang, Z.P., Lu, B., Zhu, C., Wu, G.H. and Vetter, W. 2007. Occurrence of organochlorine pollutants in the eggs and dropping-amended soil of Antarctic large animals and its ecological significance. Sci. China Ser. D-Earth Sci. 50(7): 1086–1096.

Zhang, L., Dickhut, R., DeMaster, D., Pohl, K. and Lohmann, R. 2013. Organochlorine pollutants in Western Antarctic Peninsula sediments and benthic deposit feeders. Environ. Sci. Technol. 47(11): 5643–5651.

10

International Regulatory Responses to Global Challenges in Marine Pollution and Climate Change

Yubing Shi[1,*] and *Dazhen Zhang*[2]

Introduction

Marine pollution, also referred to as 'pollution of the marine environment', may occur as a result of different activities. Examples are land-based activities, vessel-related activitiese, dumping at sea, atmospheric and offshore hydrocarbon exploration, seabed mining, and so on. As discussed in Chapter 4, these types of marine pollution are often transboundary in nature and are harmful to human health and marine ecosystem. Similarly, climate change is a global issue involving the interests of all States. The Fifth Assessment Report of the Intergovernmental Panel on Climate Change (IPCC), finalized and published in 2014, has further confirmed the existence of global warming when compared with the previous IPCC reports. It indicates that climate change has negatively affected natural and human systems on all continents and across the oceans, and asserts that

---

[1] Associate Professor, Center for Oceans Law and the China Seas, South China Sea Institute, Xiamen University.
[2] Australian National Centre for Ocean Resources and Security (ANCORS), University of Wollongong.
E-mail: dz907@uowmail.edu.au
* Corresponding author: shiyubing@hotmail.com

substantial and sustained reduction of greenhouse gas (GHG) emissions would contribute to the tackling of climate change.[1] International issues need international responses. Both the marine pollution and climate change are issues with international dimensions, and thus require the global regulation by the international community.

This chapter focuses on the international regulation that has been developed at the global and regional levels for the purpose of the prevention, control and reduction of marine pollution and climate change. It consists of three main parts. The first part highlights the challenges in regulating marine pollution and climate change. The second and third parts introduce the regulatory initiatives of the international community in combating marine pollution and climate change respectively. It is worth emphasizing that there have been comprehensive international regulations on marine pollution and climate change. However, deficiencies remain in these regulations.

Marine Pollution and Climate Change as Global Regulatory Challenges

Marine pollution as a regulatory challenge

An overview of marine pollution

Marine pollution is not a new phenomenon, and there are various interpretations on this concept from different disciplines. From the international law perspective, there are two typical definitions on marine pollution. One is the definition provided by the Joint Group of Experts on Scientific Aspects of Marine Pollution (GESAMP) in 1969. Based on the views of the GESAMP, 'marine pollution' refers to

> "the introduction by man, directly or indirectly, of *substances* into the marine environment (including estuaries) *resulting in* such deleterious effects as *harm to living resources*, hazards to human health, hindrance to *marine activities including fishing*, impairment of quality for use of sea water and reduction of amenities."[2] [emphasis added]

---

[1] Intergovernmental Panel on Climate Change (IPCC), 'Climate Change 2014 Synthesis Report: Summary for Policymakers', available at < http://www.ipcc.ch/pdf/assessment-report/ar5/syr/AR5_SYR_FINAL_SPM.pdf> accessed 14 October 2015, p 8.

[2] Qing-nan Meng, *Land-based Marine Pollution: International Law Development* (Graham and Trotman 1987), at 4; Joint Group of Experts on Scientific Aspects of Marine Pollution (GESAMP), 'Report of the First Session (London, UN Doc.GESAMP I/11, 1969)', at 5.

Another definition of marine pollution can be seen from the 1982 United Nations Convention on the Law of the Sea (LOSC).[3] In accordance with Article 1(4) of the LOSC, 'pollution (of the marine environment)' refers to

> "the introduction by man, directly or indirectly, of *substances or energy* into the marine environment, including estuaries, which *results or is likely to result in* such deleterious effects as *harm to living resources and marine life*, hazards to human health, hindrance to *marine activities, including fishing and other legitimate uses of the sea*, impairment of quality for use of sea water and reduction of amenities."[4] [emphasis added]

The second definition provided by the LOSC represents an advance when compared with the first definition. The italicized term 'results or is likely to result in' indicates that this definition has adopted the precautionary approach.[5] Based on the 15th principle of the 1992 Rio Declaration on Environment and Development, this approach requires decision-makers not to use the 'lack of full scientific certainty' as a reason to postpone cost-effective measures to prevent environmental degradation.[6] Furthermore, the LOSC definition classifies marine pollution into 'substances or energy' whereas the GESAMP definition only refers to 'substances'. The types of harm and hindrances listed by the LOSC also add extra 'marine life' and 'other legitimate uses of the sea'. This broader formation of marine pollution better reflects the development of international law, and makes the LOSC definition widely accepted by the international community.[7]

The definition of marine pollution helps to judge whether a human activity leads to marine pollution, which also enables the application of marine pollution related international treaties to this activity. To date the issue of marine pollution has been comprehensively regulated, and the sources of marine pollution include land-based activities, shipping, dumping at sea, seabed activities, and so on.

Challenges in regulating marine pollution

Marine pollution has negative impacts on the oceans, which has been broadly discussed by various literatures and regulated by a number of

---

[3] *United Nations Convention on the Law of the Sea* (LOSC), 10 December 1982, 1833 UNTS 3, art 1(4).

[4] *Ibid.*

[5] Md Saiful Karim, *Prevention of Pollution of the Marine Environment from Vessels: the potential and Limits of the International Maritime Organization* (Springer 2015) 4.

[6] *Rio Declaration on Environment and Declaration*, 31 ILM 874 (14 June 1992) principle 15.

[7] Patricia Birnie, Alan Boyle and Catherine Redgwell, *International Law and the Environment* (Oxford University Press, 3rd, 2009) 189.

international treaties. In practice, it may take various forms, such as sewage, petroleum, metallic effluents, chlorinated hydrocarbons (organo-chlorines) and radionuclides.[8] These substances or energies, once discharged into the ocean, may lead to serious damage to the marine environment. Furthermore, the significant role of the ocean as a bottomless carbon sink might also be challenged by marine pollution.[9] Indeed the definition of the marine pollution provided by the LOSC has clearly revealed the significant harm of this type of pollution.

Given the threat from marine pollution, the international community has sought to monitor and regulate marine pollution through a number of specific international agreements. This process has been accelerated particularly after the establishment of the International Maritime Consultative Organization (IMCO) in 1958. The IMCO was renamed as the International Maritime Organization (IMO) in 1982. Although marine pollution has been comprehensively regulated, challenges remain as to the further regulation of this issue due to the development of international seaborne trade and climate change.

Firstly, it is challenging to properly balance the interests of large developed flag States and developing flag States. Traditionally the IMO has been dominated by main developed flag States despite of the 'no more favourable treatment' (NMFT) principle incorporated in treaties adopted by the IMO.[10] The differing regulatory interests between developed and developing States hinder or postpone the adoption of many treaties. For example, developed States push the adoption of regularly upgrading shipping technologies in negotiating treaties such as Annex VI to International Convention for the Prevention of Pollution from Ships (MARPOL 73/78)[11] and International Convention for the Control and Management of Ships' Ballast Water and Sediments (BMW Convention).[12] This type of regulation actually serves as a trade barrier for developing flag States, and is thus opposed by many developing States.

---

[8]   Christopher C. Joyner, 'The Southern Ocean and Marine Pollution: Problems and Prospects', (1985)17(165) *Case Western Reserve University's Journal of International Law* 165, 167.

[9]   Ibid 168.

[10] Md Saiful Karim, *Prevention of Pollution of the Marine Environment from Vessels: the potential and Limits of the International Maritime Organization* (Springer 2015) 19; see, e.g., *International Convention for the Prevention of Pollution from Ships (MARPOL 73/78)*, signed 2 November 1973, 12 ILM 1319, as amended by the 1978 Protocol to the 1973 Convention, 1341 UNTS 3, 17 ILM 546 (entered into force 2 October 1983) art 5(4).

[11] *International Convention for the Prevention of Pollution from Ships (MARPOL 73/78)*, signed 2 November 1973, 12 ILM 1319, as amended by the 1978 Protocol to the 1973 Convention, 1341 UNTS 3, 17 ILM 546 (entered into force 2 October 1983).

[12] *International Convention for the Control and Management of Ships' Ballast Water and Sediments*, opened for signature 13 February 2004, IMO Doc. BWM/CONF/36 (not yet in force) ('*BWM Convention*').

Secondly, global regulation of climate change has also imposed new challenges on marine pollution. On the one hand, combating climate change requires the reduction of greenhouse gas (GHG) emissions from international shipping. To regulate GHG emissions from international shipping and include this issue into an amended Annex VI to MARPOL 73/78, whether CO_2 is a type of marine pollution aroused fierce debate.[13]

Currently it is still open to debate whether CO_2 is a type of marine pollution. On the other hand, climate change alters the marine environment of the Arctic and makes a new route for international shipping possible. Current research indicates that black carbon emitted by ships engaged in international shipping, an aerosol with potent climate forcing capacity, has been regarded as the second largest contributor to climate change in the Arctic.[14] Moreover, black carbon is an important component of Particulate Matter (PM) which often causes heart disease, lung cancer, respiratory illness, low birth weight and other health problems.[15] Although black carbon has been identified as a type of marine pollution, the international community has been slow in addressing this matter. At the 68th Marine Environment Protection Committee meeting of the IMO in May 2015, the IMO approved the definition on black carbon but also noted that it was impossible to consider control measures to reduce black carbon emissions from international shipping at this stage.[16] It is still technically difficult to regulate black carbon emissions from ships, which serves as a partial reason for this non-regulation.[17]

Some new forms of marine pollution are emerging with the rapid growth of international seaborne trade. Underwater noise pollution is one of these pollutions. Anthropogenic sources of noise in the marine environment, including those from commercial shipping, oil and gas exploration, dredging and fishing, can generate acoustic pollution that can travel considerable distances.[18] Marine mammals and other forms of marine

---

[13] See, e.g., *Report of the Marine Environment Protection Committee on Its Sixtieth Session*, MEPC 60th Session, Agenda Item 22, Doc MEPC 60/22 (12 April 2010) Annex 4, at 2; *Report of the Marine Environment Protection Committee on Its Sixty-First Session*, MEPC 61st Session, Agenda Item 24, Doc MEPC 61/24 (6 October 2010) Annex 3, at 2.

[14] Laura Boone, 'Reducing Air Pollution from Marine Vessels to Mitigate Arctic Warming: Is it Time to Target Black Carbon', (2012) 1 *Carbon & Climate Law Review* 13: 14–15.

[15] Ibid 16.

[16] IMO, 'Report of the Marine Environment Protection Committee on Its Sixty-eighth Session', IMO Doc MEPC 68/21 (29 May 2015) para 3.29.

[17] See, e.g., Kate Deangelis, 'Black Carbon: the Most Important Ignored Contributor to Climate Change', (2011) 26 *Maryland Journal of International Law* 239: 254–255.

[18] Donald R Rothwell and Tim Stephens, *The International Law of the Sea* (Oxford and Portland, Oregon 2010) 341–342.

life have been threatened by this type of marine pollution.[19] However, it is very challenging to reach a consensus in regulating noise pollution globally.

Climate change as a regulatory challenge

An overview of climate change

In recent years climate change has attracted mounting attention from the international community, and has been regarded as the 'mother of all issues'.[20] Both the United Nations Framework Convention on Climate Change (UNFCCC) and the IPCC provide definitions on climate change. According to the UNFCCC, 'climate change' refers to "a change of climate which is attributed directly or indirectly to human activity that alters the composition of the global atmosphere and which is in addition to natural climate variability observed over comparable time periods".[21] This can be regarded as a legal definition on climate change. IPCC Fourth Assessment Report defines climate change as "a change in the state of the climate that can be identified by changes in the mean and/or the variability of its properties, and that persists for an extended period, typically decades or longer", which may result from either natural internal processes and 'external forcings' or anthropogenically-induced activities.[22] This definition emphasizes the technical aspects of climate change.

The IPCC was jointly established by the United Nations Environment Program (UNEP) and the World Meteorological Organization (WMO) in 1988 to provide knowledge and assessment of climate change.[23] Through the efforts of thousands of scientists around the world, IPCC has published five assessment reports (1990, 1995, 2001, 2007 and 2014) which serve as a bridge between the scientists and policy-makers. The conclusions revealed in these reports underpin the adoption of laws and regulations by the governments of world. For instance, the IPCC Third Assessment Report

---

[19] Jeremy Firestone and Christina Jarvis, 'Response and Responsibility: Regulating Noise Pollution in the Marine Environment', (2007) 10 *Journal of International Wildlife Law and Policy* 109, 109.

[20] Milke Hulme, 'The Idea of Climate Change' (2010) 19(3) *GAIA: Ecological Perspectives for Science & Society* 171, 171.

[21] *United Nations Framework Convention on Climate Change*, opened for signature 9 May 1992, 31 ILM 848 (entered into force 21 March 1994) ('*UNFCCC*') art 1(2).

[22] IPCC, 'Fourth Assessment Report' (2007) <http://www.ipcc.ch/publications_and_data/ar4/syr/en/contents.html> accessed 17 November 2013, Appendix Glossary. 'External forcing' refers to a forcing agent outside the climate system causing a change in the climate system, and some of its examples include volcanic eruptions, solar variations and anthropogenic changes in the composition of the atmosphere and land-use change.

[23] IPCC, *Organization* <http://www.ipcc.ch/organization/organization.shtml> accessed 15 October 2015.

in 2001 stated that "most of the observed warming over the last 50 years *is likely to* have been due to the increase in GHG concentrations".[24] The IPCC Fourth Assessment Report reinforced this assessment, stating that "most of the observed increase in global average temperatures since the mid-20th century *is very likely* due to the observed increase in anthropogenic GHG concentrations".[25] The IPCC Fifth Assessment Report provides that "[GHGs are] *extremely likely to* have been the dominant cause of the observed warming since the mid-20th century".[26] The italicized terms of above expressions clearly point out that climate change mainly results from excessive GHG emissions from human activities. It is thus important to reduce GHG emissions from all relevant industries.

Challenges in regulating climate change

Climate change has caused negative impacts on natural and human systems on all continents and across the oceans.[27] In the context of marine environment, GHG emissions has gradually led to the rise of ocean temperature and altered the dynamics of the marine environment. Research indicates that the rise of ocean temperature may change the species distribution, polar systems, and global and regional weather patterns,[28] and the carbon stored in the form of methane hydrates from the seabed may also be released accordingly.[29] Climate change may also lead to sea-level rise, which may cause certain coastal hazards, such as flooding of coastal land, storm surges, erosion, destruction of infrastructure, settlements and facilities. Furthermore, sea level rise may influence maritime jurisdictional claims by coastal States due to the changes in their baselines.[30] Ocean acidification is another environmental hazard that climate change may bring to the ocean. It occurs when the ocean has absorbed too much CO_2. Ocean acidification may have profound negative impacts on the marine

---

[24] IPCC, 'IPCC Third Assessment Report' (2001), available at <http://www.grida.no/publications/other/ipcc_tar/> accessed 15 October 2015, Synthesis Report, p 51.

[25] IPCC, 'IPCC Fourth Assessment Report' (2007), available at <http://www.ipcc.ch/publications_and_data/ar4/syr/en/contents.html> accessed 17 November 2014, Synthesis Report, p 37.

[26] IPCC, 'IPCC Fifth Assessment Report' (2014), available at <http://www.ipcc.ch/report/ar5/syr/> accessed 15 October 2015, Synthesis Report, p 47.

[27] Ibid 49.

[28] Duncan E.J. Currie and Kateryna Wowk, 'Climate Change and CO_2 in the Oceans and Global Oceans Governance' (2009) 3(4) *Carbon & Climate Law Review* 387, 389.

[29] Ibid.

[30] Clive Schofield, 'Shifting Limits? Sea Level Rise and Options to Secure Maritime Jurisdictional Claims' (2009) 3(4) *Carbon & Climate Law Review* 405, 405.

ecosystems and biodiversity.[31] Particularly ocean acidification reduces the ocean's capacity to absorb manmade CO_2, leads to economic loss and engenders food security.[32] It appears that reducing CO_2 emissions is 'the only' realistic mitigation option to address ocean acidification.[33]

Given the growing contribution of GHG emissions from human activities to global climate change, the international community has realized the importance and urgency of reducing GHG emissions and has made continuous regulatory efforts. The adoption of the UNFCCC and its Kyoto Protocol is one of these examples. The global regulatory outcomes that have been achieved are discussed later in this chapter. Nevertheless, this regulatory process is not straightforward but rather lengthy and complex.

Firstly, the international community has pledged to limit the global climate change to two degrees Celsius which will be the increase in the global average temperature by 2100.[34] The two degrees Celsius goal was first put forward by the G-8 in 2009, and later agreed in the Copenhagen Accord. In 2010 this goal was formally incorporated into the UNFCCC process and later explicitly written in the 2015 Paris Agreement. However, it is projected that this goal will be difficult to achieve. A report by the Asian Development Bank concluded that an increase of two degrees Celsius by 2050 is 'almost unavoidable'.[35] Now that the specific reduction targets and time frame for achieving this goal have not yet been agreed under the UNFCCC process, it is still challenging for the international community to have further regulation on climate change.

Secondly, no consensus has been reached as to whether the Common But Differentiated Responsibility (CBDR) principle should be incorporated and how to interpret the differentiation in furthering the regulation of global climate change. The 1992 Rio Declaration on Environment and Development formulated the CBDR principle for the first time.[36] Essentially, this principle

---

[31] See, e.g., Cheryl Logan, 'A Review of Ocean Acidification and America's Response' (2010) 60(10) *BioScience* 819: 821–823; Wim H. Van der Putten, Mirka Macel and Marcel E. Visser, 'Predicting Species Distribution and Abundance Responses to Climate Change: Why It is Essential to Include Biotic Interactions across Trophic Levels' (2010) 365 *Philosophical Transactions of the Royal Society B: Biological Sciences* 2025, 2025.

[32] The International Geosphere-Biosphere Programme, The Intergovernmental Oceanographic Commission and The Scientific Committee on Oceanic Research, 'Ocean Acidification Summary for Policymakers—Third Symposium on the Ocean in a High-CO_2 World' (International Geosphere-Biosphere Programme, Stockholm, Sweden, 2013) 1.

[33] Ibid.

[34] Lavanya Rajamani, 'The Cancun Climate Agreements: Reading the Text, Subtext and Tea Leaves' (2011) 60(2) *The International and Comparative Law Quarterly* 499, 501.

[35] Michael Westphal, Gordon Hughes and Jorn Brommelhorster (eds.). *Economics of Climate Change in East Asia* (Asian Development Bank 2013) executive summary, xvi.

[36] *Rio Declaration on Environment and Development*, 31 ILM 874, 14 June 1992 ('Rio Declaration') principle 7.

requires both developed and developing States to tackle environmental issues, but underscores that developed States should take the primary responsibility. This differentiated arrangement is based on the mutual recognition that the current environmental issues were primary caused by the historical contribution of developed States, and developed and developing States have differentiated capability in combating these issues.[37] Although the CBDR principle was fully incorporated by the UNFCCC and its Kyoto Protocol, this principle has been weakened in the negotiations of recent years. During the negotiations of a new climate agreement, developing States, India as an example, claim that the CBDR principle should continue to be incorporated and the differentiated responsibility should be interpreted as different central obligations.[38] Developed States, the US as an example, assert that a new climate agreement should be applicable to all States, and the CBDR principle may be reflected in an appropriate manner.[39] This conflict reflects the differing regulatory interests of developed and developing States in addressing the issue of climate change, and it needs to be addressed if a universal climate agreement is to be adopted.

Interaction between the regulation for Marine Pollution and for Climate Change

Marine pollution and climate change are not two completely-isolated issues. Instead, they are relevant from an international law perspective. In practice the global regulation for marine pollution often interacts with the regulation for climate change.

Firstly, both marine pollution and climate change are issues with international dimensions, and both of them have negative impacts on marine ecosystem and the marine environment. Therefore, they should be regulated globally in order to relieve their negative impacts on the environment. Furthermore, global regulation on these issues should cover all the major industries so as to avoid 'carbon leakage' or 'pollution leakage'. 'Carbon leakage' generally refers to differentiated carbon policies and their

---

[37] Yubing Shi, 'Greenhouse Gas Emissions from International Shipping: the Response from China's Shipping Industry to the Regulatory Initiatives of the International Maritime Organization', (2014) 1(29) *International Journal of Marine and Coastal Law* 77, 85.

[38] Rakesh Kamal, 'What to Expect from India's INDCs', September 2015, available at <http://www.downtoearth.org.in/news/what-to-expect-from-india-s-indcs-51181> accessed 17 October 2015.

[39] The US White House, 'U.S.-China Joint Presidential Statement on Climate Change', September 2015, available at <https://www.whitehouse.gov/the-press-office/2015/09/25/us-china-joint-presidential-statement-climate-change> accessed 16 October 2015, para 3.

subsequent impacts on the effectiveness of GHG emissions reduction.[40] Suppose that a carbon tax scheme has been applied to most sectors of transport but the shipping sector stay unregulated, investors may flow into this sector to avoid the tax. Similarly, 'pollution leakage' may occur if some types of marine pollution stay unregulated.

Secondly, regulations on marine pollution may also be helpful for the combating of climate change. As discussed earlier, it is arguable that CO_2 is a type of marine pollution. Indeed CO_2 has been regulated as a type of air pollution in the US.[41] If this is the case, global regulation on CO_2 emissions will definitely contribute to the tackling of climate change. Another fact is that the discharge or emission of certain forms of marine pollution is often simultaneous. That means that GHGs are also emitted when a ship is emitting SOx and NOx during an international voyage. Given that the reduction of GHGs contributes to addressing the climate change issue, the regulation of marine pollution is often beneficial for the mitigation of climate change, and vice versa.

International Regulatory Responses to Marine Pollution

Traditional international law did not, in general, concern pollution at sea, it merely imposed duties on States to regulate marine pollution. The global awareness of marine pollution changed since severe oil spill accidents happened in the 1960s and 1970s. As a result, policy-makers, legislators and the public increased the alertness of disastrous damages to the marine environment brought by the accidents and other marine pollutions. International conventions and regional agreements concerning the control and management of marine pollution from a variety of sources are well developed. In this part international regulatory responses to marine pollution are examined from the perspective of global regulation and several regional agreements.

Global regulation

General regulatory framework

In 1956, the United Nations convened its first Conference on the Law of the Sea (UNCLOS I) at Geneva, Switzerland. It is at UNCLOS III that the

---

[40]   Larry Parker and John Blodgett, "Carbon Leakage" and Trade: Issues and Approaches' (19 December 2008) <http://www.fas.org/sgp/crs/misc/R40100.pdf> accessed 10 October 2015.

[41]   *Clean Air Act of the United States of America*, Pub L No 108–201, Stat, 42 USC §7401 et seq. (1970, as amended in 1977 and 1990) Sec. 103(g)(1).

respect to the protection of marine environment reached a significant stage—an environmental law framework was established under the 1982 United Nations Convention on the Law of the Sea (LOSC). The prevention, reduction and control of marine pollution are the primary objectives of the LOSC. To achieve such an objective, the LOSC establishes rules on information, scientific research, monitoring, environmental assessment, enforcement[42] and liability.

Part XII of the LOSC specifically addresses the 'protection and preservation of the marine environment' providing core provisions relevant to this issue. Section 1 of Part XII sets up the general provisions. The primary obligation of all States is 'to protect and preserve marine environment'. The LOSC regulates that 'States have the sovereign right to exploit their natural resources pursuant to their environmental policies and in accordance with their duty to protect and preserve the marine environment'.[45] To further amplify this general obligation, a distinction is drawn between the duty to protect the environment and the responsibility not to cause damage by pollution to other States and their environment.[46] Article 1(4) of the LOSC provides a definition on pollution of the marine environment. This definition is inclusive of all sources of marine pollution encompassing not only the traditionally concerned vessel-source pollution but and pollution from land-based activities and the atmosphere. Article 194(3) further elaborates the obligation to prevent pollution damage by addressing particular sources of pollution respectively. While under Article 194(2), States are required not to cause damage by pollution, being directed to:

> 'take all measures necessary to ensure that activities under their jurisdiction or control are so conducted as not to cause damage by pollution to other States and their environment, and that pollution arising from incidents or activities under their jurisdiction or control does not spread beyond the areas where they exercise sovereign rights in accordance with this Convention.'

---

[42] This includes developing rules in relation to enforcement by coastal-States and port states.

[43] *LOSC* arts 21(1)(f), 42(1)(b) and 54.

[44] In exercising their rights, coastal States are to "have due regard to the rights and duties of other States and shall act in a manner compatible with the provisions of the Convention". (Article 56(2)) The rights of other States include freedoms of navigation in the EEZ (Article 58(1)).

[45] *LOSC* art 192.

[46] Philippe Sands, Jacqueline Peel, *Principles of International Environmental Law* (Cambridge University Press, 3rd ed., 2012) 351.

States parties must not transfer damage or hazards, or transform one type of pollution into another, and must limit the use of technologies or the introduction of alien or new species which may cause significant and harmful changes to the marine environment.[47] These general obligations the LOSC sets forth serve as the basis for more detailed standards in relation to marine pollution. Supplementary obligations regulated in other sections under Part XII cover the requirements of global and regional cooperation in notification of damage, developing contingency plans against pollution and exchange of information and data about marine pollution.[48]

The IMO is a UN specialised agency dealing with international shipping, shipping casualties and marine environment. The Marine Environment Protection Committee (MEPC) is IMO's senior technical body on marine pollution related matters. IMO's original mandate was principally concerned with maritime safety. The Organization Assembly assumed responsibility for pollution issues for the 1954 International Convention for the Prevention of Pollution of the Sea by Oil (OILPOL Convention) in 1959. Until the late 1960s, major oil spill accidents such as the Torrey Canyon in 1967 and the Amoco Cadiz in 1978 triggered international efforts and led to the adoption under IMO's auspices of the 1969 International Convention and the 1971 Oil Pollution Fund Convention (now 1992 Fund Convention).[49] Following the 1972 Stockholm Declaration, the Convention on the Prevention of Marine Pollution by Dumping of Wastes and Other Matter (1972 London Convention) was adopted. In 1973, IMO adopted the International Convention for the Prevention of Pollution from Ships (MARPOL 1973) which was amended many times to also include requirements addressing pollution from chemicals, other harmful substances, garbage, sewage and air pollution and emissions from ships. IMO also carried out other international instruments in regulating oil pollution preparedness, response and co-operation (OPRC Convention and its 2000 OPRC-HNS Protocol), control of harmful anti-fouling systems on ships (AFS Convention), prevention of the potentially devastating effects of the spread of invasive harmful aquatic organisms carried by ships' ballast water (BWM Convention), safe and environmentally sound recycling of ships (Hong Kong Convention), etc. By developing regulations to promote the effective control of all sources of marine pollution, the IMO's regulatory framework is acting to assist States to take practicable steps to prevent pollution of the sea.

---

[47] *LOSC* arts 195, 196.

[48] *LOSC* arts 198–200.

[49] Philippe Sands, Jacqueline Peel, *Principles of International Environmental Law* (Cambridge University Press, 3rd ed., 2012) 348.

Regulation on land-based and atmospheric pollution

Pollution from land-based sources contributes most to ocean pollution. It is estimated that 80% of the pollution entering the marine environment comes from land-based sources.[50] The primary source of land-based pollution is substances and energy entering the marine environment by run-off from land, rivers, pipelines and other outfall structures.[51] It may also arise from or through the atmosphere, generated principally from land-based activities but also from ships and aircraft.[52]

Article 207 of the LOSC requires States to 'prevent, reduce and control pollution of the marine environment from land-based sources, including rivers, estuaries, pipelines and outfall structures'. States must take into account: internationally agreed rules, standards and recommended practices and procedures; characteristic regional features; the economic capacity of developing countries and their need for economic development; and the need 'to minimise, to the fullest extent possible, the release of toxic, harmful or noxious substances, especially those which are persistent, into the marine environment'.[53]

Land-based and atmospheric marine pollution has been subject to some attention in proceedings before international courts and tribunals.[54] In the MOX Plant case[55], International Tribunal for the Law of the Sea (ITLOS) issued an important provisional measures order, which required the two States to cooperate in order to develop measures to prevent marine pollution from a nuclear fuel reprocessing plant "is a fundamental principle in the prevention of pollution of the marine environment".[56] In Land Reclamation by Singapore in and Around the Straits of Johor,[57] ITLOS repeated its statement in the MOX Plants case[58] under the LOSC. It observed that it could not be excluded that Singapore's land reclamation works may arouse adverse effects on the marine environment (including within the territorial sea of Malaysia) and directed Singapore not to conduct land reclamation that

---

[50] UNEP, 'About the GPA', available at: http://unep.org/gpa/About/about.asp (accessed 7 October 2015).

[51] Philippe Sands, Jacqueline Peel, *Principles of International Environmental Law* (Cambridge University Press, 3rd ed., 2012) 373.

[52] Ibid.

[53] *LOSC* arts 207 (1), (4) and (5).

[54] Donald Rothwell and Tim Stephens, *The International Law of the Sea* (Hart Publishing 2010) 379.

[55] *The MOX Plant Case (Ireland v United Kingdom)* (2001) 47 ILM 405; ITLOS, Order of 3 December 2001 on Provisional Measures.

[56] ibid.

[57] *Case concerning Land Reclamation by Singapore in and around the Straits of Johor (Malaysia v. Singapore)* (2003) ITLOS Provisional Measures.

[58] Above n 55.

might cause serious harm to the marine environment. In 1973, in response to French atmospheric nuclear testing in the Pacific that had negative impacts on the regional marine environment, Australia and New Zealand commenced proceedings against France in the ICJ.[59] Both applicants resisted that such nuclear testing generating radioactive fall-out constituted an infringement of freedom of the high seas by interfering with the freedom of navigation and overflight, and by interfering with the freedom to explore and exploit marine living and non-living resources. Therefore, Australia and New Zealand sought interim orders with regards to potential pollution on each respective metropolitan land mass, and also ocean space. However, the issue of marine pollution was not considered. The Court ordered France to refrain from nuclear tests causing radioactive substances discharge on Australian or New Zealand territory. In 1995, New Zealand made efforts to reopen the case due to the French resumption of testing and claimed that France was polluting the marine environment which violated international law. As with the proceedings in the 1970s, the ICJ found that there was no basis for the proceedings, as the underground testing did not disturb the original decision which concerned atmospheric testing.

The 1995 Global Programme of Action for the Protection of the Marine Environment from Land-Based Activities (GPA),[60] and a Declaration, were adopted by 108 States and the EU at a conference held in Washington 1995. The GPA was designed based on relevant provisions of Chapters 17, 33 and 34 of Agenda 21, the Rio Declaration on Environment and Development, and the 1985 Montreal Guidelines on the Protection of the Environment Against Pollution from Land-based Sources (1985 Montreal LBS Guidelines).[61] It aims at 'preventing the degradation of the marine environment from land-based activities by facilitating the realisation of the duty of States to preserve and protect the marine environment', and is designed to assist states 'in taking actions individually or jointly within their respective policies, priorities and resources, which will lead to the prevention, reduction, control and/or elimination of the degradation of the marine environment, as well as to its recovery from the impacts of land-based activities'.[62] The GPA recommends actions at the State, regional and international levels to address the problem of marine pollution from land-based activities. Nationally, these recommendations relate to the identification and assessment of problems, the establishment of priorities for action, setting management objectives for priority problems, identifying, evaluating and selecting strategies and measures to achieve objectives and

---

[59] *Nuclear Tests Case* (*Australia v France*) (*Interim Protection*) (1974) ICJ Reports 253.
[60] The GPA is administered by a UNEP-led GPA Co-ordination Office.
[61] 24 May 1985, Doc UNEP/GC/DEC/13/1811.
[62] 5 December 1995, Doc UNEP (OCA)/LBA/IG.2/7,7.

developing criteria to assess the effectiveness of strategies and measures.[63] Regionally, States are encouraged to strengthen the participation in and the effective functioning of regional and subregional arrangements to support effective national action, strategies and programmes.[64] Internationally, the GPA seeks to develop institutional arrangements, and facilitate capacity-building and the mobilisation of financial resources.[65] The GPA records agreement on the need for international action to develop a global, legally binding instrument dealing with persistent organic pollutants.[66] Lastly, Chapter V of the GPA provides specific guidance to States, regional and international organizations concerning recommended objectives and actions for addressing particular sources of land-based pollution, including sewage, persistent organic pollutants (POPs), radioactive substances mobilization, heavy metals, oil (hydrocarbons), nutrients, sediment, litter, and physical alteration and destruction of habitats.[67] One particular weakness of the GPA is ocean acidification; besides is one of the most serious impacts, it has not been included in any of the targets set in the discourse of the GPA.[68]

Regulation on vessel-source pollution

Shipping introduces many pollutants to the marine environment, as a result of the normal operation of vessels including some harmful substances discharged from the routine working of vessels, the seafaring practices as well as from ships' ballast water. Accidents at sea, mostly the collision and maritime casualties, can also bring significant problems to the marine environment and cause grave consequences. The likelihood as such has brought the adoption of a group of international conventions to address the bulk of vessel-source pollution.

Operational vessel-source pollution

MARPOL 73/78 is the main international convention regulating pollution from vessels. This convention mainly regulates jurisdiction and powers of enforcement and inspection, the detailed anti-pollution regulations are contained in its annexes. MARPOL 73/78 establishes specific international

---

[63]   Ibid., Chapter II.
[64]   Ibid., Chapter III.
[65]   Ibid., Chapter IV.
[66]   Ibid., paras 86, 88.
[67]   Ibid., Chapter V.
[68]   Donald Rothwell and Tim Stephens, *The International Law of the Sea* (Hart Publishing 2010) 374.

regulations to implement the objective of completely eliminating international pollution of the marine environment by oil and other harmful substances and minimizing accidental discharges.[69] It sets out a framework for the adoption of the regulations in the annexes and provides basic definitions. MARPOL 73/78 applies to ships that are entitled to fly the flag of a Party or operate under the authority of a Party, but it does apply to warships or other ships operated by a State and used only on governmental non-commercial services.[70] The Parties must prohibit and sanction violations and accept certificates required by the regulations which are issued by other Parties as having the same validity as their own certificates.[71] A ship which is in the port or offshore terminal of a Party may be subject to an inspection to verify the existence of a valid certificate unless there are 'clear grounds for believing that the condition of the ship or its equipment does not correspond substantially with the particulars of that certificate'.[72] The inspection Party must ensure that the ship does not sail 'until it can proceed to sea without presenting an unreasonable threat of harm to the marine environment'. The MARPOL Convention requires Parties to apply the Convention to ships of non-Parties so as to ensure that 'no more favourable treatment is given to such ships'.[73] The Convention also provides for the detection of violations and enforcement, reporting requirements on incidents involving harmful substances, the communication of information to the IMO, and technical cooperation.[74]

The six annexes to MARPOL 73/78 contain pollution control standards in relation to the particular pollutants. Annexes I and II bind all parties, whereas the four other annexes are optional that a State may declare it does not accept when first becoming a Party to the Convention or may subsequently accede to.[75] Annex I addresses technical, technological and operational practices. Annex II sets out measures for the control of pollution by noxious liquid substances in bulk. The 2007 revised Annex II establishes a four-category system: the most hazardous (category X) substances subject to a complete prohibition on discharge; less hazardous (categories Y and Z) substances may be discharged in limited quantities in certain

---

[69] The term 'harmful substances' is defined in Article 2 (2) of MARPOL 73/78 as: "any substance which, if introduced into the sea, is liable to create hazards to human health, to harm living resources and marine life, to damage amenities or to interfere with other legitimate uses of the sea."

[70] *MARPOL 73/78* art 3(1), (3).

[71] *MARPOL 73/78* art 5(1), (2).

[72] *MARPOL 73/78* art 5(2).

[73] *MARPOL 73/78* art 5(4).

[74] *MARPOL 73/78* arts 6, 8, 11 and 17. Protocol I sets out detailed Provisions Concerning Reports on Incidents Involving Harmful Substances.

[75] *MARPOL 73/78* art 14.

circumstances; 'other substances' (harmless products) may be discharged when cleaning tanks, or releasing bilge and ballast water.[76] Annex III regulates the prevention of pollution by harmful substances carried by sea in packaged form. It is implemented through the IMO International Maritime Dangerous Goods Code, which includes standards concerning packaging, marketing, labelling, documentation, stowage and quantity limits.[77] No jettisoning of harmful substances in packaged form from vessels is permitted under Annex III except for safety reasons.[78] Regulations in relation to the prevention of pollution by sewage from ships are contained in Annex IV. The regulations address such matters as surveys and certification,[79] and facilities for reception of sewage.[80] Discharge of sewage into the sea is prohibited unless the sewage complies with disinfection requirements or the ship has an approved sewage treatment plant, or is situated in the waters of State imposing less stringent requirements.[81] Annex V regulates the prevention of pollution by garbage from ships. It applies to all ships and regulates different types of garbage, subject to rules of special application, special areas and exceptions. The disposal from ships into the sea of all plastics is prohibited;[82] dunnage, lining and packing materials that will float are not to be disposed within 25 nautical miles of land; disposal of food waste and all other garbage is prohibited within 12 nautical miles of land.[83] For special areas such as the Mediterranean and the Gulfs area, disposal of all garbage except for food wastes is prohibited.[84] The regulations for the prevention of air pollution from ships are contained in Annex VI to MARPOL 73/78. The Annex sets limits on sulphur oxide and nitrogen oxide emissions from ship exhausts, prohibits deliberate emissions of ozone depleting substances and regulates the emissions of volatile organic compounds.[85] It provides provisions for the establishment of special SOx Emission Control Areas (SECAS) with more stringent controls on sulphur emissions by ships in these areas.[86] Annex VI also prohibits the incineration on board ships of certain products, such as contaminated packing materials and polychlorinated biphenyls (PCBs).[87]

---

[76] *MARPOL 73/78* Annex II Regulations 1–18.
[77] *MARPOL 73/78* Annex III Regulations 2–6.
[78] *MARPOL 73/78* Annex III Regulation 7.
[79] *MARPOL 73/78* Annex IV Regulations 3–7.
[80] *MARPOL 73/78* Annex IV Regulations 10–12.
[81] *MARPOL 73/78* Annex IV Regulation 11.
[82] *MARPOL 73/78* Annex V Regulations 3(1) (a).
[83] *MARPOL 73/78* Annex V Regulations 3(1) (b) and (c).
[84] *MARPOL 73/78* Annex V Regulation 5.
[85] *MARPOL 73/78* Annex VI Regulations 12–15.
[86] *MARPOL 73/78* Annex VI Regulation 14. The Baltic Sea and North Sea are designated as SOx Emission Control Areas under the protocol.
[87] *MARPOL 73/78* Annex VI Regulation 16.

With regard to GHGs, in 2011 an extra Chapter 4 was added to Annex VI which covers mandatory technical and operational energy efficiency measures aimed at reducing GHG emissions from ships. Some of the most significant measures are the mandatory Energy Efficiency Design Index (EEDI) for new ships and the Ship Energy Efficiency Management Plan (SEEMP) for all ships.[88] The amended MARPOL Annex VI in 2011 represents the first legally binding instrument dealing with GHG emissions reduction from international shipping.

The issue caused by the introduction of invasive species through vessels' ballast water reveals to be another operational vessel-source pollution that initiates major problem for marine environmental management.[89] The 2004 BWM Convention aims at establishing a global legal regime to control the discharge of ballast water and in particular the discharge of invasive species into the sea. It contains 22 Articles, and one Annex with 5 sections (A to E). The general obligation is for State Parties to take steps to 'prevent, reduce and if possible eliminate the transfer of harmful aquatic organisms and pathogens through the control and management of ships' ballast water and sediments'.[90] The Convention requires cooperation amongst Parties, especially in the same region, continued research, development of programs, not to cause further or greater harm and to enforce laws or regulations to avoid the actual uptake of ballast water with potentially harmful organisms.[91] All ships are obliged to implement a 'Ballast Water and Sediments Management Plan' and to carry a 'Ballast Water Record book',[92] and all vessels are to be surveyed and certified and may be inspected by the port State.[93] The Annex to the Convention contains details for the management and control requirements for ships, standards for ballast water management and survey and certification requirements for ballast water management. Since its adoption, 14 Technical Guidelines have been developed to support port State authorities, shipmasters and owners, equipment manufacturers and class societies, and to encourage the harmonized implementation of the different requirements.[94]

---

[88]   *MARPOL 73/78 Annex VI Regulations 20 and 21.*

[89]   Michael White, *Australasian Marine Pollution Laws* (Federation Press, 2nd ed., 2007) 52.

[90]   *BWM Convention* art 2(1).

[91]   *BWM Convention art 2(7), (8) and (9).*

[92]   *BWM Convention* art 5.

[93]   *BWM Convention* art 9.

[94]   BIO-UV, 'Ballast Water Management Convention (IMO)', 2015, available at: http://www. ballast-water-treatment.com/reglementation/237-2 (accessed 7 October 2015).

Accidental vessel-source pollution

i. Safety of Shipping—SOLAS

In response to the Titanic disaster, the first International Convention for the Safety of Life at Sea (SOLAS) was adopted in 1914. It was revised four times before a thoroughly updated SOLAS[95] was adopted in 1974. The primary purpose of SOLAS is to set down standards for the construction, equipment and operation of ships that will promote their seaworthiness.[96] SOLAS is structured of a 13-article convention including general provisions with an Annex divided into 12 chapters. The main objective of SOLAS is the responsibility of flag States to ensure that ships under their flag comply with the requirements of the convention and annex.[97] SOLAS requires flag States to inspect and survey ships and issue certificates of seaworthiness.[98] Every ship in port is subject to the control of port State authorities to verify the validity of issued certificates.[99] Certificates, if valid shall be accepted unless there are clear grounds for believing that the condition of the ship or its equipment does not correspond substantially with the certificate.[100] Given the codified circumstances, the port State is to take steps to ensure that the ship does not sail until it can processed to sea, or leave the port for the purpose of proceeding to a repair yard, without danger to the ship or its crew.[101]

ii. Qualifications and Working Conditions for Seafarers—STCW

The 1978 International Convention on Standards of Training, Certification and Watchkeeping for Seafarers (STCW)[102] (revised in 1995) is the international instrument that sets forth the requirements for issuing certificates to individuals who serve as masters, chief mates, officers of the navigational watch, and engineer officers in charge of a watch on seagoing ships. The Convention requires flag States to certify that masters and crew meet defined training and qualification standards, and specifies the level of

---

[95] *International Convention for the Safety of Life at Sea*, opened for signature 1 November 1974, 1184 UNTS 2 (entered into force 25 May 1980) ('*SOLAS*').

[96] Donald Rothwell and Tim Stephens, *The International Law of the Sea* (Hart Publishing 2010) 360.

[97] *SOLAS* art 1.

[98] *SOLAS* Annex, Chapter I, Regulation 6.

[99] *SOLAS* Annex, Chapter I, Regulation 19 (a).

[100] *SOLAS* Annex, Chapter I, Regulation 19 (b).

[101] SOLAS Annex, Chapter I, Regulation 19 (c).

[102] *International Convention on Standards of Training, Certification and Watchkeeping for Seafarers*, opened for signature 7 July 1978, 1361 UNTS 2 (entered into force 28 April 1984), as amended by the 1995 Protocol, 1969 UNTS (entered into force 1 February 1997) ('*STCW*').

watchkeeping to be maintained. Flag States have the major responsibility for ensuring that these standards are met. Port State control is also utilized in the STCW allowing port States to prevent ships from sailing where it is found that there are serious deficiencies which pose a danger to persons, property or the environment.[103]

iii. Preventing Collisions at Sea—COLREG

The 1972 Convention on the International Regulations for the Prevention of Collisions at Sea (COLREG)[104] concerns maritime safety relating to the prevention of collisions which updates and replaces the 1960 Collision Regulation. It specifies rules regarding matters as look out, safe speed, use of radar and other equipment to reduce risk of collisions, action to be taken to avoid collisions, navigation through narrow channels, conduct during poor visibility and rules concerning lights, shapes and signals. One of the most important innovations of COLREG relates to traffic separation schemes which provide guidance in determining safe speed, the risk of collision and the conduct of vessels operations in or near traffic separation schemes. Traffic separation schemes and other systems for routeing ships have been adopted by the IMO as mandatory regime in most major congested shipping areas, resulting in significantly fewer collisions and casualties.

Liability for vessel-source pollution

The oil spills and other pollution incidents may cause long-lasting, and in some cases permanent, environmental damage. Currently IMO has played a major part in establishing the insurance schemes to address liability and compensation issues. There are two IMO conventions that work as a scheme to pay compensation and to reimburse the clean-up costs and damages by oil spills that occur from oil tankers, which are the 1992 Civil Liability Convention (CLC 1992) and the 1992 Fund Convention. The scheme has been gradually extended from persistent oil spills by tankers to spills by non-tankers' bunkers, which is running under the 2001 Bunker Convention.

The original 1969 CLC was adopted following the accident involving the Liberian-registered Torrey Canyon. The most recent protocol of 1969 CLC is the 1992 Liability protocol. With the entry into force of the 1992 protocol, the Convention is known as the International Convention on Civil Liability for Oil Pollution Damage 1992. The 1992 CLC establishes the liability of the owner of a ship for oil pollution damage from the ship

---

[103] *STCW* art X.
[104] *Convention on the International Regulations for Preventing Collisions at Sea*, opened for signature 20 October 1972, UKTS 77 (entered into force 15 July 1977) ('COLREG').

as a result of an incident in the territory of a Party, and applies preventive measures to minimise such damage.[105] The 1992 CLC amended the definition of 'pollution damage'.[106] In order to bring a claim for environmental damage, the 1992 definition requires preventive measures taken to be 'reasonable' and to have actually been undertaken or to be undertaken.[107] The 1992 CLC establishes joint and several liability for damage which is not 'reasonably separable'.[108] This allows a series of exceptions, which are war and hostilities and insurrection; intentional acts involving a third party; any governmental negligence; or it extinguishes all other claims for compensation.[109] The 2000 amendments (entered into force on 1 November 2003) of the Convention increased the compensation limits for the owner to 4.51 million SDRs[110] for ships not exceeding 5,000 units of gross tonnage and 631 SDRs for each additional unit of tonnage to a maximum, at 140,000 units of tonnage, of 89.77 million SDRs.[111] The owner must maintain insurance or other financial security to cover and limit its liability, establish a fund for the total sum of liability with the court or other competent authority of any one of the Contracting States in which action is brought.[112] Claims may be brought before the courts of any Party or Parties in which the pollution damage has occurred or the preventive measures have been taken, and judgments are recognizable and enforceable in courts of all parties.[113] The court of the State in which a fund is constituted is exclusively competent to apportion and distribute the fund.[114]

---

[105] *1992 CLC* arts II and III (i). The Convention does not apply to warships or other ships owned or operated by a State and being used at the time of the incident for non-commercial purposes: Article 3 of the 1992 Protocol extended the application of the Convention to pollution damage caused in the EEZ of a Party or, if the Party has not declared an EEZ, to the area extending to no more than 200 nautical miles from the baseline from which its territorial sea is measured.

[106] Article I (6) of the 1992 CLC defines 'pollution damage' as: (a) loss or damage caused outside the ship by contamination resulting from the escape or discharge of oil from the ship, wherever such escape or discharge may occur, provided that compensation for impairment of the environment other than loss of profit from such impairment shall be limited to costs of reasonable measures of reinstatement actually undertaken or to be undertaken; (b) the costs of preventive measures and further loss or damage caused by preventive measures.

[107] *1992 CLC art* 1(7).

[108] *1992 CLC art* IV.

[109] *1992 CLC* arts III (2) (3) and IV.

[110] Special Drawing Rights: The daily conversion rates for Special Drawing Rights (SDRs) can be found on the International Monetary Fund website at http://www.imf.org/

[111] *1992 CLC* art V(1).

[112] *1992 CLC* arts V(3), VI and VII.

[113] *1992 CLC* arts IX(1) and X. Article X provides for the circumstances that a Court's judgment is not recognized, where the judgment was obtained by fraud, or the defendant was not given reasonable notice and a fair opportunity to present his case.

[114] *1992 CLC art* IX(3).

Supplementary to the 1992 CLC, the 1992 Fund Convention was adopted to provide additional compensation for victims of oil pollution and to transfer some of the economic consequences to the owners of the oil cargo, as well as the shipowner. It adopts the same definitions as the 1992 CLC.[115] The 1992 Fund Convention, which establishes an International Oil Pollution Compensation Fund (IOPC Fund), provides a regime to compensate the pollution damage victims for inadequate or unavailable compensation processed under the 1992 CLC.[116]

The 1992 Fund Convention limits the obligation of the Fund in certain situations, including war, lack of evidence that the damage resulted from an incident involving one or more ships, damage by warships or state-operated non-commercial ships, and contributory negligence.[117] The 1992 Fund Convention increased the maximum liability to SDRs 135 million per incident or for certain natural damage, and to SDRs 200 million for any period when there are three parties to the Convention where the combined relevant quantities of contributing oil received by persons in the territories of those parties equalled or exceeded 600 million tons in the preceding year.[118]

In 2001, the IMO adopted the International convention on Civil Liability for Bunker Oil Pollution Damage (2001 Bunker Oil Convention),[119] filling a gap left by previous oil pollution conventions to ensure that victims be able to enjoy adequate, prompt and effective compensation from oil spill damage caused by ship's bunkers.[120] The 2001 Convention is largely based on the 1992 CLC, which makes shipowners strictly liable for fuel spills,[121] but also allows States to limit liability in accordance with national or international regimes, but not exceeding an amount calculated in light with the 1976 Convention on Limitation of Liability for Maritime Claims.[122] The 2001 Bunker Oil Convention relies on the same approach to addressing environmental damage as the 1992 CLC, limiting compensation for environmental damage to "reasonable measures of reinstatement".[123] The Convention requires ships over 1,000 gross tonnage to maintain compulsory insurance or other financial security to cover the liability for pollution

---

[115] *1992 Fund Convention* art 1.

[116] *1992 Fund Convention* art 2(1).

[117] *1992 Fund Convention* art 4(2),(3).

[118] *1992 Fund Convention* art 6(1).

[119] *IMO International Convention on Civil Liability for Bunker Oil Pollution Damage*, opened for signature 27 March 2001, 40 ILM 1493 (entered into force 21 November 2008) ('*2001 Bunker Oil Convention*').

[120] Philippe Sands, Jacqueline Peel, *Principles of International Environmental Law* (Cambridge University Press, 3rd ed., 2012) 755.

[121] *2001 Bunker Oil Convention* art 3.

[122] *2001 Bunker Oil Convention* art 6.

[123] *2001 Bunker Oil Convention* art 1(9)(a).

damage. For ships not exceeding 2,000 gross tonnage, the limit of liability for claims for loss of life or personal injury is SDRs 2 million, whereas the limit of liability for property claims is SDRs 1.51 million.

Regulation on dumping at sea

The 1972 London Convention is an instrument of global application to all marine waters (other than internal waters), which has attached the support of 87 parties, more than half of which are developing countries. In 1996, the Protocol was adopted to further modernize the Convention. To date the 1996 Protocol has been ratified by 45 States.[124] The Convention remains in force and thus the Convention and the protocol are parallel regimes, with different parties to each agreement.

The objective of the 1972 Convention is to 'prevent the pollution of the sea by the dumping of waste and other matter that is liable to create hazards to human health, to harm living resources an marine life, to damage amenities or to interfere with other legitimate uses of the sea', and to encourage the development of regional agreements.[125] The definition of 'dumping'[126] under this Convention include incidental disposal of waste.[127] Under Article III, 'wasters or other matters' are broadly defined as 'material and substance of any kind, form or description'. The core rules of the 1972 London Convention are the ones that prohibit or regulate the dumping of waste. The Convention divided waste into three categories, each of which is subject to specific obligations: (1) the 'black list' under Annex I[128] such as oil and high-level nuclear waste, the dumping is prohibited, except in emergency situations and after consultation with countries likely to be affected, and with the IMO;[129] (2) the 'grey list' under Annex II 'special carer' substances and wastes requires a prior 'special permit';[130] and (3) other substances that may be dumped under a permitting system which served

---

[124] The Protocol supersedes the Convention between those parties to the Protocol that are also parties to the Convention. Article III (3) of 1996 Protocol.

[125] *1972 London Convention* arts I,VIII.

[126] 'Dumping' means: 1. any deliberate disposal into the sea of wastes or other matter from vessels, aircraft, platforms or other man-made structures at sea; 2. any deliberate disposal into the sea of vessels, aircraft, platforms or other man-made structures at sea; 3. any storage of wastes or other matter in the seabed and the subsoil thereof from vessels, aircraft, platforms or other man-made structures at sea; and 4. any abandonment or toppling at site of platforms or other man-made structures at sea, for the sole purpose of deliberate disposal. 1972 *London Convention* art III.

[127] 1972 *London Convention* art III(1)(a)(b).

[128] The prohibition does not apply to some Annex I substances.

[129] *1972 London Convention* Annex I art IV(1)(a).

[130] *1972 London Convention* Annex II art IV(1)(b).

only to provide a level of supervision of the dumping industry. 'Special' and 'general' permits are granted by designated national authorities, for matter intended for dumping loaded in its territory, or loaded by vessel or aircraft registered in its territory, or flying its flag when the loading occurs in the territory of a non-Party State to the Convention.[131] The grant of 'special' and 'general' permits must comply with additional criteria,[132] and the designated authorities must keep detailed records of all matter permitted to be dumped, and monitor the condition of the sea individually or in cooperation with other Parties, which should be reported together with other information to the IMO.[133] The convention also requires collaboration between parties on training, research and monitoring and methods for disposal and treatment of waste, to develop procedures to assess liability and the settlement of disputes, and the promotion of measure to protect the marine environment against pollution from specific sources (such as hydrocarbons and radioactive pollutants).[134]

The 1996 Protocol is a major revision to the Convention reaching the zenith of the reform process initiated in the framework of the 1972 Convention.[135] It sets a broader objective than the 1972 Convention, aiming to 'protect and preserve the marine environment from all sources of pollution'.[136] To this end, the Protocol generally prohibits all forms of dumping at sea, except for possibly acceptable substances on its 'reverse list'. Within this 'reverse list' approach, Annex 1 only considers dumping, with a permit, of the certain substances.[137] The Parties to this Protocol are required to take effective measures to prevent, reduce, and where practicable, eliminate marine pollution caused by dumping or incineration at sea.[138] The protocol incorporates the polluter pays principle and the precautionary approach with respect to environmental protection from dumping of wastes or other matter.[139] The Protocol expressly prohibits incineration of wastes at sea and the export of wasted or other matter to other countries for dumping or incineration at sea.[140] In a move designed

---

[131] *1972 London Convention* art VI (1)(a)(b) and (2).
[132] *1972 London Convention* art VI (3) and Annex III to the Convention.
[133] *1972 London Convention* art VI (1)(c) and (d).
[134] *1972 London Convention* arts IX, X, XI and XII.
[135] Application to internal waters is voluntary. Article 7 (2) of the 1996 Protocol.
[136] 1996 Protocol art 6.
[137] This includes dredged material; sewage sludge; fish waste, or material resulting from industrial fish processing operations; vessel and platforms or other man-made structures at sea; inert, inorganic geological material; organic material of natural origin; bulky items and similarly harmless materials; and carbon dioxide streams from carbon dioxide capture processes for sequestration. Annex 1.
[138] *1996 Protocol* art 2.
[139] *1996 Protocol* art 3(1).
[140] *1996 Protocol* art 6.

to facilitate the deployment of carbon capture and storage technologies for the mitigation of climate change, amendments made to the protocol in 2006 allow the storage of carbon dioxide under the seabed from 10 February 2007. The amendments add 'CO_2 streams from CO_2 capture processes for sequestration' to Annex I.[141] However CO_2 stream may only be considered for dumping if: disposal of into a sub-seabed geological formation; they consist overwhelmingly of carbon dioxide; no wastes or other matter are added for the purpose of disposing of them.[142] Advocated by some as a climate change mitigation measure, on the other hand, dumping in the context of iron ocean fertilization practices has not been accepted under the London Convention and its 1996 Protocol. In 2008, parties to these agreements decided, taking into consideration the precautionary approach, against allowing activities whose principal intention is stimulating primary productivity in the oceans.[143] Parties admit ocean fertilization for legitimate scientific research and in 2010 established an assessment framework in which to assess scientific research proposals involving ocean fertilization.[144] The Protocol includes extended technical co-operation and assistance provisions,[145] as well as a commitment to develop procedures for assessing and promoting compliance with the Protocol.[146] In 2007, the meeting of the contracting parties to the Protocol adopted a set of compliance procedures and mechanisms and established a new subsidiary body—the London Protocol Compliance Group.[147]

IMO adopted the Hong Kong International Convention for the Safe and Environmentally Sound Recycling of Ships (the Hong Kong Convention) in 2009, aiming at ensuring that ships, when being recycled after reaching the end of their operational lives, do not pose any unnecessary risks to human health, safety and to the environment. The Hong Kong Convention intends to address all the issues around ship recycling, including the fact that ships sold for scrapping may contain environmentally hazardous substances such as asbestos, heavy metals, hydrocarbons, ozone-depleting substances

---

[141] International Maritime Organazation (IMO), Ocean Fertilization under the LC/LP <http://www.imo.org/ en/OurWork/ Environment/LCLP/ EmergingIssues/ geoengineering/ OceanFertilizationDocumentRepository/ OceanFertilization/ Pages/default.aspx>.

[142] The second meeting of contracting parties in November 2007 adopted "Specific Guidelines for Assessment of Carbon Dioxide Streams for Disposal into Sub-seabed Geological formations".

[143] Resolution LC-LP.1 (2008) on the regulation of ocean fertilisation, 31 October 2008.

[144] Resolution LC-LP.2 (2010) on the assessment framework for scientific research involving ocean fertilisation, 14 October 2010.

[145] *1996 Protocol* art 13.

[146] *1996 Protocol* art 11.

[147] "Compliance Procedures and Mechanisms Pursuant to Article 11 of the 1996 Protocol to the London convention 1972", 9 November 2007.

and others.[148] It also addresses concerns raised about the working and environmental conditions at many of the world's ship recycling locations.[149] Ships to be sent for recycling will be required to carry an inventory of hazardous materials, which will be specific to each ship. An appendix to the Convention provides a list of hazardous materials the installation or use of which is prohibited or restricted in shipyards, ship repair yards, and ships of Parties to the Convention. Ships will be required to have an initial survey to verify the inventory of hazardous materials, additional surveys during the life of the ship, and a final survey prior to recycling. Ship recycling yards will be required to provide a 'Ship Recycling Plan',[150] specifying the manner in which each ship will be recycled, depending on its particulars and its inventory.[151] Parties will be required to take effective measures to ensure that ship recycling facilities under their jurisdiction comply with the Convention.[152] The Convention has not been triggered by the condition of entry into force.[153]

Regional regulation/treaties

Article 197 of the LOSC requires States to cooperate regionally in 'formulating and elaborating international rules, standards and recommended practices and procedures consistent with this Convention'. The regional basis for the protection and preservation of the marine environment calls for efforts concerning ocean governance put at regional level. Regional agreements have gradually incorporated the international environmental principles that emerged in international treaties. An overview of the principal regional agreements from the sea areas of Northeast Atlantic and the Baltic Sea is provided here.

---

[148] *2009 Hong Kong Convention* Annex, Regulation 20.
[149] *2009 Hong Kong Convention* Annex, Regulation 22.
[150] *2009 Hong Kong Convention* Annex, Regulation 9.
[151] *2009 Hong Kong Convention* Annex, Regulation 20.
[152] *2009 Hong Kong Convention* Annex, Regulation 18.
[153] Article 17 of the 2009 Hong Kong Convention states that: This Convention shall enter into force 24 months after the date on which the following conditions are met: 1. not less than 15 States have either signed it without reservation as to ratification, acceptance or approval, or have deposited the requisite instrument of ratification, acceptance, approval or accession in accordance with Article 16; 2. the combined merchant fleets of the States mentioned in paragraph 1.1 constitute not less than 40 per cent of the gross tonnage of the world's merchant shipping; and 3. the combined maximum annual ship recycling volume of the States mentioned in paragraph 1.1 during the preceding 10 years constitutes not less than 3 per cent of the gross tonnage of the combined merchant shipping of the same States.

Northeast Atlantic—the 1992 OSPAR Convention

By seeking to regulate all sources of marine pollution in a single instrument, the 1992 OSPAR Convention adopts a new comprehensive yet simplified approach to the protection of the marine environment. The five Annexes to the Convention regulate pollution from land-based sources, dumping and incineration, on offshore sources, on the assessment of the quality of the marine environment and on the protection and conservation of the ecosystems and biological diversity of the 'maritime area'.[154] The Convention adopts significant legal developments including a commitment to 'sustainable management'[155] (rather than sustainable development), the incorporation of the precautionary principle and the polluter pays principle,[156] and the creation of a new Commission with powers to take legally binding decisions and participate in compliance.[157] Pollution is to be eliminated and degraded areas should be restored 'so as to safeguard human health and to conserve marine ecosystems and, when practicable, restore marine areas which have been adversely affected'.[158] The OSPAR Commission adopted a Strategy for the Protection of the Marine Environment of the North-East Atlantic 2010–2020 based on its previous strategic and action plans. Maintaining its previous priority objectives, the new Strategy sets ecosystem approach at the core of OSPAR's objectives and encourages international cooperation with regional organisations (such as EU) and other relevant instruments and organisations.[159]

Baltic Sea—the 1992 Helsinki Convention

The 1974 Convention on the Protection of the Marine Environment of the Baltic Sea Area (1974 Baltic Convention) failed to fulfil its aims, and did not prevent massive pollution of the Baltic Sea.[160] The 1974 regime was superseded by the 1992 Convention on the Protection of the Marine Environment of the Baltic Sea Area (1992 Helsinki Convention).[161] The 1992

---

[154] *1992 OSPAR Convention* arts 3–7.

[155] 'Sustainable management' is defined in the Preamble of the Convention as "the management of human activities in such a manner that the marine ecosystem will continue to sustain the legitimate uses of the sea and will continue to meet the needs of present and future generations".

[156] *1992 OSPAR Convention* art 2(2)(a)(b).

[157] *1992 OSPAR Convention* art 10.

[158] *1992 OSPAR Convention* art 2(1)(a)(b).

[159] Strategy of the OSPAR Commission for the Protection of the Marine Environment of the North-East Atlantic 2010–2020, OSPAR Agreement 2010-3, Bergen: September 2010.

[160] Philippe Sands, Jacqueline Peel, *Principles of International Environmental Law* (Cambridge University Press, 3rd ed., 2012) 362.

[161] Helsinki, 9 April 1992, enforced on 17 January 2000.

Convention enlarges the application area by including the internal waters. It amends the six Annexes to the 1974 Convention and adds Annex VII on the prevention of pollution from offshore activities.[162] Parties must, individually or jointly, take all legislative, administrative or other measures to 'prevent and eliminate pollution in order to promote the ecological balance'.[163] Under the Convention, parties must apply the precautionary principle and the polluter pays principle, promote the use of Best Environmental Practice and Best Available Technology.[164] It requires Parties to ensure the measurements and calculations of emissions to water and air, and use best endeavours to ensure that implementation of the Convention does not cause transboundary pollution in areas beyond the Baltic Sea or lead to other 'unacceptable environmental strains'.[165] The parties are required to prevent and eliminate pollution by harmful substances from all sources under Annex I.[166] Pollution from land-based sources is to be prevented and eliminated in accordance with Annex III,[167] and pollution from ships is subject to the measure required by Annex IV.[168] Incineration is prohibited, as is dumping, subject to exemptions for dredged material and safety.[169]

International Regulatory Responses to Climate Change

In 1987 a report entitled *Our Common Future* was released, which aroused worldwide attention to the global issues of development and environment. The issue of climate change was also highlighted in this report. In the same year the UN General Assembly discussed this report and put the issue of climate change to the political agenda under the auspices of the WMO and United Nations Environment Programme (UNEP). In 1988 the IPCC was established. The above global institutional developments facilitated the international legal responses to climate change. In a broad

---

[162] Current list of Annexes under the 1992 Convention is: Annex I Harmful substances; Annex II Criteria for the use of Best Environmental Practice and Best Available Technology; Annex III Criteria and measures concerning the prevention of pollution from land-based sources; Annex IV Prevention of pollution from ships; Annex V Exemptions from the general prohibition of dumping of waste and other matter in the Baltic Sea Area; Annex VI, Prevention of pollution from offshore activities; ANNEX VII Response to pollution incidents.
[163] *1992 Helsinki Convention* art 3(1).
[164] *1992 Helsinki Convention* arts 3 (2), (3) and (4).
[165] *1992 Helsinki Convention* art (5) and (6).
[166] *1992 Helsinki Convention* art 5.
[167] *1992 Helsinki Convention* art 6.
[168] *1992 Helsinki Convention* art 8.
[169] *1992 Helsinki Convention* arts 10, 11.

sense, international legal responses to climate change constitute various global and regional treaties and non-binding political agreements to tackle climate change by States or through intergovernmental organisations. For the purpose of this chapter, only the regulatory developments under the UNFCCC-Kyoto Protocol regime are examined here, and an analysis of regulating climate change in the European Union (EU) is provided.

Global regulation

The UNFCCC and its Kyoto Protocol

The international climate change regime, also referred to as the UNFCCC-Kyoto Protocol regime, was established in 1992 when the UNFCCC was adopted, culminated in 1997 when the Kyoto Protocol was signed, and is currently under development.[170] This regime mainly consists of the 1992 UNFCCC, its 1997 Kyoto Protocol, and its Conferences of the Party (COPs) and COPs serving as the Meeting of the Parties to the Kyoto Protocol (CMPs).[171]

The UNFCCC is generally regarded as the 'Constitution' for the international climate change regime.[172] To date the UNFCCC has 196 parties including 195 States and the EU.[173] However, the bodies under the UNFCCC have been changing due to the termination of some temporary bodies mandated by the convention for certain periods. For example, the Ad Hoc Working Group on Long-term Cooperative Action (AWG-LCA) and the Ad Hoc Working Group on Further Commitments for Annex I Parties under the Kyoto Protocol (AWG-KP) were terminated in 2012 as mandated. Figure 10.1 provides the current structure of the UNFCCC bodies. This figure reveals that except for the Subsidiary Body for Scientific and Technological Advice (SBSTA) and the Subsidiary Body for Implementation (SBI) which are permanent subsidiary bodies under the UNFCCC, most of the other subsidiary bodies are temporary.

The UNFCCC aims to stabilize GHG concentrations 'at a level that would prevent dangerous anthropogenic interference with the climate

---

[170] Daniel Bodansky, 'The History of the Global Climate Change Regime' in Urs Luterbacher and Detlef F. Sprinz (eds.), *International Relations and Global Climate Change* (Cambridge, MA: MIT Press, 2001) 23.

[171] See Ronald D. Brunner, 'Science and the Climate Change Regime' (2001) 34(1) *Policy Sciences* 1, 1.

[172] D.M. Bodansky, 'The Emerging Climate Change Regime' (1995) 20(1) *Annual Review of Energy and the Environment* 425, 426.

[173] UNFCCC, *Status of Ratification of the Convention* <http://unfccc.int/essential_background/convention/status_of_ratification/items/2631.php> accessed 27 October 2015.

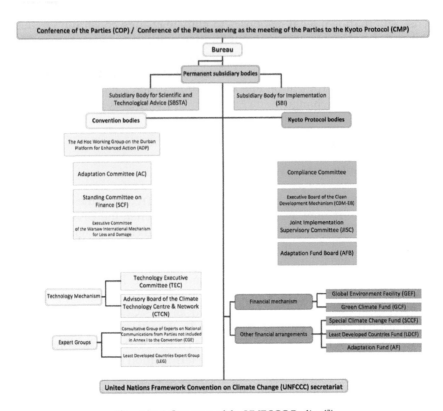

Figure 10.1. Structure of the UNFCCC Bodies.[174]

system' rather than reverse GHG emissions.[175] However, the convention does not define what is the safe level rather it regulates that it should be 'within a time frame sufficient to allow ecosystems to adapt naturally to climate change, to ensure that food production is not threatened and to enable economic development to proceed in a sustainable manner'.[176] It appears that the international community would be tolerant with slow increase of GHG concentrations.[177] This regulation was acceptable when it was adopted in 1992. However, this goal became too general when the subsequent IPCC reports revealed the urgency of tackling climate change.

---

[174] UNFCCC, *Bodies* (2015) <http://unfccc.int/bodies/items/6241.php> accessed 27 October 2015.

[175] *UNFCCC* art 2.

[176] *UNFCCC* art 2.

[177] Birnie, Patricia W., Alan E. Boyle and Catherine Redgwell, *International Law and the Environment* (Oxford University Press, 3rd ed., 2009) 358.

The two degrees Celsius goal was thus incorporated into the UNFCCC process in 2010 to improve this regulation.

The convention also provides several guiding principles for the parties to achieve the above objective. For example, Article 3 lists the principle of equity, the precautionary principle, the CBDR principle, the sustainable development principle, and the promotion of a supportive and open international economic system. Among these principles, the CBDR principle has been generally regarded as an important principle of international climate change regulation as well as a principle of international environmental law. Based on the CBDR principle, the UNFCCC classifies States into Annex I States (OECD States and economies in transition), Annex II States (OECD States only) and non-Annex I States (mostly developing States). All parties have 'common responsibility', that is to develop national inventories of anthropogenic emissions, to promote sustainable management and to exercise the reporting obligations.[178] Regarding the 'differentiated responsibilities', Annex I parties have the obligations of returning emissions to 1990 levels by 2000, whereas Annex II Parties must provide financial assistance and promote the transfer of technologies to developing States.[179] However, Annex I Parties are entitled to choose their preferred starting points, resources, economies and other individual circumstances without a uniform requirement.[180] The UNFCCC are thus asserted by some scholars as 'neither strong nor clear'.[181]

The UNFCCC continued the 'framework treaty' model for international environmental agreements which was first established by the 1979 Geneva Convention on Long-Range Transboundary Air Pollution[182] and the 1985 Vienna Convention for the Protection of the Ozone Layer.[183] This convention provides an important forum for its member parties to further the negotiations of combating global climate change. The first Conference of the Parties of the UNFCCC (COP 1) in 1995 adopted the 'Berlin Mandate', which treated the commitments of States under the UNFCCC as 'not adequate' and hereby launched new negotiations on a protocol or another legal instrument.[184] This instrument, together with the 'Geneva Ministerial

---

[178] *UNFCCC* art 4(1).

[179] *UNFCCC* art 4(2)(3)(4)(5).

[180] *UNFCCC* art 4(2).

[181] Birnie, Boyle and Redgwell, above n 169, 360.

[182] *Convention on Long-Range Transboundary Air Pollution*, opened for signature 13 November 1979, 18 ILM 1442 (entered into force 16 March 1983).

[183] *Vienna Convention for the Protection of the Ozone Layer*, opened for signature 22 March 1985, 26 ILM 1529 (entered into force 22 September 1988).

[184] *The Berlin Mandate*, Report of the Conference of the Parties on its First Session, FCCC/ CP/1995/7/Add.1 (28 March–7 April 1995) preamble.

Declaration' adopted by the UNFCCC (COP 2) in 1996, eventually led to the adoption of the Kyoto Protocol in 1997. The Kyoto Protocol serves as the only protocol of the UNFCCC. However, the entry into force of this protocol witnessed a lengthy process until 2005.

The 1997 Kyoto Protocol only listed six types of GHGs from the legal perspective, namely CO_2, CH_4, N_2O, HFCs, PFCs and SF_6, but a seventh type of GHG, NF_3 was added to the category in the Durban Climate Change Conference in 2011.[185] Compared with the UNFCCC, the Kyoto Protocol divides its parties into Annex I States and non-Annex I States, or generally developed States and developing States. Kyoto Protocol was criticized by some developed States (e.g., the US) for its exclusion of mandatory GHG emissions limitation requirements for non-Annex I States such as large GHG emitters China, India and Brazil. However, it is the incorporation of the CBDR principle that enabled the adoption of this protocol by consensus. The Kyoto Protocol stipulates legally binding emissions commitments for Annex I States for the first commitment period from 2008 to 2012, and these targets are more ambitious in aiming for real reduction from 1990 emissions levels if compared to the UNFCCC.[186] As such, the Kyoto Protocol has been regarded as the culmination of international efforts to tackle the climate change problem.[187] Furthermore, a system of differentiated targets within the rolling time scale was agreed as Annex B to the Protocol, which reflects the differing circumstances between the main industrial actors.[188]

Joint Implementation (JI), the Clean Development Mechanism (CDM) and Emissions Trading (ET) are three innovative market-based mechanisms established under the Kyoto Protocol. Article 6 of the Kyoto Protocol regulates the JI which allows Annex I States to trade emission reduction units (ERUs) among themselves. Implementing cooperative projects or establishing GHG sinks are commonly-used methods to obtain ERUs.[189] Article 12 of the Kyoto Protocol stipulates the CDM, based on which Annex I States may invest actual GHG emission reduction projects in non-Annex I States and thus receive the generated Certified Emission

---

[185] *See Outcome of the Work of the Ad Hoc Working Group on Further Commitments for Annex I Parties under the Kyoto Protocol at its Sixteenth Session*, Decision 1/CMP.7, Report of the Conference of the Parties serving as the Meeting of the Parties to the Kyoto Protocol on its Seventh Session, FCCC/KP/CMP/2011/10/Add.1 (Mar. 15, 2012).

[186] Rafael Leal-Arcas, 'Kyoto and the COPs: Lessons Learned and Looking Ahead' (2010) 23 *Hague Yearbook of International Law* 17, 24.

[187] Hodgkinson, D.I. and Garner, R. Global Climate Change: Australian Law and Policy (LexisNexis Butterworths 2008) 34–64.

[188] Ong, D.M. 'International Legal Efforts to Address Human-induced Global Climate Change' in Fitzmaurice, M., Ong, D.M. and Panos Merkouris (eds.). *Research Handbook on International Environmental Law* (Edward Elgar Publishing Limited, 2010) 450, 456.

[189] Ong, above n 180, 456.

Reductions (CERs). CERs may be achieved through financial support or the transfer of technologies. It is projected that through the implementation of CDMs Annex I Parties can meet their emissions reduction targets while the non- Annex I Parties will benefit from these projects.[190] However, as the only mechanism of the three options available to developing States, CDM has not achieved satisfying outcomes.[191] Article 7 of the Kyoto Protocol introduces the EI, which allows Annex I Parties to purchase emissions credits from other Annex I Parties in order to fulfil their commitments provided that such trading is supplemental to their domestic actions.[192] However, States like Russia and Ukraine have a large surplus due to their economic collapse and emissions in these States may be much lower than their targets. In this case, the transfers through the ET would 'do little or nothing' to help slow climate change.[193] Moreover, due to the conservative reduction targets, it is arguable that the Kyoto Protocol attempts to address a long-term climate change problem with 'unfeasible short-term measures'.[194]

Post-Kyoto efforts and developments

The UNFCCC and its Kyoto Protocol are the most significant international regulatory responses to global climate change. However, deficiencies also remain. The subsequent COPs and CMPs continue to work on this issue to address these deficiencies. After the adoption of the *Kyoto Protocol* in 1997, as of November 2016, 22 COPs and 12 CMPs had been held. A broad range of matters have been discussed and various decisions made during these conferences. Based on distinct missions and achievements, the development of the climate change regime under the post-Kyoto era can be divided into three stages as follows.

Stage 1: from 1998 to 2004 (COP 4–COP 10)

During the first stage, various unresolved issues within the UNFCCC and its Kyoto Protocol were first raised by the *Buenos Aires Plan of Action* (BAPA) at COP 4 in 1998 and then discussed and supplemented in the subsequent COPs. Seven important issues were raised in the BAPA and

---

[190] Ibid 457.
[191] Rudi M. Lof, 'Addressing Market Failures in the CDM: A Funding-based Approach' (2009) *Carbon & Climate Law Review* 25, 25.
[192] Birnie, Boyle and Redgwell, above n 169, 367.
[193] Leal-Arcas, above n 178, 45.
[194] Ibid 48.

most of them were required to be finished before COP 6 in 2000.[195] To address these matters, the *Bonn Agreements* adopted in COP 6 stipulated the 'core elements for the implementation of the BAPA', providing specific approaches and requirements for such implementation.[196] However, the tasks set in the BAPA were not accomplished until COP 10 in 2004 due to the diverse interests from developed and developing States. During this process, the *Marrakesh Accords* adopted in COP 7 made key contributions in successfully drafting detailed rules, procedures, technical guidelines and work programs.[197] The post-Kyoto cycle of policy-making launched by the BAPA was thus basically fulfilled with only minor matters supplemented by the subsequent three COPs.

Stage 2: from 2005 to 2010 (COP 11/CMP 1–COP 16/CMP 6)

The second stage commenced in 2005 when the *Kyoto Protocol* entered into force and the 'twin track' Convention and Protocol negotiations were launched. One of the focuses of the work in this stage was to establish a second commitment period by means of a new Protocol, an amendment to the *Kyoto Protocol*, or a new climate change agreement after the first commitment period indicated in the *Kyoto Protocol* expired on 31 December 2012.

In this stage, most of the COPs and CMPs have been working along with four subsidiary bodies: the AWG-LCA which was launched in Bali (COP 13) in 2007 and terminated in Doha (COP 18) in 2012, the AWG-KP which was established in Montreal (CMP 1) in 2005 and terminated in Doha (CMP 8) in 2012, and the SBSTA and SBI.

The AWG-LCA was established as a subsidiary body under the Convention at COP 13 and CMP 3 of the *UNFCCC* process in 2007, working for long-term cooperative action under the *UNFCCC*. The AWG-LCA organized 15 sessions from March 2008 to December 2012.[198] As the

---

[195] *The Buenos Aires Plan of Action*, Decision 1/CP.4, Report of the Conference of the Parties on its Fourth Session, FCCC/CP/1998/16/Add.1 (25 January 1999) ('*BAPA*'). These seven issues include financial mechanisms, technology transfer, adverse effects of climate change and implementation of response measures, activities implemented jointly, flexibility mechanisms, and the preparation for future COPs/CMPs.

[196] *The Bonn Agreements on the Implementation of the Buenos Aires Plan of Action*, Decision 5/CP.6, Report of the Conference of the Parties on the Second Part of its Sixth Session, FCCC/CP/2001/5 (25 September 2001).

[197] See *The Marrakesh Accords*, Decisions 2–14/CP.7, Report of the Conference of the Parties on its Seventh Session, FCCC/CP/2001/13/Add.1 (21 January 2002).

[198] UNFCCC, *International Bunker Fuels under the AWG-LCA* <http://unfccc.int/methods/emissions_from_intl_transport/items/6141.php> accessed 29 October 2015.

two mechanisms worked in parallel, the AWG-KP worked for a second commitment for Annex B Parties of the *Kyoto Protocol* beyond the end of the first commitment period in 2012,[199] while the AWG-LCA primarily worked for long-term cooperative action under the *UNFCCC*.

It was not until the Bali Climate Change Conference in 2007 that the development of a post-2012 climate change legal framework began, although the establishment of the AWG-KP at the CMP 1 in 2005 launched the negotiations for the next phase of the *Kyoto Protocol*. This was not only because of the establishment of the twin-track negotiation process,[200] but also due to the substantial contributions from the *Bali Road Map*. The *Bali Road Map* constitutes a set of decisions that represent the work to be done under various negotiating 'tracks'. A number of outcomes have been achieved in the climate change conferences following the Bali conference in 2007. Examples in this stage include the *Copenhagen Accord*,[201] and the *Cancun Agreements*.[202] The *Copenhagen Accord* committed developed States to US$ 30 billion fast-starting financing for adaptation and mitigation in developing States for the period 2010–2012 and decided to establish the Copenhagen Green Climate Fund and a Technology Mechanism. The *Cancun Agreements* established new mechanisms for the Measurement, Reporting and Verification (MRV) of mitigation efforts and support for both developed and developing States. Furthermore, these agreements also created a framework for addressing deforestation in developing States.

Stage 3: from 2011 to present (COP 17/CMP 7–COP 22/CMP 12/or later)

The main tasks of the third stage are the adoption of an agreement to establish a second commitment period under the Kyoto Protocol beginning 2013, and the adoption of a globally uniform Paris Climate Agreement in December 2015 through the work of the Ad Hoc Working Group on the Durban Platform for Enhanced Action (ADP).

---

[199] *Consideration of Commitments for Subsequent Periods for Parties Included in Annex I to the Convention under Article 3, Para 9 of the Kyoto Protocol*, Decision 1/CMP.1, Doc FCCC/KP/CMP/2005/8/Add.1 (2006) art 1.

[200] The twin-track negotiation process refers to the simultaneous negotiations under the COPs and CMPs of the *UNFCCC*.

[201] *Copenhagen Accord*, Decision 2/CP.15, Report of the Conference of the Parties on its Fifteenth Session, FCCC/CP/2009/11/Add.1 (30 March 2010) ('*Copenhagen Accord*').

[202] *The Cancun Agreements*, Decisions 1-2/CMP.6, Report of the Conference of the Parties serving as the Meeting of the Parties to the Kyoto Protocol on its Sixth Session, FCCC/KP/CMP/2010/12/Add.1 (15 March 2011); Decision 1/CP.16, Report of the Conference of the Parties on its Sixteenth Session, FCCC/CP/2010/7/Add.1 (15 March 2011) ('*Cancun Agreements*').

The *Durban Package* adopted in 2011 has been regarded as an advance to the climate regime on the grounds that: it ended the uncertainty of the future of the *Kyoto Protocol* by extending it for a second commitment period, established a roadmap for adopting a post-2020 climate regime applicable to all and fulfilled the promise of the *Cancun Agreements*.[203] However, the *Durban Package* decisions do not contain a reference to the CBDR principle or even 'equity'.[204] It is thus argued that the weakened role of the CBDR principle, in particular the interpretation of 'differentiation', in the above decisions or statements represents 'a shift towards greater parallelism between developed and developing countries'.[205] Or in other words, 'differentiated responsibility' might be replaced by 'symmetry' as a guide for a future climate regime.[206]

The 2012 *Doha Climate Gateway* adopted the Doha Amendment to the Kyoto Protocol, which includes new commitments for Annex I Parties to the Kyoto Protocol who agreed to take on commitments in a second commitment period (1 January 2013–31 December 2020). Meanwhile it also transferred some issues to be considered by the SBSTA and SBI, such as MRV and Reducing emissions from deforestation and forest degradation in developing countries, and the role of conservation, sustainable management of forests, and enhancement of forest carbon stocks (REDD+).

The 2013 Warsaw Climate Conference established the Warsaw international mechanism for loss and damage associated with climate change impacts, as well as the Warsaw REDD+ framework. This meeting also adopted an ADP decision that, *inter alia*, invited parties to initiate or intensify domestic preparations for their intended nationally determined contributions (INDCs). The adopted *Lima Call for Climate Action* in 2014 sets in motion the negotiations towards a 2015 agreement, including the process for submitting and reviewing INDCs. At the Geneva Climate Change Conference in February 2015, States agreed on the 'Negotiating Text' for the Paris Climate Agreement (Nov/Dec 2015). The specific texts of this draft agreement were under further negotiations of States in the subsequent ADP meetings convened in June, August-September, and October 2015 in Bonn, Germany. Eventually the Paris Agreement was adopted in December 2015 representing a significant progress in the development of global climate change regime.

---

[203] Lavanya Rajamani, 'The Durban Platform for Enhanced Action and the Future of the Climate Regime' (2012) 61(2) *International and Comparative Law Quarterly* 501, 515.

[204] Ibid 507.

[205] Lavanya Rajamani, 'The Climate Regime in Evolution: The Disagreements that Survive the Cancun Agreements' (2011) 5(2) *Carbon & Climate Law Review* 136, 144.

[206] Rajamani, above n 195, 502.

2011 and 2014 amendments to MARPOL Annex VI

GHG emissions from international shipping stayed unregulated when the Kyoto Protocol was adopted in 1997. However, the shipping industry also has growing contribution to global climate change. In 2012, CO_2 emissions from international shipping were 2.2% of global CO_2 emissions.[207] The average CO_2 emissions from international shipping were 2.6% of global CO_2 emissions during the period 2007–2012.[208] There is a decline of CO_2 emissions from international shipping from 2.7% of global CO_2 emissions in 2007 to 2.2% of global CO_2 emissions in 2012. However, maritime emissions are projected to increase significantly due to the growth of international seaborne trade.[209]

Both the international climate change regime and the IMO have attempted to regulate the reduction of GHG emissions from international shipping. The UNFCCC started to negotiate the allocation of shipping GHG emissions to States in 1995 and then its SBSTA and the subsequent AWG-LCA continued to work on addressing this GHG issue. However, no outcomes were achieved under the UNFCCC-Kyoto Protocol regime. The IMO received its GHG mandate from Article 2(2) of the 1997 Kyoto Protocol. In the same year the IMO commenced its regulatory efforts on this issue when Resolution 8 on 'CO_2 emissions from ships' was adopted by the International Convention for the Prevention of Pollution from Ships (MARPOL) conference. This resolution requested the IMO to conduct a study on GHG emissions and consider strategies for CO_2 reduction.[210] In 2003 the IMO adopted another resolution on "IMO policies and practices related to the reduction of greenhouse gas emissions from ships" which formally urged the IMO to develop mechanisms to address this issue. After 14 years' lengthy negotiations and discussions within the IMO, the issue of GHG emissions from international shipping was formally regulated through amendments to Annex VI to MARPOL 73/78. Through adding a new Chapter 4 to Annex VI, MARPOL 73/78 introduces a mandatory Energy Efficiency Design Index (EEDI) for new ships and the Ship Energy Efficiency Management Plan (SEEMP) for all ships. Meanwhile, during the

---

[207] Smith, T. W. P.; Jalkanen, J. P.; Anderson, B. A.; Corbett, J. J.; Faber, J.; Hanayama, S.; O'Keeffe, E.; Parker, S.; Johansson, L.; Aldous, L.; Raucci, C.; Traut, M.; Ettinger, S.; Nelissen, D.; Lee, D. S.; Ng, S.; Agrawal, A.; Winebrake, J.J.; Hoen, M.; Chesworth, S.; Pandey, A., 'Third IMO GHG Study' (International Maritime Organization (IMO), 2014) executive summary, Table 1.

[208] Ibid.

[209] Ibid para 5.1.

[210] IMO, 'Main Events in IMO's Work on Limitation and Reduction of Greenhouse Gas Emissions from International Shipping' (2011) <http://www.imo.org/MediaCentre/resources/Pages/Greenhouse%20gas%20emissions.aspx> accessed 16 October 2015, p 3.

period 2000 to 2013 market-based measures (MBMs), as a supplementary measure to the adopted EEDI and SEEMP, were also discussed by States and non-governmental organizations within the IMO. Currently MBMs have not been regulated yet.

EEDI

The EEDI is the main technical measure regulated by the 2011 amendments to Annex VI to MARPOL 73/78. It imposes binding obligations to reduce GHG emissions from ships by a formula prescribed in Regulation 21 of the Annex for different ship types and size segments. This formula can be seen as a technological threshold based on which a ship can be measured whether it has met the EEDI requirement. As such the EEDI is a 'performance-based' non-prescriptive measure. Ship designers and shipbuilders are entitled to choose the most cost-efficient technologies provided that the EEDI requirements are met. In this way the EEDI provides a strong incentive for the shipping industry to improve ship fuel consumption with new technological developments. Non-compliant ships which have not obtained the International Energy Efficiency Certificate may not be able to trade in that port States that are parties to Annex VI are required to verify whether there is such certificate on board when a ship calls at its port.[211]

The 2011 amendments to Annex VI to MARPOL 73/78 regulate that the EEDI requirements are applicable to new ships, new ships which have undergone a major conversion and new or existing ships which have undergone a major conversion that is so extensive that the ship is regarded by the Administration as a newly-constructed ship.[212] Indeed the EEDI under the 2011 amendments to Annex VI only applies to seven types of ships, namely bulk carriers, gas carriers, tankers, container ships, general cargo ships, refrigerated cargo ships and combination carriers.[213] The emissions from these types of new ships are estimated to cover 70% of emissions from all new ships.[214] Ships with diesel-electric propulsion, turbine propulsion and hybrid propulsion are excluded from the EEDI requirement.[215] These exemptions were mainly because of the technical difficulty of incorporating them into the EEDI formulae as regulated in Regulation 21 of the Annex. In 2014, MARPOL Annex VI was amended again to extend the application of EEDI to include an extra five types of ships. They are Liquefied Natural

---

[211] *MARPOL 73/78 Annex VI (2011 amendments)* reg 10(5).

[212] *MARPOL 73/78 Annex VI (2011 amendments)* regs 2.3.23, 2.3.24.

[213] *MARPOL 73/78 Annex VI (2011 amendments)* reg 21, Table 1, 2.

[214] IMO, above n 210, 12.

[215] *MARPOL 73/78 Annex VI (2011 amendments)* reg 19.3.

Gas (LNG) carrier, roll-on/roll-off (ro-ro) cargo ship (vehicle carrier), ro-ro cargo ship, ro-ro passenger ship, and cruise passenger ship having non-conventional propulsion.[216] In this way, the effectiveness of the EEDI was further strengthened.

The IMO regulations are usually adopted by consensus. However, the 2011 amendments to Annex VI to MARPOL 73/78 were adopted by a majority vote. Some leading developing States such as China, Brazil, Chile and Saudi Arabia voted against the regulation. It is thus arguable that the lack of consensus reveals a challenge for the future implementation of the Annex VI.[217]

SEEMP

The SEEMP is the operational measure regulated by the 2011 amendments to Annex VI to MARPOL 73/78. It introduces a mechanism for improving the operational energy efficiency of all new and existing ships. The SEEMP aims to minimize shipping GHG emissions by means of reducing fuel consumption, and the energy efficiency operational indicator (EEOI) is often utilized as a monitoring tool and to establish benchmarks related to the ships' energy efficiency.[218] Regulation 22 of Annex VI briefly regulates the SEEMP, which provides that '[e]ach ship shall keep on board a ship specific Ship Energy Efficiency Management Plan (SEEMP). This may form part of the ship's Safety Management System (SMS).'[219]

In order to make the SEEMP regulation enforceable, the IMO developed the 2012 Guidelines for the Development of a Ship Energy Efficiency Management Plan (2012 SEEMP Guidelines) for ships to assist with the preparation of the SEEMP. The 2012 SEEMP Guidelines adopt a four-step approach to improve a ship's energy efficiency, namely planning, implementation, monitoring, and self-evaluation and improvement. Furthermore, it introduces specific procedures and measures at each stage. The SEEMP does not set any targets for energy efficiency and the 2012 SEEMP Guidelines are not legally binding on shipowners and ship

---

[216] *MARPOL 73/78 Annex VI (2014 amendments)* reg 21.

[217] James Harrison, 'Recent Developments and Continuing Challenges in the Regulation of Greenhouse Gas Emissions from International Shipping' (2012) *University of Edinburgh Research Paper Series* <http://ssrn.com/abstract=2037038> accessed 1 June 2013, p 19.

[218] The EEOI can be applied to almost all new and existing ships and is generally used to measure ships energy efficiency at each voyage or over a certain period of time. It enables ship operators to measure the fuel efficiency of a ship in operation and to gauge the effect of any changes in operation. Currently, the EEOI is circulated to encourage shipowners and ship operators to use it on a voluntary basis.

[219] *MARPOL 73/78 Annex VI (2011 amendments)* reg 22.

operators. Currently the EEOI as a performance indicator for the SEEMP remains voluntary. These deficiencies need to be addressed so as to improve the effectiveness of the SEEMP.[220]

Regional regulation

To date many States have incorporated their obligations under the international climate change regime into their domestic legislation. Regarding the regional regulatory initiatives, the EU has made significant efforts and achieved some outcomes. Here the EU's emissions trading system and its newly-adopted Monitoring, Reporting and Verification (MRV) regulation are examined.

The EU Emissions Trading System (EU ETS)

The ET mechanism was first regulated by the Kyoto Protocol and is utilized in the EU. This system employs a 'cap-and-trade' approach to reduce GHG emissions from industries and companies. It imposes a limit/cap on overall emissions from high-emitting industry sectors, and companies may buy and sell emission allowances as needed. This limit/cap is reduced over time to enable the decrease of total emissions. The EU ETS has applied since 1 January 2005 as the world's largest company-level 'cap-and-trade' system. As of 30 October 2015, 28 EU member States as well as Iceland, Liechtenstein and Norway have participated in this scheme which covers more than 11,000 power stations and manufacturing plants.[221] Aviation operators flying within and between most of these States are also covered. It is estimated that 45% of total EU emissions are limited by the EU ETS.[222]

The EU has very ambitious targets for the reduction of GHG emissions. Compared to 2005 levels, the EU has committed to reduce emissions from sectors covered by EU ETS by 21% lower in 2020, and this percentage will be 43% lower by 2030.[223] Accordingly, the EU ETS has adopted a phased approach to reach its reduction goals. The period 2005–2007 is its first trading

---

[220] Yubing Shi, 'The Challenge of Reducing Greenhouse Gas Emissions from International Shipping: Assessing the International Maritime Organization's Regulatory Response' (2012) 23 *Yearbook of International Environmental Law* 131, 153.

[221] Directorate-General for Climate Action (DG CLIMA), 'The EU Emissions Trading System (EU ETS)' (27 October 2015) available at: <http://ec.europa.eu/clima/policies/ets/index_en.htm> accessed 30 October 2015.

[222] Ibid.

[223] Ibid.

period. In this stage, the EU ETS successfully developed as the world's biggest carbon market, whereas its number of allowances was proved to be excessive which made the price of the first-phase allowances drop to zero in 2007.[224] The second phase ranges from 2008 to 2012. During this period, the number of allowances was reduced by 6.5%, which increased the demand due to the unexpected emissions reduction by economic downturn.[225] The period 2013–2020 is its third trading period. A major revision approved in 2009 has been effective on 1 January 2013. This reform introduced an EU-wide cap on emissions (reduced by 1.74% each year) as well as a shift towards auctioning of allowances rather than free allocation in the first and second phases.[226] On 22 January 2014, the European Commission released its proposals for the EU 2030 Energy and Climate Framework.[227] This package proposed to establish a new GHG target for 2030 and the implementation of a market stability reserve as of 2021.[228]

It appears that the establishment of the EU ETS has effectively reduced the emissions below the cap since 2009.[229] Furthermore, it is arguable that the EU ETS has also successfully pushed the global regulation on climate change. On 1 January 2012 the EU included the emissions from the international aviation industry into the EU ETS because of slow progress within the International Civil Aviation Organization (ICAO). In the next month, China, the US, Russia, India, and representatives from 19 other countries signed the Moscow Joint Declaration to demonstrate their opposition to the EU's unilaterally action. In December 2012 the EU suspended this policy either due to the improved performance from the ICAO or because of strong pressure from other States. In the same year, the EU published a consultation document seeking views on how best to reduce GHG emissions from ships so as to finally include GHG emissions from international shipping in an EU ETS. Presently the further regulation of GHG emissions from international shipping and aviation is progressing

---

[224] EU, 'The EU Emissions Trading System (EU ETS)' (October 2013) available at <http://ec.europa.eu/clima/publications/docs/factsheet_ets_en.pdf> accessed 30 October 2015, p 4.

[225] Ibid.

[226] Ibid.

[227] *Communication from the Commission to the European Parliament, the Council, the European Economic and Social Committee and the Committee of the Regions: A Policy Framework for Climate and Energy in the Period from 2020 to 2030*, COM (2014) 015.

[228] Commission Proposal for a Decision of the European Parliament and of the Council concerning the Establishment and Operation of a Market Stability Reserve for the Union Greenhouse Gas Emission Trading Scheme and amending Directive 2003/87/EC, COM(2014)20/2.

[229] Benedikt Günter, 'In the Market: Reforming the EU ETS Revisited' (2014) 1 *Carbon and Climate Law Review* 65, 65.

under the auspices of the IMO and ICAO respectively. Nevertheless, the deficiencies remain in the EU ETS. This regional initiative is currently under the reform due to changing economic situations and global regulation on climate change.

EU's MRV regulation

In order to accomplish its ambitious targets for the reduction of GHG emissions, the EU has been very keen on making its unilateral regulations when there is a lack of international regulation. Regarding the regulation of reducing GHG emissions from maritime transport, the EU welcomed the adoption of the 2011 amendments to Annex VI to MARPOL 73/78. However, it also recognized that the EEDI and SEEMP requirements will not lead to absolute emissions reduction due to the growth of international seaborne trade. The EU thus supported the development and implementation of a global data collection system for CO_2 emissions from shipping. This system may serve as the first step, while the ultimate goal is to reach a global agreement that may be achieved under the auspices of the IMO.[230] In June 2013 the European Commission developed a proposal for an EU Regulation on Monitoring, Reporting and Verification (MRV) and submitted it to the European Parliament and the Council. Consequently, the MRV Regulation 2015/757 was adopted and came into force on 1 July 2015.

The EU's MRV Regulation is a significant step in furthering the reduction of shipping GHG emissions. It requires ships exceeding 5,000 Gross Tonnages to monitor and report their CO_2 emissions on all voyages to, from and between EU ports from 1 January 2018. It is projected that the introduction of an EU MRV system will lead to emissions reduction and to energy savings, and also contribute to the removal of market barriers and facilitate international trade.[231] The European Commission has assessed that compared with a 'business as usual' situation, the MRV system will reduce CO_2 emissions from the journeys covered by up to 2%.[232] Meanwhile, this system will also reduce ship-owners' net costs by up to €1.2 billion per year in 2030.[233]

---

[230] *Development of a Global Data Collection System for Maritime Transport*, submitted by Austria, Belgium, Bulgaria, Croatia, Cyprus, the Czech Republic, Denmark, Estonia, Finland, France, Germany, Greece, Hungary, Ireland, Italy, Latvia, Lithuania, Luxembourg, Malta, the Netherlands, Poland, Portugal, Romania, Slovakia, Slovenia, Spain, Sweden, the United Kingdom and the European Commission, MEPC 68th Session, Agenda Item 4, IMO Doc MEPC 68/4/1 (3 March 2015) para 10.

[231] Ibid para 12

[232] *PR Newswire US* (1 July 2015), 'Verifavia Launches First Dedicated Shipping Verification Service as EU 'MRV' Rules Come Into Force'.

[233] Ibid.

Discussion on a similar mechanism, namely a data collection system for global application, had been held within the IMO before the adoption of the EU MRV Regulation in July 2015. IMO had obtained general support from its member States and agreed that it was imperative to develop a data collection system for adoption by the Organization for global application.[234] However, it was agreed at the IMO in May 2015 that the development of a data collection system for ships should employ a three-step approach (data collection, data analysis and decision making), and it is premature to decide whether this system should be voluntary or mandatory.[235] Even though, the Data Collection System for Fuel Oil Consumption of Ships was adopted at the 70th Meeting of the MEPC within the IMO in October 2016 in the form of an amendment to MARPOL Annex VI. It is arguable that this achievement was partially facilitated by the adoption of the EU's MRV regulation.

Conclusion

It is concluded that marine pollution and climate change are two significant issues with international dimensions. The international community has recognized the negative impacts of marine pollution and climate change on the marine environment and human health. To date both the two issues have been regulated at the international and regional levels. However, challenges remain as to the further regulation of the two issues.

Regarding the regulation of marine pollution, how to properly balance the interests of large developed flag States and developing flag States constitutes a key barrier. Meanwhile, there is a lack of political will and technical certainty as to the global regulation of some new types of marine pollution, such as underwater noise pollution from commercial shipping. Concerning the regulation of climate change, it is increasingly challenging to limit the global climate change to two degrees Celsius the increase in the global average temperature by 2100. Some scholars assert that 'it is already apparent' that this goal is 'impossible to achieve'.[236] Furthermore, the lack of mandatory reduction targets for any States in the 2015 Paris Agreement

---

[234] Ibid para 4.7.

[235] *Report of the Marine Environment Protection Committee on Its Sixty-eighth Session*, MEPC 68th Session, Agenda Item 21, IMO Doc MEPC 68/21 (29 May 2015) paras 4.8, 4.10.

[236] Sherry P. Broder and Jon M. Van Dyke, 'The Urgency of Reducing Air Pollution from Global Shipping', in Aldo Chircop, Norman Letalik, Ted L. McDorman and Susan J. Rolston (eds.). *The Regulation of International Shipping: International and Comparative Perspectives* (Martinus Nijhoff Publishers 2012) 249.

makes the implementation of this agreement questionable. Regulations for marine pollution interact with the regulations for climate change. On the one hand, regulations for marine pollution may be helpful for the combating of climate change. On the other hand, tackling climate change requires the reduction of GHG emissions from international shipping which may be regarded as a type of marine pollution. Additionally, the 'carbon leakage' can also be effectively avoided provided that both issues are comprehensively regulated at the global and regional levels.

Index

Printed and bound by CPI Group (UK) Ltd, Croydon, CR0 4YY

01/11/2024

01782624-0008